To my freshman comp teacher, Ian Osborne,
who did not so much teach me how to write as that I could write,
and
To all the mathematicians (mostly safely dead) whose work appears here.
I, like Newton, stand on the shoulders of giants.

ZAHLVERGNUGEN
by Martin Christensen

Chapter 1   Computational Shortcuts

If you have watched enough late-night TV, you have surely seen the "program-length commercial" in which a boy, about 8 years old, is asked to multiply 103 by 107. Immediately, he answers correctly with 11,021. "How did he do that?," you probably wondered; after a while, you might have come to the conclusion that he was told the answer beforehand. This is not the case, however. The boy merely applied one of the many computational shortcuts provided in the book or videotape promoted during the commercial and offered at the end of it (and probably many other times as well) for, say, $49.95. For a fraction of that price, the book that you are now reading contains all these shortcuts plus their mathematical justifications, as well as much more mathematics, from algebra and geometry through trigonometry and calculus into such advanced topics as linear algebra, group theory, combinatorial theory, and fractal geometry, all on a level, it is hoped, accessible to the average adult with little mathematical knowledge beyond 9 x 9 = 81, or that all-too-common soul who did learn some "advanced" math in high school or college but has now forgotten everything except $A = \pi r^2$, the formula for the area of a circle.

This book is called "Zahlvergnugen" as an imitation of Volkswagen's "Fahrvergnugen" series of commercials. In German, *Zahl* means number and *vergnugen* means joy, so *Zahlvergnugen* should mean "the pleasure of numbers." I have slightly generalized this meaning to "the fun of mathematics." As you will see, math is not necessarily boring; learning it can be fun.

1.1 Quick Multiplication

The commercial example in which 103 x 107 = 11,021 displays one of the simplest rules of quick multiplication. By dissecting 11,021 into three parts, 1-10-21, the nature of the rule is apparent; since 10 = 3 + 7 and 21 = 3 x 7, the rule is to take the amounts by which the two numbers to be multiplied (which are called "factors") exceed 100, and first add them together, then multiply them together, and finally write in order 1, the result of the addition, and the result of the multiplication. Thus 105 x 109: 5 + 9 = 14, 5 x 9 = 45, 105 x 109 = 1-14-45 = 11,445. And 113 x 106: 13 + 6 = 19, 13 x 6 = 78, so 113 x 106 = 11,978. But the rule needs modification for a few cases, as can be seen from 102 x 104. 2 + 4 = 6, 2 x 4 = 8, but certainly 102 x 104 is not 168. This exception is taken care of by requiring that both the addition and multiplication have a two-digit result; a zero is added in front if necessary. With this provision in effect, 2 + 4 is not just 6 but 06, and 102 x 104 = 10,608. Looking at 111 x 114 shows another problem: with 11 x 14 being 154, the product appears to be 125,154, but that should seem a little large, since 111 x 1,000 is only 111,000. In view of the two-digit rule previously stated, the source of the problem should be the three-digit product 154. The problem is fixed by another provision that says when the product has three or more digits, all digits to the left of the last two (here, the 1 of 154) are to be added to the result of the addition. Thus, in 114 x 111, 14 + 11 = 25, 14 x 11 = 154, 25 + 1 = 26, so the answer is 12,654. Finally, considering, say, 143 x 138, we must deal with 43 x 38, which seems almost as tough as our beginning problem. This tells us that this rule is only useful for numbers that are close to 100. However, there are many other quick multiplication rules, as will be shown in the next few pages. Some of them may apply to 143 x 138; most will not. A few may offer a partial solution, reducing the problem to a slightly easier one. For some problems, such as 784 x 623, the old-fashioned "long" way may be the best available. But the chance of this is low, and is even lower in the two-digit times two-digit problems that crop up most often.

1.1.1   The Magical Number 37

We begin this section by noting that 37 x 3 = 111. On the surface, this may not seem like much – an easily remembered, but isolated, result. However, this product is the starting point for shortcuts involving multiplication by 37, 74 (37's double), or any number close to them. To see why, notice that not only does 37 x 3 equal 111, but 37 x 6 = 222, 37 x 9 = 333, etc. This is true because multiplication is "associative"; in multiplying several numbers together, we can start with any two we want. Thus, 37 x 12 = 37 x (3 x 4) since 3 x 4 = 12. With the associative property, we rewrite this as (37 x 3) x 4, which is the same as 111 x 4 which is obviously 444. To go past 37 x 27 = 999, we use the rule: Write down the first and last digits of what you get when the other factor is divided by 3, then write the sum of these two digits *twice* in between. (This can be used only if that sum is less than ten.) So, 37 x 69: 69 is 23 times 3. Writing two 5's (2 + 3) between the 2 and the 3 produces the answer, 2,553.

To multiply by 74, we break this number down into 37 x 2 and use the associative property again: 74 x 12 = (37 x 2) x 12 = 37 x (2 x 12) = 37 x 24 = 37 x (3 x 8) = (37 x 3) x 8 = 111 x 8 = 888. Of course, these intermediate steps are not written down in practice. To multiply 74 x 12, you notice that 74 is twice 37, so you double the 12, divide the answer 24 by 3, and write down three copies of the resulting 8 to form the product.

You have probably noticed that all the numbers that have been multiplied by 37 or 74 have been exact multiples of three. You are probably wondering (or snickering) about what happens when 37 or 74 is multiplied by another number, say 37 x 22. In this case, the "distributive" property of multiplication is used. This property is more important than the associative property because in involves addition (or subtraction) as well. To calculate 37 x 22, first split 22 into a multiple of 3 and either a 1 or a minus 1. Here 22 = 21 + 1 = (3 x 7) + 1, so 37 x 22 = 37 x (21 + 1) = (37 x 21) + (37 x 1) (the distributive property in action; the 37 is distributed to both the 21 and the 1) = 777 + 37 = 814. Watch what happens in 74 x 43; the 43 is doubled to get 86, the 86 is dissected to get 87 – 1, and the 87 broken down into 3 x 29. Since 2 and 9 add up to more than nine, we cannot write the sum twice between the 2 and the 9. Instead we write, from left to right, one more than the two, one more than the ones digit of 2 + 9, the exact ones digit of 2 + 9, and the nine itself for 3,219. From this, we must subtract 37 (since we had 87 – 1, not just 87, to deal with), and get a final answer of 3,182 for 74 x 43.

Next, we consider multiplying by numbers close to 37 or 74. For 48 x 38, write 38 as 37 + 1 and 48 as 16 x 3. The problem becomes (16 x 3) x (37 + 1) = 16 x (3 x 37) + 48 = 1,776 + 48 = 1,824. 81 x 73 becomes (81 x 74) – 81, which becomes [(3 x 37) x 54] – 81 or 5,994 – 81 which equals 5,913. The 54 is twice 27, which in turn is one-third of 81. For 39 x 76, write the 76 as 74 + 2. Thus, 39 x 76 = (39 x 74) + (39 x 2) = 26 x (3 x 37) + 78 = 2,886 + 78 = 2,964.

For a number that is not a multiple of 3 multiplied by a number that is close to, but not exactly, 37 or 74, the distributive property must be used twice. To obtain 43 x 38, we decompose 38 into 37 + 1 and 43 into 42 + 1. Now, (37 x 1) + (42 + 1) = (37 + 1) x 42 + (37 + 1) x 1, which in turn is expanded to (37 x 42) + (1 x 42) + (37 x 1) + (1 x 1). Our original product has become four separate products, but three of them are trivial and the fourth is a multiple of three times 37. The 37 x 42 equals 1,554 (1 + 4 twice between 1 and 4), and the sum of the other products is 80. Therefore, the answer to 43 x 38 is 1,634. With this, it is now possible to multiply 143 x 138, the problem left unanswered in the previous section. With 43 + 38 = 81 and their product 1634, 143 x 138 = 19,734. The 97 is the sum of 81 and the underlined 16, all but the last two digits of 1634.

It is interesting to evaluate some products in two or more different ways to see which one is the easiest. For example, 36 x 75. This is 2,700, a result quickly found by the shortcuts involving aliquot parts of 100 (here 75 is three-fourths of 100), but it may also be calculated as (37 – 1) x 75 = (37 x 75) – 75 = (37 x 3) x 25 – 75 = 2,775 – 75, or as 36 x (74 + 1) = (36 x 74) + 36 = 24 x (3 x 37) + 36 = 2,664 + 36. The answer, of course, comes out the same in all three calculations.

It is not necessary that numbers be very close to 37, 74, or multiples of three to use this method, just that the difference between the target number and the actual factor be easy to multiply by. This can be seen in 69 x 47 = 69 x (37 + 10). One-third of 69 is 23, so 69 x 37 = 2,553, to which is added 690 (69 x 10) to get a product of 3,243. (The 690 addition, by the way, can be further reduced by adding 700 and subtracting 10.) In 82 x 74, instead of splitting 82 into 81 + 1, it can be broken down into 72 + 10. Then one-third of 72 is 24, and double that is 48. Since 4 + 8 is 12, the technique at the top of this page gives 5,328 to which is added 740 for a total of 6,068. Splitting 82 into 81 + 1 instead results in 5,994 + 74 which is also 6,068. The extra technique is here avoided, in exchange for an addition that could give some trouble. However, additions such as this can themselves be done quickly. For 5,994 + 74, notice that 5,994 is 6,000 – 6. This transforms the sum into 6,000 + (74 – 6) which can be done immediately. In 497 + 368, 497 = 500 – 3, which immediately produces 865 for the sum.

There are other "magic numbers" besides 37, but they come up much less often. A fairly useful result is 7 x 11 x 13 = 1,001. Pairing two of the factors in each of the three possible ways results in three two-factor equations: 7 x 143 = 11 x 91 = 13 x 77 = 1,001. This makes 143, 91 and 77 all somewhat "magic." The distributive and associative properties may be applied here as well, to give answers to "not quite" problems. Thus 56 x 91 = [(5 x 11) + 1] x 91 = 5 x (11 x 91) + 91 which results in 5,096. The 143 appearing as a factor of 1,001 gives another way to calculate 143 x 138: 143 x (140 – 2) = (143 x 7) x 20 – 143 x 2 = 20,020 (*not* 20,002) – 286 = 19,734. Note also that 143 x 5 = 715, the "magic number" for Henry Aaron, and you will see that baseball is mixed up with mathematics, too,

Other "magic" results include 73 x 137 = 10,001 and 41 x 271 = 11,111. Using the first of these, 143 x 138 = [(2 x 73) – 3] x (137 + 1) = (2 x 73 x 137) + (2 x 73 x 1) – (3 x 137) – (3 x 1) or 20,002 (*not* 20,020) + 146 – 411 – 3 = 20,148 – 414 = 19,734. The 411 is simple to derive; it is (3 x 100) + (3 x 37), which takes us back to the beginning of this section.

It is obvious that 78 x 10 = 780, 23 x 100 = 2,300, 9,876 x 1,000 = 9,876,000, etc. What is not so obvious is that the distributive property can be used to make multiplication by numbers close to 10, 100, 1,000, etc., almost as easy as multiplication by those powers of 10 themselves. For example, 9 can be replaced by the equivalent value (10 – 1). Thus, 58 x 9 = (58 x 10) – (58 x 1) = 580 – 58 = 522. Larger numbers work the same way: 3,807 x 9 = 38,070 – 3,807= 34,263. The associative property can be used as well to facilitate multiplying by 18, 27, and so forth. To multiply 283 x 27, first split up 27 into 3 x 9, then multiply the 283 by the 3 to get 849. Then 849 x 9 = 8,490 – 849 = 7,641.

Multiplying by 11 is even easier since only addition is required. Writing 11 as (10 + 1), 78 x 11 becomes 780 + 78 = 858, and 3,183 x 11 = 31,830 + 3,183 = 35,013. Moving to higher powers of 10, 97 (for example) can be written as 100 – 3. Therefore, 157 x 97 = 15,700 – (157 x 3). The latter product is equal to 471, so 157 x 97 = 15,229. Splitting 104 into 100 + 4 makes multiplying this number by 82 quite easy: 82 x 104 = 82 x 100 + 82 x 4 = 8,200 + 328 = 8,528. The more distant number 89 can be split into 100 – (10 + 1); thus, 47 x 89 = 4,700 – (470 + 47) = 4,700 – 517 = 4,183. Note here that the addition (inside the parentheses) must be done first. Calculating the answer by first subtracting 470 from 4,700, then adding the 47, will get you the incorrect 4,277 (which is actually 47 x 91), and subtracting 470 and then subtracting another 47 from the result will usually take longer, although it does produce the correct result.

There are also more mechanistic ways to multiply by 9, 11, and some other numbers. Jaime Escalante (of "Stand and Deliver" fame) demonstrated one way to come up with the basic products involving 9. The first step is to write the products, from 9 x 1 to 9 x 10, in a column. Next, write the numbers

9 x 1 =	0	0 9	0 through 9 next to the equal signs, so that each
9 x 2 =	1	1 8	number is one less than the factor that is being
9 x 3 =	2	2 7	multiplied by 9, immediately to its left. The last
9 x 4 =	3	3 6	step is to write the same numbers, 0 to 9, from
9 x 5 =	4	4 5	bottom to top to form the ones' digits of all the
9 x 6 =	5	5 4	answers. Presto! Each product from 9 x 1 = 09
9 x 7 =	6	6 3	to 9 x 10 = 90 is complete. This simple list of
9 x 8 =	7	7 2	9 multiplication facts contains many patterns, and
9 x 9 =	8	8 1	mathematics is to a large extent the science of
9 x 10 =	9	9 0	patterns.

The most important pattern is that each of the digit pairs that make up the products has a sum of 9. That is, 9 x 2 = 18 and 1 + 8 = 9, 9 x 3 = 27 and 2 + 7 = 9, and so on. This is true in general in the form that the sum of the digits of any multiple of 9 is also a multiple of 9. Example: 3,807 x 9 = 34,263 (see above) and 3 + 4 + 2 = 6 + 3 = 9, so that the sum of all five digits is 18 or 9 x 2. When applied to multiplying 9 by single digits, this pattern produces a simple rule: subtract 1 from the factor to get the tens' digit, and then subtract the tens' digit from 9 to produce the ones' digit. Thus 9 x 7: 7 – 1 = 6, 9 – 6 = 3, so 9 x 7 = 63. Equivalently, we can proceed in the other direction: subtract the factor from 10 to get the ones' digit, then subtract the ones' digit from 9 to get the tens' digit: 10 – 7 = 3, 9 – 3 = 6, 9 x 7 = 63 again.

The pattern involving the sum of the digits of a multiple of 9 can be proven by considering place value. Recall that 45,678 means 4 x 10,000 + 5 x 1,000 + 6 x 100 + 7 x 10 + 8. For any digit, say the 5, we form the difference between that digit and its value, that is, (5 x 1,000) – 5 or 5 x (1,000 – 1) since 5 = 5 x 1 and the distributive property works backwards as well as forwards. Now 1,000 – 1 = 999 is obviously a multiple of 9. In general, one less than any power of 10 is a number consisting exclusively of nines and therefore a multiple of 9. For the ones' digit, (8 x 1) – 8 = 0 and we adopt the convention that zero = 9 x 0 is a multiple of 9 (or any number at all). Thus, for any number, the difference between it and the sum of its digits is a multiple of 9. If the number is one as well, then the digit sum must be also.

The old rule of "casting out nines" comes directly from the last paragraph. Since the difference between any number and the sum of its digits is a multiple of 9, it follows that the remainder when you divide the sum of a number's digits by 9 is the same as the remainder when the whole number is divided by 9. It is also true that the remainder of the sum (difference, product) of two remainders is the same as the remainder of the sum (difference, product) of the numbers that gave rise to the remainders in the first place. (See Chapter 4 for a fuller discussion of the related topic of modular arithmetic.) This can be used as a check on arithmetic results. Back on the first page, 105 x 109 was calculated to be 11,445. Now 105 = (9 x 11) + 6 and 109 = (9 x 12) + 1, so the remainder when 11,445 is divided by 9 must be 6. The sum of the digits of 11,445 is 15, and 1 + 5 = 6, so the remainder is indeed 6 as required. In 103 x 107 = 11,021, the remainders are 4, 8, and 5 respectively. 4 x 8 is not 5, so something is wrong? Not really. 4 x 8 = 32, and 32 = (3 x 9) + 5, so the final remainders do match. In all cases, the computation must be continued until a single number between zero and eight is obtained. (A "remainder" of nine is the same as a remainder of zero.)

To multiply a two-digit number by 11, just add the digits together and write the answer between those digits. Thus, 35 x 11 = 385, 42 x 11 = 462, 85 x 11 = 8135 (Whoops!). When the sum goes over 9, we must "carry" the one into the first digit of the answer, just as ones are carried in the addition of 46 + 38 = 84 rather than 714. Therefore, 85 x 11 = 935, 48 x 11 = 528, and 97 x 11 = 1,067. (No "whoops!" on that last answer, even though it is in four digits; the 9 plus the carry gives the two-digit number 10 as the first "digit.") To multiply eleven by larger numbers, we expand the rule above: each digit of the answer is equal to the corresponding digit in the factor, plus the one to its immediate right (with any carry added in, of course). If all the digits in the factor are small, no carries will be needed: 42,531 x 11 = 467,841, where 6 = 4 + 2, 7 = 2 + 5, 8 = 5 + 3, and 4 = 3 + 1. Larger digits tend to create carries, as in 59,748 x 11 = 657,228. The 6 is 5 plus the carry from 5 + 9, the first two is the ones' digit of 7 + 4 plus the carry from 4 + 8, and so on.

Multiplying 9 by large numbers mechanically involves the same type of computation, but with subtracting instead of adding, and hence borrowing instead of carrying. The rule: Each digit in the answer is found by subtracting the corresponding digit in the factor from the one to its right, and then adjusting for any borrow from the preceding digit's calculation. If the result is less than zero, subtract how much less it is from 10 to get the answer digit, and you have a borrow to count against the next digit to the left. For the first (ones') digit in the answer, just subtract the ones' digit in the factor from 10, and there is an automatic borrow for the next (tens') digit. (The exception is a ones' digit of zero in the factor, which gives you a zero ones' digit in the product and no borrow.) For 48,726 x 9, the calculations go 10 – 6 = 4 with the automatic borrow, 6 – 2 – borrow = 3, 2 – 7 is five less than zero, and 10 – 5 = 5 with a borrow, 7 – 8 – borrow is two less than zero, and 10 – 2 = 8 with another borrow, 8 – 4 – borrow = 3, and 4 – 0 = 4, for an answer of 438,534. Note that in the calculation of the last (hundred-thousands') digit, there is no corresponding digit in the factor, so a zero was subtracted. The calculation is not done until there is no digit to the immediate right of the space corresponding to the next digit of the product to be calculated.

Multiplying by 99 and 101 can be mechanized as well. The procedures are the same as in multiplying by 9 and 11, except that two-digit groups replace single digits. Thus, for 4,523 x 101, since 45 + 23 = 68, the answer is 456,823. The division into groups must be carried out from right to left, and a single digit left over at the end is treated as a two-digit group beginning with a zero. In multiplying by 99, the first group is subtracted from 100, not 10.

1.1.3 Aliquot Parts of 100

This section of shortcuts was emphasized in 19th-century arithmetic texts, because numbers of this type come up frequently in commercial applications. The old-fashioned term "aliquot" means roughly "factor." Thus, 50 ,25, 20, 33 1/3, 16 2/3, and 12 ½ are all aliquot parts of 100, being respectively 1/2 , ¼, 1/5, 1/3, 1/6, and 1/8 of 100. Their multiples such as 75 (3/4 of 100) and 83 1/3 (5/6 of 100) are also considered aliquot parts. (For those of you who shudder at the sight of fractions, Chapter 2 explains them as special combinations of "ordinary" numbers.)

To multiply a number by an aliquot, simply apply the fraction it represents to the other factor, and multiply the result by 100 (just tack on two zeros). Thus, 36 x 75 from Sec. 1.1.1: 75 is ¾ of 100, and ¾ of 36 is 27, so the answer is 2,700. Calculating three-fourths of 36 is done by dividing 36 by four and multiplying the resulting 9 by three. Similarly, 60 x 65 = 3,900: 60 = 3/5 of 100, 65/5 = 13, and 13 x 3 = 39. Admittedly, numbers like 33 1/3 and 62 1/2 rarely appear as factors, but 34 and 62 do show up, and the distributive and associative properties can be used to convert 33 1/3 to 34 or 62 ½ to 62. Example: 34 x 252. Since 34 = 33 1/3 + 2/3, we can write 34 x 252 = (33 1/3 x 252) + (2/3 x 252) = 8,400 + 168 = 8.568. Here, 252/3 = 84 and 2 x 84 = 168. In 62 x 196, (62 ½ x 196) – (1/2 x 196) = (5/8 x 100) x 196 – 98 = 5 x 2450 - 98 = 12,152. 196/8 in this problem is 24 ½, which when multiplied by 100 becomes 2,400 plus one-half of 100. Also, 5 x 2,450 = (1/2 x 10) x 2,450, and 2,450/2 = 1,225. Five is an aliquot part of 10, and can be treated as such when multiplying.

A similar technique can be used to multiply by 51, 76, 79, etc. For 79 x 237, write (80 – 1) x 237 = [(4/5 x 100) x 237] – (1 x 237) = (189 3/5 x 100) – 237 = 18,723. The division of 237 by 5 produces 47 2/5, which multiplied by 4 results in 188 8/5 = 189 3/5, then 3/5 of 100 is 60.

1.1.4 Multiplying Close-Together Numbers

This section is one of the most useful, for answers determined here come out faster than just about anywhere else. To multiply 58 x 62, subtract 2 x 2 from 60 x 60 to get 3,596. In this example, 60 is the number halfway between 58 and 62, and 2 is the difference between 58 or 62 and 60. Mathematically, we write 58 = 60 – 2, 62 = 60 + 2, and use that all-important distributive property twice: (60 – 2) x (60 + 2) = [60 x (60 + 2)] – [2 x (60 + 2)] = (60 x 60) + (60 x 2) – (2 x 60) – (2 x 2). Since multiplication is commutative, the + 60 x 2 and – 2 x 60 cancel each other out, and we are left with (60 x 60) – (2 x 2). This shortcut can only be used if the difference between the two factors is even, and is most useful if the two ones' digits add up to ten. 33 x 37 (which can be calculated by the magic-number shortcut as it involves 37) has these properties: 33 = 35 - 2, 37 = 35 + 2, so 33 x 37 = (35 x 35) – (2 x 2) = 1,225 – 4 or 1,221. (Multiplying 35 by itself can be done instantly by using the trick in Sec. 1.2.1.) In 92 x 108, 100 is the in-between number: 92 x 108 = (100 x 100) – (8 x 8) = 9,936. Now, suppose we wish to multiply two numbers whose ones' digits do not add up to ten. The price for this is that the in-between number will not end in zero or five, and therefore will be a little tougher to multiply by itself; however, it may still fall under one of the other squaring shortcuts in the next section, and if it does, a quick answer to the original problem is still likely. Thus, in 66 x 48 = (57 + 9) x (57 – 9), the 57 falls under another squaring shortcut that holds for any number between 41 and 59, and here produces 3,249. Subtracting 81 (9 x 9) from this gives 66 x 48 = 3,168. (This problem can also be calculated from the aliquot-parts shortcut: 48 = 50 – 2, and thus 66 x 48 = 3,300 – 132 = 3,168.) The same squaring shortcut yields 43 x 49 = (46 x 46) – (3 x 3) = 2,116 – 9 = 2,107.

With numbers that are a small odd distance apart, it is sometimes worthwhile to rewrite one of the factors so that this shortcut can be applied. With 78 x 83, the 83 can be split into 82 + 1 so that a neat, round 80 is the in-between number. Thus, 78 x 83 = (78 x 82) + 78 = 6,396 + 78 = 6,474. With 74 x 79, we split 79 into 76 + 3, and the product turns into (74 x 76) + (74 x 3) = 5,624 + 222 = 5,846. The 74 x 3, of course, comes from the first section on 37, showing how more than one technique can be applied to a problem that does not become transparent after the first shortcut is applied.

1.2 Squaring

If the shortcut methods presented in the previous pages seem like a great help in what would otherwise be drudgery of routine multiplication, the techniques shown here will make multiplication seem positively easy when the two factors involved happen to be identical. To "square" a number is to multiply it by itself. To show that a number has been squared, a little ^{2}, called an exponent, is written above and to the right of the number. (2 refers to the number of factors used.) Thus, 8^2 = 8 x 8 = 64, 47^2 = 47 x 47 = 2,209, and 707^2 = 707 x 707 = 499,849.

1.2.1 Numbers Ending in 5

To square a number ending in 5, multiply the part before the 5 by one more than itself, and then write 25 after that answer. Thus, 65^2 = 4,225 since 6 x 7 = 42, and 135^2 = 18,225 since 13 x 14 = 182. The justification for this trick is as follows: Write 65 as (10 x 6) + 5, then 65^2 = [(10 x 6) + 5] x [(10 x 6) + 5] =[(10 x 6) x (10 x 6)] + [5 x (10 x 6)] + [(10 x 6) x 5] + (5 x 5), which, with a little more rearranging, can be written as [(100 x 6 x (6 + 1)) + 25. Any number could have been used instead of the 6, so the rule must hold for all numbers ending in 5. Larger numbers can be used, with shortcuts from previous pages taking care of the remaining multiplication. Thus 245^2 = 60,025 (24 x 25 = 600, aliquot parts), and 365^2 = 133,225 (36 x 37 = 1,332, magic number 37).

1.2.2 Numbers between 41 and 59

These can be squared by the following rule: Subtract 25 from the number to get the first two digits, then square the difference between the number and 50 to get the last two. Therefore, 57^2 = 3,249 since 57 – 25 = 32 and 7 x 7 = 49, and 41^2 = 1,681 since 41 – 25 = 16 and 9 x 9 is 81. (The 9 is 50 – 41.) For 48^2, 48 – 25 = 23 and 2 x 2 = 4; using the basic principle that a single digit where two are needed is to be preceded by a zero, we arrive at 2,304.

Mathematically, any number between 41 and 59 can be written as 50 + a, where a is between -9 and +9. (See Chapter 3 for arithmetic with letters.) Then $(50 + a)^2$ = 2,500 + (100 x a) + a^2 = 100 x (25 + a) + a^2. Our original number is 50 + a, so the first two digits are indeed 25 less than the number we started with, whether a is positive or negative. Also, since a is between -9 and +9, a^2 = a x a is between 0 and 81 and hence has no more than two digits. The rule of signs guarantees that a^2 is positive irrespective of the sign of a itself.

This rule can be used in conjunction with the one above it to facilitate the squaring of 415, 425, etc. Thus 535^2: 53 x 54 = 2,809 + 53 = 2,862, so 535^2 = 286,225. It can also be extended to numbers between 469 and 531 as well. These squares have six digits; the first three are the number less 250, and the last three are the square of the difference between the number and 500. Thus $484^2 = 234{,}256$ since $484 - 250 = 234$, $500 - 484 = 16$, and $16 \times 16 = 256$. Similarly, $509^2 = 259{,}081$.

1.2.3 Numbers Near Powers of Ten

Squaring these numbers is almost as easy as using the 41-59 rule. The first half of their squares is a number twice as far from the power of 10 as the original number is, and the second half is the square of the distance between the original number and the power of 10. Mathematically, $(10^n + a)^2 = 10^n \times (10^n + (2 \times a)) + a^2$. The little n is an exponent, just like the ^{2}, and counts the number of 10's that are multiplied together: $10^2 = 10 \times 10 = 100$; $10^3 = 10 \times 10 \times 10 = 1{,}000$; $10^5 = 10 \times 10 \times 10 \times 10 \times 10 = (10 \times 10) \times (10 \times 10 \times 10) = 10^2 \times 10^3 = 100{,}000$. Note, in the last example, that $10^5 = 10^2 \times 10^3$ and $5 = 2 + 3$. This is an example of the first law of exponents (Sec. 3.1.3). In general, the exponent on 10 is equal to the number of zeros that follow the 1.

Now, a few numerical examples. $13^2 = 169$; 16 is twice as far from 10 as 13 is, and 9 is 3×3. $8^2 = 64$; 6 is twice as far from 10 as 8 is, and 4 is the square of 2. $93^2 = 8{,}649$ since $100 = 93 + 7$ and $93 - 7$ is 86 and $7 \times 7 = 49$. $108^2 = 11{,}664$ since $108 = 100 + 8$, $108 + 8 = 116$, and $8 \times 8 = 64$. $123^2 = 15{,}129$. Here, $123 = 100 + 23$, $123 + 23 = 146$, and $23^2 = 529$; the 5 of 529 is added to the 146. $978^2 = 956{,}484$; $1000 - 978 = 22$, $978 - 22 = 956$, and $22^2 = 484$. $1{,}008^2 = 1{,}016{,}064$; $1{,}008 - 1{,}000 = 8$, $1{,}008 + 8 = 1{,}016$, and $8 \times 8 = 64$. The questions about when to insert zeros (as here) and when to carry something into the first half of the square are answered by the following rule: Squaring numbers starting with a 9 (or 4, 5, 6, 7, or 8) doubles the number of digits, and squaring numbers starting with a 1 (or 2) comes one short of doubling the number of digits. Hence, 123^2 must have five digits since $(3 \times 2) - 1 = 5$, while 978^2 must have six digits. (Squaring numbers beginning with 3 can either produce a full doubling or come up one short, as is apparent from $31^2 = 961$ and $32^2 = 1{,}024$.) For $1{,}008^2$, the square must have $(4 \times 2) - 1 = 7$ digits, therefore, three digits must follow the 1016.

1.2.4 Numbers Close to Easy-to-Square Numbers

The mathematical formula here is $(a + b)^2 = a^2 + b \times ((2 \times a) + b)$. Here, a represents the (relatively large) number whose square can be easily calculated, and b is the (small) difference between a and the number actually given to be squared. The nonmathematical, or at least nonsymbolic, rule is: Write down the easily calculated square. Then double the easy-to-square number, add the difference, and multiply the sum by that difference. Since the difference is small, this will be an easy multiplication. Add this product to the easy square, and you have your answer. Thus $76^2 = (75 + 1)^2 = 5{,}625 + 151$ or $5{,}776$. The 151 is twice 75, plus 1, times 1. Note that 76^2 ends with 76; it is one of only two two-digit numbers (the other is 25) to have that property. This means that the American revolutionaries picked one of the two best years in the 18th century, and even the second millennium, due to the first seven in 5,776, to declare independence. No wonder the British couldn't beat them!

If your number is less than an easy-to-square number, replace "add" with "subtract" in the rule (and $+$ with $-$ in the formula). Thus $69^2 = (70 - 1)^2 = 4{,}900 - 139 = 4{,}761$. Note that $139 = (2 \times 70) - 1$, not $(2 \times 70) + 1$.

Your number does not have to be particularly close to an easy-to-square number, so long as the arithmetic involved in applying the rule is easy. Thus, 878^2: $878 = 978 - 100$, so $878^2 = 956{,}484$ (from previous section) $- 100 \times (1{,}956 - 100) = 956{,}484 - 185{,}600 = 770{,}884$. Similarly, $163^2 = (103 + 60)^2 = 10{,}609 + (60 \times (206 + 60)) = 10{,}609 + 15{,}960 = 26{,}569$. The multiplication of 60×266 is eased by thinking of 266 as $266\ 2/3 - 2/3 = (800 \times 1/3) - 2/3$, so $60 \times 266 = 16{,}000 - 40$. 163^2 can also be attacked by writing it as $(160 + 3)^2 = 25{,}600 + (3 \times 323) = 26{,}569$. This method is easier provided that 160^2 is considered "easy."

1.2.5 Squaring Any Number

No matter how extensive the list of shortcuts, you will eventually come across a number that does not lend itself to any of them. However, there is still a method of squaring that is roughly twice as fast as conventional multiplication. It is based on the conventional method, but takes advantage of the fact that, say in $7{,}654 \times 7{,}654$, you will not only have to multiply the first 6 by the second 4, but also the second 6 by the first 4. Both products have the same place value, so why not add both of them at once? This is the insight on which the method is based.

At left, the square of 7,654 is calculated. The partial products are determined much as they are in conventional multiplication, with the key exception that when digits in different columns are multiplied, the product is *doubled* before it is added in. Also, digits are multiplied only by ones to their left or by themselves, and each new row is started two, rather than one, columns to the left of the start of the previous row. The first row, 6 1 2 1 6, is calculated as follows: 4 x 4 = 16 (same digit, so no doubling); write down 6, carry the 1. 5 x 4 = 20, doubled is 40, plus the

$$
\begin{array}{r}
7\ 6\ 5\ 4 \\
\hline
6\ 1\ 2\ 1\ 6 \\
7\ 6\ 2\ 5 \\
8\ 7\ 6 \\
4\ 9 \\
\hline
5\ 8\ 5\ 8\ 3\ 7\ 1\ 6
\end{array}
$$

1 yields 41; write 1, carry 4. 6 x 4 = 24, doubled is 48, + 4 = 52; write 2, remember 5. 7 x 4 = 28, doubled is 56, + 5 is 61. Since we are out of digits, the entire 61 is put down to complete the row.

If there is a 5 in the number you are squaring, the row corresponding to that 5 can be calculated quickly. The last two digits in the row are 25, and the rest of the row is a copy of all the digits to the left of the 5. Application of this shortcut-within-a-shortcut produces the second row, 7 6 2 5.

For the third row, 8 7 6, 7 x 6 = 42, doubled is 84, plus 3 from 36 (= 6 x 6) gives 87. The last row, corresponding to the digit 7, is easy. Since there are no digits to the left of the 7, all we need write is the square of the 7 itself. Once all the rows are calculated, they are totaled as in ordinary multiplication.

If the number to be squared contains several large digits like 9 or 8, you will find yourself going over 100 when you double some products. Don't worry; you will just have to carry a number bigger than 9 to the next step. The most you can carry is 17, from 9 x 9 doubled = 162 plus a large carry from the previous step.

1.3 Division Shortcuts

There are fewer methods of quick division than of quick multiplication. Most of them consist of improvements on the rather inefficient conventional algorithm. The method called "peasant division" works as follows: The divisor is repeatedly doubled until the result exceeds the dividend. Next to the sequence of doubled numbers, the corresponding powers of 2 are written. Then, the largest number in the sequence that is smaller than the dividend is subtracted from the dividend, and the associated power of 2 is circled. The remainder from the subtraction then replaces the dividend, and the process is repeated until a remainder smaller than the divisor is obtained. The powers of 2 that have been circled are then added to obtain the quotient.

On the left, the division of 7,497 by 83 is illustrated. The sum of the boldfaced figures is 90, which is the quotient, and the last subtraction gives the remainder, 27. This method requires the writing of more digits than the conventional method, but uses no "division" in the sense that 74/8 is not used to estimate the first digit of the quotient. (Note that if the dividend had been 7,450, the estimate provided by 74/8 would have been wrong.) Nor does the peasant method require the re-multiplication of the divisor by each successive digit of the quotient, which is where most errors are made. Once the sequence of doubled numbers is finished, all multiplication is finished.

83	1	7497
166	**2**	5312
332	4	2185
664	**8**	1328
1328	**16**	857
2656	32	664
5312	**64**	193
10624	128	166
		27

If doubling is too much, a modification of this method replaces it by simple addition. In this method, the second row is still double the divisor, but all succeeding rows are the sum of the two immediately above them. Thus, 1079 = 664 + 415, and 1743 = 1079 + 664. The associated parts of the quotient are calculated in the same way, giving rise to what is known as the "Fibonacci sequence." (See Sec. 4.2 for a detailed study of this very important, and entertaining, mathematical object.) This method will require more additions than the peasant method, but it is likely to require fewer subtractions, since it never requires the use of two consecutive rows as numbers to be subtracted. (If it did, then the next row under could be used instead, as it is the sum of those two.)

83	1	7497
166	2	7387
249	3	110
415	5	83
664	8	27
1079	13	
1743	21	
2822	34	
4565	55	
7387	**89**	
11952	144	

One class of true computational shortcuts for division applies when the divisor is 5, 25, 50, or a few other numbers. Since 5 = 10/2, any number can be divided by 5 by multiplying by 2 and then dividing by 10. Thus, 7,195/5: 7,195 x 2 = 14,390 and therefore 7,195/5 = 1,439. Similarly, 25 = 100/4, so to divide by 25, multiply by 4 and then divide by 100.

In some cases, it may be profitable to "break up" a division into two or more parts that correspond to factors of the divisor. Thus, to divide by 56, divide by 7 and then divide the quotient by 8 (since 56 = 7 x 8). Similarly, a divisor of 176 can be split into 11 and two 4's. This split-up can be combined with the previous rule to make division by 35, 55, or other numbers easier. To divide by 35, multiply by 2 and divide by 70; to divide by 75, multiply by 4 and divide by 300.

If only a close approximation is desired, good results can be achieved with divisors of 9, 11, 98, 103, etc. Since 9 is one-tenth less than 10, it stands to reason that dividing something by 9 gives a result one-tenth more than dividing the same number by 10. Thus, 473 divided by 9: 473/10 is about 47, and one-tenth of that is about 5. 47 + 5 is 52, so 473/9 should be about 52. Actually, 473/9 is 52 with a remainder of 5, so this method is a good approximation. It must be stressed, though, that this method will not give an exact answer. This is made clearer by using a larger number: 71,483/9 is approximately 7,148 + 715 = 7,863 but the actual value is 7,942 with a remainder of 5. The true result, in symbolic form, is $n/9 = (n/10) + (n/10)/9$. We still have a division by 9 to deal with. But this in turn can be approximated, leading to an endless sequence of ever-better approximations. Ultimately, the formula becomes $n/9 = (n/10) + (n/100) + (n/1,000) + (n/10,000)$ plus an endless series of ever-smaller terms. (See Sec. 7.10 for a full treatment of infinite series.) For now, you can stop adding when the terms become small enough to ignore, say, less than one. Thus, 71,483/9 can be approximated as 7,148 + 715 + 72 + 7 = 7,942. Round-off peculiarities can induce an error of one or two in the final answer, but, in general, the more steps are taken, the better your approximation becomes. For division by 11, the formula is the same, except that a minus sign precedes every other term, beginning with the $(n/100)$ term. Thus, 71,483/11 is approximately 7,148 – 715 + 72 – 7 = 6,498. The actual value is 6,498 with a remainder of 5.

Similar reasoning applies to numbers near 100. For example, 98 is two hundredths smaller than 100, so a number divided by 98 should give a result two hundredths greater than the same number divided by 100; and in reality, the quotient is two 98ths greater. The mathematical formula becomes $n/98 = (n/100) + (2 \times n/10,000) + (4 \times n/1,000,000)$, etc.

1.4 Cube Root Extraction

How about this for "How did you do that?" responses: You tell a friend to enter a two-digit number into a calculator, multiply it by itself, and then multiply by that number again. He or she then reads off the calculator's answer, and you almost immediately give him the number he started with! Sounds impossible? In reality, it is quite easy.

The process of multiplying three copies of a number together is called *cubing*, and the result is the *cube* of the original number. The word "cube" is used due to an analogy with the "square" of a number as being two copies of that number multiplied together; a cube can be thought of as a three-dimensional square. The mathematical symbol for cubing is a little ^{3}; like the ^{2} for squaring, the number counts the times the base number is multiplied together. Thus 6^3: 6 x 6 x 6 = 36 x 6 = 216. (The same principle works for larger numbers of factors as well: $8^4 = 8 \times 8 \times 8 \times 8 = 4,096$ and $7^6 = 7 \times 7 \times 7 \times 7 \times 7 \times 7 = 117,649$.)

Any mathematical process has an *inverse* process that essentially undoes the work of the first one. Subtraction and addition are inverses: 4 + 5 = 9, and subtracting 5 from 9 takes us back to 4. So are multiplication and division: 7 x 3 = 21, and 21/3 takes us back to 7. The inverse process for raising a number to some power, e. g., squaring or cubing, is called *root extraction*. In particular, the inverse of squaring is square root extraction, and the inverse of cubing is cube root extraction. Since $7^2 = 49$, the square root of 49 is 7, and because $6^3 = 216$, the cube root of 216 must be 6.

Now, a peculiarity of our number system makes the extraction of cube roots particularly easy. Examining the ones' digits of the cubes in the adjoining table shows that each possible ones' digit from

Number	Cube
1	1
2	8
3	27

zero to nine occurs exactly once. Furthermore, the ones digit of the cube is the same as the ones' digit of the number being cubed (the cube root) unless it is 2, 3, 7, or 8, in which case it is the difference between that digit and ten. For the extraction of cube roots, this means that the ones'

4	64
5	125
6	216
7	343
8	512
9	729
10	1,000

digit of the root is the same as the ones' digit of the cube, unless it is 2, 3, 7, or 8. If the ones' digit does fall in that group, subtract it from 10 to obtain the ones' digit of the cube root. Thus, given 343, the ones' digit is 3, which is part of the special group. 10 – 3 is 7, so the cube root of 343 is 7. Given 729, which has a ones' digit of 9, we see that 9 is not in the special group, so it is the cube root of 729.

The astute reader will notice that if I was given 723 and 349 instead of 343 and 729, the method would still have produced 7 and 9 respectively. This extraction method works only on genuine cubes, and 723 and 349 are not genuine cubes. These numbers still have cube roots, but they are not whole or even rational numbers. In Chapter 6, which discusses "real" numbers, we will see in what way numbers like 723 have cube roots.

At the beginning of this section, I promised easy extraction of two-digit cube roots. Here is how it is done: The cubes from which these roots are extracted have four, five, or six digits. The last three are ignored for a moment, and the others are compared to the table above, which must be memorized. Given 140,608, for instance, we ignore the 608 and concentrate on the 140. Since this number is between $125 = 5^3$ and $216 = 6^3$, the tens' digit of the cube root is 5. Now the ones' digit of 140,608 is tested. Since 8 is in the special group, the ones' digit of the cube root is 10 – 8 or 2, and therefore the cube root of 140,608 is 52 (provided that 140,608 is a genuine cube). The answer can be checked by calculating 52^3: 52 x 52 = 2,704 (squaring trick for 41-59), and 2,704 x 52 = (2,704 x 50) + (2,704 x 2) = 135,200 + 5,408 = 140,608. In the last calculation, 2,704/2 = 1,352, which provides the 135,200. So 140,608 is a genuine cube, and 52 is indeed its cube root. Given 117,649: 117 is between 64 and 125, and 9 is not in the special group (2, 3, 7, or 8), so the cube root of 117,649 is 49. Note that $49 = 7^2$, and above I claimed that 117,649 was 7^6. Therefore, $(7^2)^3 = 7^6$ and 2 x 3 = 6. This is an example of the second law of exponents (the first law appeared in Sec. 1.2.3). Both of these laws and some others are explained in Chapter 3.

This method can be extended to three-digit cube roots (7, 8, or 9-digit cubes), although a little more work is required. The hundreds' and ones' digits are derived in the same way as the tens' and ones' digits are in the case of two-digit roots. For the middle (tens') digit, a technique called *interpolation* is used. The digits in the millions (which determine the hundreds' digit of the root) are not just between one reference value and the next, but are some definite fraction of the way along. This fraction is then used to estimate the tens' digit of the root. Thus, given 77,308,776, we first determine the hundreds' digit. It is 4 because 77 is between 64 and 125. The ones' digit is 6 because it is not in the special group. For the tens' digit, 77 is 13 more than 64. From 64 to 125 there is a difference of 61, so 77 is a little more than one-fifth of the way from 64 to 125. Therefore the tens' digit must be 2, and the cube root of 77,308,776 must be 426 (again, provided that 77,308,776 is a genuine cube; but if you can't trust an amateur mathematician, who can you trust?)

The interpolation may not be accurate for small three-digit cube roots, so the cube table from 11 to 20 must be memorized as well. Note that the cube does not get halfway from 1,000 to 8,000 until the root is two-thirds of the way from 10 to 20.

Number	Cube
11	1,331
12	1,728
13	2,197
14	2,744
15	3,375
16	4,096
17	4,913
18	5,832
19	6.859
20	8,000

Chapter 2   Fractions Made Easy

This chapter is for all those people who panic at the sight of a fraction, like $\frac{2}{3}, \frac{3}{4}, \frac{5}{7}, \frac{13}{8}, \frac{99}{100}, \ldots$ (Relax! They aren't as tough as they seem.) and whose memories are of pie wedges and the queer fact that making the number on the bottom *bigger* somehow made the fraction *smaller*, then later on having to find the "lowest common denominator" to add $\frac{1}{2}$ and $\frac{1}{3}$, rather than just adding the tops and bottoms to get $\frac{2}{5}$.

Fractions are to be explained as a new type of "number" built up from simpler numbers. This is something that mathematicians love to do: make complex objects up from combinations of simple ones. Whole numbers are the starting blocks, then negative numbers are created from positive ones, fractions from integers. Later on in this book, you will see "real" numbers built up from fractions, then "complex" numbers from real numbers. And mathematicians do this type of thing with objects other than numbers, too. They create "groups" and "fields," "functions" and "vectors" out of simpler objects. Many of their creations go beyond the scope of this book, but the ones mentioned just above do not. Now we will discuss one of the simpler built-up objects, the *ordered pair*.

2.1 Ordered Pairs

An ordered pair of numbers is a collection of two of them, distinguished from others not only by the numbers themselves but also by the order in which they are written. In the ordered pair (4, 6) the first member is 4 and the second member is 6. I stress that this is not the same as the ordered pair (6, 4) even though both pairs contain a 4 and a 6. In (4, 6) the first member is 4; in (6, 4) the first member is 6. Two ordered pairs are considered the same if *and only if* their first members are identical *and* their second members are identical. The numbers in an ordered pair do not have to be distinct: (5, 5) and (3, 3) are valid ordered pairs. The concept of ordered pair can be generalized to ordered triples, quadruples, ..., "*n*-tuples" which contain three, four, ..., *n* numbers. Also, ordered pairs do not have to contain only numbers; they can contain any mathematical object, including other ordered pairs. For the purposes of fractions, however, the ordered pair can be viewed as containing only two numbers.

2.2 What Is a Fraction?

A fraction is *defined* as an ordered pair of numbers, with the restriction that the second member of the ordered pair cannot be zero. Since a fraction is to be considered a number in its own right, we must also define when two fractions are to be considered equal, when a fraction is to be considered equal to an ordinary number (*integer*), and, of course, how to add, subtract, multiply, and divide fractions. The following rules will do this. (Letters are used to stand for arbitrary numbers.)

1. Equality of Fractions
$(a, b) = (c, d)$ if and only if $a \times d = b \times c$. In particular, $(a \times k, b \times k) = (a, b)$ provided, of course, that k is not zero.

2. Equality of Fractions with Integers
$(a, 1)$ is equal to and identified with the integer a. By rule 1, $(a \times k, k)$ is also equal to a.

3. Addition of Fractions
$(a, b) + (c, d) = (a \times d + b \times c, b \times d)$. If $b = d$, then $(a, b) + (c, b) = (a + c, b)$.

4. Subtraction of Fractions
$(a, b) - (c, d) = (a \times d - b \times c, b \times d)$. If $b = d$, then $(a, b) - (c, b) = (a - c, b)$.

5. Multiplication of Fractions
$(a, b) \times (c, d) = (a \times c, b \times d)$.

6. Division of Fractions

$(a, b) \div (c, d) = (a \times d, b \times c)$, provided that c is not zero. If c is zero, then $(0, d) = (0 \times d, d) = 0$ by rule 2, and division by zero is not allowed with fractions any more than it is with ordinary integers.

Since zero is not allowed as the second member of a fraction's ordered pair, neither b nor d can be zero in rules 1-6, and therefore the product $b \times d$ in rules 3, 4, and 5 is non-zero. Therefore, rules 3-6 define arithmetic for all fractions.

The notation $\frac{a}{b}$ for fractions is functionally identical with the notation (a, b). Since ordered pairs have many other uses in mathematics, the ordered pair notation for fractions will be used only in this chapter. The first member in the ordered pair, like the top number in the $\frac{a}{b}$ notation, is called the numerator of the fraction, and the second member, like the bottom number, is called the denominator.

2.2.1 Numerical Examples

No book on mathematics can be complete without lots of examples to illustrate each point, and this book is no exception.

Rule 1. $(4, 6) = (6, 9)$ because $4 \times 9 = 6 \times 6 = 36$. If all the numbers a, b, c, and d are positive then (a, b) is less then, equal to, or greater than (c, d) in the same way that $a \times d$ is less than, equal to, or greater than $b \times c$. This explains the strange result about increasing the denominator decreasing the fraction mentioned at the beginning of this chapter: $(4, 7)$ is less than $(4, 6)$ because 4×6 is less than 7×4.

Rule 1 also emphasizes the importance of order in ordered pairs: $(5, 8)$ is not equal to $(8, 5)$ because 5×5 is not equal to 8×8.

The second part of Rule 1 is exemplified by $(2, 3)$ and $(4, 6)$. Since $4 = 2 \times 2$ and $6 = 3 \times 2$, the $(4, 6)$ can be rewritten as $(2 \times 2, 3 \times 2)$ which is defined to be equal to $(2, 3)$. Also, $2 \times 6 = 3 \times 4 = 12$, so the first half of Rule 1 also brands the two fractions as equal.

Rule 2: $(7, 1)$ is the same as 7. The process works both ways, so the integer 8 can be written as the fraction $(8, 1)$. This is how integers are combined arithmetically with fractions: they are turned into fractions themselves by this rule, and then Rules 3-6 are applied.

Also, $(14, 7) = (2 \times 7, 7) = 2$ by the second half. The process of figuring out the 2, given 14 and 7, is precisely the division process ($14/7 = 2$), so a fraction is sometimes referred to as an "indicated division," that is, a division that is to be carried out. Of course, in a fraction like $(8, 3)$ there is no integer that we can multiply by 3 to arrive at 8, and this means that $8 \div 3$ is not an integer. This is why fractions were invented; to give answers to problems like $8 \div 3$.

Rule 3: $(1, 2) + (1, 3) = (1 \times 3 + 2 \times 1, 2 \times 3) = (5, 6)$ and $(1, 8) + (3, 8) = (1 + 3, 8) = (4, 8)$, which is also $(1 \times 4, 2 \times 4)$ and therefore equal to $(1, 2)$. The relative complexity of this rule and of the related Rule 4 have given many sixth-graders nightmares. Part of the problem is that teachers start with the simpler second half of the rule, when the two denominators are the same, and then say that when they are not we must make them so, giving rise to the complex first half of the rule and its "least common denominator" variant. This approach stems from the pie-wedge view of a fraction and the desire to present children with something "concrete" to enable them to deal with what is really a powerful abstraction.

Rule 4: $(4, 6) - (6, 9) = (4 \times 9 - 6 \times 6, 6 \times 9) = (0, 54) = $ zero. This shows that $(4, 6) = (6, 9)$ since subtracting one from the other results in zero. The general rule confirms Rule 1 for equality.

Rule 5. $(5, 9) \times (0, 3) = (0, 27) = $ zero. As with integers, zero times anything is zero.

$(8, 5) \times (5, 8) = (40, 40) = 1$. Two numbers p and q that satisfy $p \times q = 1$ are called *reciprocals*. In general (a, b) and (b, a) are reciprocals, and in particular the reciprocal of a is $(1, a)$ since a can be written as $(a, 1)$.

Rule 6: $(5, 8) \div (5, 8) = (5 \times 8, 8 \times 5) = (40, 40) = 1$ again.

$(a, b) \div (c, d) = (a \times d, b \times c) = (a, b) \times (d, c)$. This is the mysterious "invert and multiply" rule given in school for dividing fractions.

$(3, 7) \div (4, 7) = (3 \times 7, 7 \times 4) = (3 \times 7, 4 \times 7) = (3, 4)$ [Rule 1]. Thus, when the denominators are equal, only the numerators need be considered. This can be turned into a derivation of "invert and multiply" by using the common-denominator principle.

2.2.2 Why the Rules Are What They Are

These rules can be derived from the notion that a fraction is an indicated division; that is, the "answer" to a division problem. Just as $21 \div 7 = 3$ means $3 \times 7 = 21$, so $4 \div 5 = (4, 5)$ means that $(4, 5) \times 5 = 4$. Thus, $(4, 5)$ is simply the number that, when multiplied by 5, becomes 4. More generally, (a, b) denotes the quantity that, when multiplied by b, becomes a.

From this, Rule 2 can be derived. $(a, 1)$ is the number that, when multiplied by 1, becomes a. Since it is obvious that this number is just a itself, we identify $(a, 1)$ with a. Similarly, $(a \times b, b)$ is the quantity that can be multiplied by b to produce $a \times b$. It is therefore obvious that $(a \times b, b) = a$.

Now we will derive Rule 1. $(a \times k, b \times k)$ is the number that, when multiplied by $b \times k$, produces $a \times k$. Thus, we can write $(a \times k, b \times k) \times b \times k = a \times k$. By the cancellation law for multiplication, the "$\times k$" can be removed from both sides of this equation, leaving $(a \times k, b \times k) \times b = a$. But we already know that $(a, b) \times b = a$. Therefore, $(a \times k, b \times k) = (a, b)$, which is the second half of Rule 1. For the first half, $(a, b) = (a \times d, b \times d)$ and $(c, d) = (b \times c, b \times d)$ by the second half. If $a \times d = b \times c$ then these expressions are certainly equal. To prove this the other way around, imagine $(a, b) = (c, d)$, that is, (c, d) may be multiplied by b to get a, or, better wet, $(c \div d) \times b = a$. Multiplying both sides of this equation by d, we get $[(c \div d) \times d] \times b = a \times d$. Since multiplication and division are inverse operations, $(c \div d) \times d = c$, so $c \times b = a \times d$, Q. E. D. (Q. E. D. is a fancy Latin abbreviation for "that is what was to be proved.")

For Rule 3, we start with the simple form $(a, b) + (c, b) = (a + c, b)$. Now, $(a, b) \times b = a$, and $(c, b) \times b = c$, so $(a, b) \times b + (c, b) \times b = a + c$. Multiplication is distributive over addition, so the left side of the last equation can be written as $[(a, b) + (c, b)] \times b$. Since this multiplication results in $a + c$, $(a, b) + (c, b) = (a + c, b)$. To add (a, b) to (c, d) we transform the fractions to $(a \times d, b \times d)$ and $(b \times c, b \times d)$ so that the simpler formula can be used. The schoolbooks' insistence on the *lowest* common denominator unnecessarily complicates the traditional algorithm. To add $\frac{1}{4}$ and $\frac{1}{6}$ conventionally, you must first find the l.c.d. (12, and finding it is not trivial), and then divide it by 4 and by 6 to find what to multiply the numerators by. It is much easier to write $\frac{1}{4} + \frac{1}{6} = \frac{1 \times 6 + 4 \times 1}{6 \times 4} = \frac{10}{24}$, or, in ordered-pair notation, $(1, 4) + (1, 6) = (1 \times 6 + 4 \times 1, 6 \times 4) = (10, 24)$.

Rule 4 is just Rule 3 with minus signs replacing plus signs. Multiplication is distributive over subtraction as well as addition.

To derive Rule 5, we begin by writing $(a, b) \times b = a$ and $(c, d) \times d = c$. Now, when we multiply these two equations together, the product of the two left sides is equal to the product of the two right sides, so $(a, b) \times b \times (c, d) \times d = a \times c$. Since multiplication is commutative, the left side of this product equation can be rearranged to get $(a, b) \times (c, d) \times b \times d = a \times c$. In other words, $(a, b) \times (c, d)$ can be multiplied by $b \times d$ to produce $a \times c$, and that is the same as saying $(a, b) \times (c, d) = (a \times c, b \times d)$.

Rule 6 can be derived from Rule 5 and the fact that division and multiplication are opposite operations. $(a, b) \div (c, d)$ is a number that, when multiplied by (c, d), is restored back to (a, b). We will denote this number by (e, f) for the moment. The condition above can be written $(e, f) \times (c, d) = (a, b)$, or, after applying Rule 5, $(e \times c, f \times d) = (a, b)$. It looks as if we have gotten nowhere, but recall Rule 1, which says that for any non-zero k, $(a, b) = (k \times a, k \times b)$. Now we have $(e \times c, f \times d) = (k \times a, k \times b)$. To match both first and second members of these two ordered pairs, k must be $c \times d$. (It must have c as a factor to match $e \times c$, and d to match $f \times d$.) The situation becomes $(e \times c, f \times d) = (c \times d \times a, c \times d \times b)$, from which it is apparent that e is $a \times d$ and f is $b \times c$, that is, $(a, b) \div (c, d) = (a \times d, b \times c)$, which is Rule 6.

2.3 Compound Fractions

Since a fraction is an ordered pair of numbers, and fractions are considered numbers in their own right, it should come as no surprise to find fractions whose members are other fractions. Such fractions could be called "matryoshka" fractions, after the Russian specialty dolls which are hollow and contain smaller dolls, which in turn contain still smaller ones, but "matryoshka" was judged too difficult to pronounce and the term "compound" fraction was used instead. A typical compound fraction is $((2, 3), (4, 5))$, or, in the conventional notation, $\frac{\frac{2}{3}}{\frac{4}{5}}$.

Compound fractions can always be reduced to ordinary ones by carrying out the indicated division that every fraction represents. Thus, $((2, 3), (4, 5)) = (2, 3) \div (4, 5) = (2 \times 5, 4 \times 3)$ by Rule 6 $= (10, 12) = (5 \times 2, 6 \times 2) = (5, 6)$ by Rule 1. Like matryoshka dolls, compound fractions may extend to deeper levels of nesting; these are handled by working from the inside out, like all expressions involving parentheses. Thus, (((2, 3), (4, 5)), ((6, 7), (8, 9))) = ((10, 12), (54, 56)) = ((5, 6), (27, 28)) = (140, 162) = (70, 81).

2.4 Mixed Numbers

A *mixed* number is defined as the indicated sum of an integer and a fraction. In conventional notation, of course, you see such symbols as $4\frac{3}{4}$ to mean $4 + \frac{3}{4}$. Since juxtaposition is used to mean multiplication in algebra, this notation is potentially confusing. The ordered pair notation 4 + (3, 4) will be used in this section.

Arithmetic with mixed numbers is performed in two ways, depending on the type of operation involved. When adding or subtracting, it is easier to leave the fractional parts alone and add the integers. Since addition is commutative, it is possible to rearrange a sum such as [6 + (1, 3)] + [4 + (2, 5)] to 6 + 4 + [(1, 3) + (2, 5)]. Then, the fractions and integers can be dealt with separately to yield 10 + (11, 15). When 6 + (2, 3) and 4 + (3, 5) are added, the resulting fraction is (19, 15). Since 19 > 15, this fraction is greater than (1, 1) = 1 and must be split up into (15, 15) = 1 and (4, 15). The 1 is added to the integer part of the sum to yield a final answer of 11 + (4, 15).

Subtraction raises the possibility of "borrowing" from the integer part of the problem to cover a deficit in the fractional part. Still, though, the problem can be rearranged to treat fractions and integers separately. To subtract [6 + (1, 3)] – [4 + (2, 5)], it must be determined whether or not borrowing will be necessary. The fraction (1, 3) is less than (2, 5) since 5 < 6 (Rule 1, Sec. 2.2.1), so a borrow is needed. This is accomplished by splitting 6 into 5 + 1, changing the 1 into (3, 3), and adding (3, 3) to (1, 3) to obtain (4, 3). These steps can be condensed and generalized to yield the rule: Subtract 1 from the integer and increase the numerator by the amount of the denominator. Thus, 6 + (1, 3) becomes 5 + (4, 3), and the whole problem becomes [5 + (4, 3)] – [4 + (2, 5)]. Now, the fractional and integral parts are separated to produce 5 – 4 + [(4, 3) – (2, 5)]. Notice the signs: the negative signs are attached to the portions of the problem that come from the mixed number to be subtracted. Applying Rule 4, the difference is computed as 1 + (14, 15).

When mixed numbers are multiplied or divided, they must be reduced to simple fractions by performing the indicated additions before further arithmetic can be carried out. This is done by converting the integers into fractions (Rule 2) and applying Rule 3. To multiply [4 + (7, 8)] x [5 + (1, 3)], the 4 + (7, 8) becomes (39, 8) and 5 + (1, 3) becomes (16, 3). Now, (39, 8) x (16, 3) = (13 x 3, 8) x (8 x 2, 3) = (13 x 3 x 8 x 2, 8 x 3) = (26 1) = 26. The 3 x 8 was canceled out pursuant to Rule 1, part 2. The conventional method would use the distributive property and calculate $4 \times 5 + 4 \times \frac{1}{3} + \frac{7}{8} \times 5 + \frac{7}{8} \times \frac{1}{3}$ or $20 + 1\frac{1}{3} + 4\frac{3}{8} + \frac{7}{24}$ and be very surprised when the fractions added up to exactly 1. Since division does not distribute (certainly 20 ÷ 5 cannot be written as 20 ÷ 4 + 20 ÷ 1!), there is no other way to divide mixed numbers other than reducing them to fractions first.

2.5 Decimals and Percents

A decimal is defined as a fraction whose denominator is a power of 10. (The Latin word for 10 is "decem," hence decade, December, and decimal.) Thus, fractions such as (6, 10), (43, 100), and (7, 1000) are decimals. The notation involving the "decimal point" such as .6, .43, and .007 is merely an abbreviation. The total number of digits to the right of the decimal point is the same as the number of zeroes in the power of 10 in the denominator, and the integer that those digits would form without the decimal point is the numerator. Thus, the three decimal-point-containing numbers above can be seen to be abbreviations for the decimal fractions preceding them. Note the zeroes in .007; they do not change the value of the number *without* the decimal point since they come in front of the 7, but they do count as digits to the right of the decimal point and thus increase the denominator of the corresponding fraction. The same definitions apply for numbers containing digits on both sides of the decimal point: 2.3 is the same as (23, 10) and 31.35 equals (3135, 100) Arithmetic with decimals is the same as with the corresponding fractions. Thus, .02 + .04 = (2, 100) + (4, 100) = (6, 100) = .06, .02 x .04 = (8, 10000) = .0008, and .02 ÷ .04 = (200, 400) = (5, 10) =.5. Note that this last fraction could have been reduced to (1, 2) as well. Some ordinary fractions can be easily converted into decimals; thus, (1, 4) = (25, 100) = .25, (1, 5) = (2, 10) = .2, and (1, 8) = (125, 1000) = .125. but most fractions cannot be so converted; (1, 7), for example. If we try to find a number k which satisfies 7 x k = 10, 100, 1000, etc., we will never find an integer value for k. But a mixed number can do; for example 142,857 + (1, 7) yields 1,000,000 when multiplied by 7. Therefore (1, 7) can be written as (142857, 1000000) + ((1, 7), 1000000) and, therefore, the decimal .142857 is very close to (1, 7).

The percent sign means (…, 100); whatever precedes the percent sign is divided by 100. To answer a question such as "What is 30% of 20?", 20 is multiplied by the fraction represented by 30%, which is (30 ,100). The result is (600, 100) or 6. To answer "6 is what percent of 30?", we divide 6 by 30, producing (naturally) (6, 30) = (1, 5)

= (20, 100) or 20%. To answer "35 is 70% of what?" we divide 35 by 70%, that is, (70, 100), to obtain (3500, 70) or 50. To solve other such problems, which may have words in different orders, start by putting the word "percent" or the % sign in the middle. The number or unknown quantity following "of" is placed last, and the remaining quantity goes in front. Then, if the "what" is in front, multiply the other two quantities; otherwise, divide the first known quantity by the second.

Chapter 3 Algebra

The first two chapters have dealt only with arithmetic, a topic that nearly all adults are familiar with. Algebra, however, raises anxieties in many people. These anxieties are ill-founded, since almost all of algebra is merely symbolic arithmetic – that is, arithmetic with letters instead of numbers.

3.1 Arithmetic with Letters

In mathematics, various letters of the alphabet are used to denote numerical quantities. A letter so used is called a *variable*.

Operations involving variables can only be indicated, not actually performed, unless we are given the numerical values that the variables represent. Therefore, we cannot do anything with the expression $a + b$ unless we know what a and b are. However, the $a + b$ can be combined with other expressions to yield a simplified result. Since variables represent numbers, they follow all the arithmetic rules that numbers do. These are summarized below.

Commutative Rules: $a + b = b + a$ and $a \times b = b \times a$.

Associative Rules: $a + (b + c) = (a + b) + c$ and $a \times (b \times c) = (a \times b) \times c$.

Identity Rules: $a + 0 = a$ and $a \times 1 = a$.

Inverse Operations: $a - b = c$ means $b + c = a$ and $a \div b = c$ means $b \times c = a$.

Distributive Rules: $a \times (b + c) = (a \times b) + (a \times c)$ and $(a + b) \times c = (a \times c) + (b \times c)$
$\qquad a \times (b - c) = (a \times b) - (a \times c)$ and $(a - b) \times c = (a \times c) - (b \times c)$.
$\qquad (a + b) \div c = (a \div c) + (b \div c)$ and $(a - b) \div c = (a \div c) - (b \div c)$.

Sign Laws: $-(-a) = a$, $(-a) \times b = -(a \times b)$, $a \times (-b) = -(a \times b)$, and $(-a) \times (-b) = a \times b$.
$\qquad a - (b + c) = (a - b) - c$ and $a - (b - c) = (a - b) + c$.

Rules for Zero: $0 \div a = 0$, $0 \times a = 0$, and $a \div 0$ has no meaning.

All of these rules hold for any values of the letters a, b, and c used in them, provided that the last rule, no division by zero, is strictly enforced. In particular, they hold even if a, b, and c are replaced with other expressions. Of course, what a is replaced with must remain the same wherever it appears in one of the above rules, and the same holds for b and c as well.

3.1.1 Notation

When several quantities are multiplied together, and no more than one of them is an ordinary number, the multiplication is indicated by juxtaposition; that is, the quantities are written right next to each other, without any space between. Hence, the notation abc means the product of a, b, and c, not a single variable. If a numeral is one of the quantities involved, it is written first, more by tradition than for any other reason. Thus, the product of x, seven, and z is written $7xz$. (It is also a tradition to write the variables in alphabetic order.) The reason for the abandonment of the multiplication sign is that it looks too much like the letter x.

3.1.2 Order of Operations

When there are no parentheses used, the first operations to be performed are multiplication and division, from left to right, and then addition and subtraction, also from left to right. Thus $4 + 5 \times 6 = 4 + 30 = 34$, and $5 - 6 \div 3 = 5 - 2 = 3$. The same rule holds for letters; $a + b \times c$ means the sum of a and the product of b and c. Note that this expression can be written $a + bc$; the juxtaposition of b and c reinforces the fact that b and c must be multiplied first before a can be added.

When parentheses are present, the operations within them must be performed first. Parentheses must be expanded from the inside out. Thus, $(a + b)c$ means the product of c with the sum of a and b, and $p - (q + r)$ means that the sum of q and r is to be subtracted from p. In $(5 + [3 \times 6] - [4 + (9 \div 3)]) - (3 + 7) \div 2$ the innermost parenthesis, $(9 \div 3)$, is expanded first. Successive evaluation steps gradually reduce the complexity of the expression until it is reduced to the single number 11. Of course, if letters are involved, we may not be able to proceed that far. But if the same variables occur in several places, we may simplify the expression considerably.

$(5 + [3 \times 6] - [4 + 3]) - (3 + 7) \div 2$
$(5 + 18 - [4 + 3]\} - (3 + 7) \div 2$
$(5 + 18 - 7) - (3 + 7) \div 2$
$(23 - 7) - (3 + 7) \div 2$
$16 - (3 + 7) \div 2$
$16 - 10 \div 2$
$16 - 5$
11

3.1.3 Exponents

Back in Chapter 1, the notion of exponents was introduced. An exponent simply counts the number of times that a certain number has been multiplied together, and this holds for variables as well. Just as $6^3 = 6 \times 6 \times 6$, so $a^3 = a \times a \times a$ or aaa. Exponents follow several rather obvious laws.

First Law of Exponents: $a^m a^n = a^{m+n}$.
Second Law of Exponents: $(a^m)^n = a^{mn}$.
Third Law of Exponents: $a^m \div a^n = a^{m-n}$ if $m > n$
$$= 1 \text{ if } m = n$$
$$= \frac{1}{a^{n-m}} \text{ if } m < n.$$

The first two of these laws are valid for any value of a, but the third does not have meaning when a equals zero. It should be obvious that multiplying the product of m a's by the product of n a's gives you the product of $(m + n)$ a's, and that multiplying together n factors, each of which is the product of m a's, produces the product of mn a's. For the third law, consider a (conventional) fraction with m a's on top and n a's on the bottom. Canceling out as many a's as possible leaves either $m - n$ on top or $n - m$ on the bottom, depending on which of m or n is greater. If $m = n$, then the top and bottom of the fraction are identical, and the fraction therefore equals one.

3.1.4 Negative Numbers and Absolute Value

It was said in grade school that 2 from 3 leaves 1, but you can't take 3 from 2. Now, mathematicians hate simple problems like $2 - 3$ that seem to have no answers. In the case of numerical problems, their solution has historically been to invent a new kind of number that can serve as the answer to the current type of "impossible" problem. For $2 - 3$, the new kind of number is a *negative* number.

Negative numbers are numbers with an arrow of direction built into them, which points backward. "Ordinary" or *positive* numbers have a forward arrow attached. When the order of the quantities in a subtraction problem is reversed, so is the arrow of the result. Formally, $(b - a) = -(a - b)$. It should be stressed that the second of the three "minus signs" in this equation is different in meaning from the others; it means "reverse the direction of the arrow (the sign) of the number that follows me." Here, the following number is the parenthesized expression $(a - b)$, in which the minus sign has its usual meaning of subtraction.

Thus, $2 - 3 = -(3 - 2) = -1$. Similarly, $3 - 7 = -4$ and $2 - 10 = -8$. The answers are to be read "negative one," "negative four," and "negative eight," not "minus one," etc.

Negative numbers, of course, are not just to be found as answers to problems. The rule for adding numbers of like sign (both positive or both negative) is: Add their absolute values (to be defined momentarily) and attach the common sign to the sum. Therefore, $(-3) + (-5) = -8$ and $(+6) + (+3) = +9$. (The numbers involved are placed in parentheses to distinguish their signs from the sign denoting the operation.) To add numbers of different signs, subtract the smaller absolute value from the larger, and attach the sign of the number with the larger absolute value to the result.

And just what is this "absolute value" that has been used in these rules? It is the number stripped of its sign. Only the length of the arrow, not its direction, is important for absolute values. The absolute values of +5, +8, -7 and -3 are 5, 8, 7, and 3, respectively. Knowing ordinary arithmetic, without signs, and the concept of absolute value enables one to apply the above rules to add any two numbers, positive or negative. The second rule is illustrated by the following examples:

$(-7) + (+4)$. The absolute values are 7 and 4, $7 - 4 = 3$, and the larger absolute value (the 7) has a minus sign attached, so $(-7) + (+4) = -3$.

$(+8) + (-6)$. The absolute values are 8 and 6. Since $8 - 6 = 2$ and the 8 has a positive sign, $(+8) + (-6) = +2$.

$(-5) + (+7) = +(7 - 5) = +2$.

$(+2) + (-9) = -(9 - 2) = -7$. Note that in these two examples, it is the second number that has the larger absolute value. That does not matter; the rules for addition do not distinguish between first and second, ensuring that signed-number addition is just as commutative as ordinary addition.

In $(-8) + (+8)$, the two absolute values are the same. There is no "larger" one, but subtracting either one from the other results in zero. By convention, $+0$ and -0 are the same thing (except to some old computers), so it does not matter which sign you put on the result. Two numbers whose sum is zero are called *additive inverses* of each other; therefore, -8 and +8 or, more generally, $-a$ and $+a$, are additive inverses. Zero is its own additive inverse.

As for subtraction, we simply define it in terms of addition: $a - b = a + (-b)$. This is another technique that mathematicians enjoy: defining complex operations in terms of simpler ones. Therefore, $(-5) - (+3) = (-5) + (-3) = -8$, and $(+6) - (-7) = (+6) + (+7) = +13$. Note that $-(-7) = +7$ (Sec. 2.1) for the same reason that two U-turns leave you traveling back in your original direction.

To multiply and divide signed numbers, the simple rule "like signs = positive result, unlike signs = negative result" is used. Thus, $(-3)(+7) = -21$ and $(-45) \div (-9) = +5$. In multiplying several signed numbers together, the result is positive if an even number of negative factors are involved (zero is even!), and negative if an odd number of such factors are present.

3.2 Solving Mathematical Mysteries

With these advanced arithmetical and notational preliminaries out of the way, we can proceed to the heart of algebra: solving mathematical mysteries.

A mathematical mystery is an equation containing one letter, such as $x + 4 = 6$. (It is customary to use letters from the end of the alphabet to stand for numbers in mystery-type equations, thanks to mathematician Rene Descartes.) Your job, when confronted by such an equation, is to determine the value or values that can be substituted for the variable so as to make the equation a statement of arithmetic fact. This is called *solving* the equation. In the example $x + 4 = 6$, it is obvious that 2 may replace x to produce the incontrovertible $2 + 4 = 6$. What is not so obvious is whether other numbers may be substituted for x as well. It will do you no good to observe that 1, 3, 4, …, 57, 58, etc. do not produce 6 when 4 is added to them; there are fractions and other types of numbers to consider as well. To prove that 2 is the *only* number that solves $x + 4 = 6$, you must transform the equation into one so simple that it is obvious that only 2 will fit. The equation $x = 2$ is such an equation. In solving mathematical mysteries, equations of this type will be our goal. (They are sometimes called *trivial* equations, so the process of arriving at one can be called "Trivial Pursuit.")

3.2.1 The Care and Feeding of Equations: The Balance Rule

In order to transform an equation such as $x + 4 = 6$ into a trivial equation, one must learn the rules for transforming equations. Equations are delicate creatures representing a balance between two quantities. The fundamental rule is that nothing can be done which might endanger that balance. Whatever is done to one side of the equation, precisely the same thing must be done to the other side.

With that all-important provision in mind, almost anything goes: we may add, subtract, multiply, or divide by any quantity – except zero. Division by zero has long since been outlawed, while multiplying by zero destroys all information contained in the equation. Adding and subtracting zero is legal, but useless, since either operation does absolutely nothing to the equation. Given, for example, the equation $y - 7 = 8$, we can add 4 to both sides of it, producing $y - 7 + 4 = 8 + 4$ or $y + (-7) + 4 = 12$, which is reduced to $y + (-3) = 12$ or $y - 3 = 12$. Or, we could multiply the equation by 5: $5(y - 7) = 5 \times 8$ or $5y - 5 \times 7 = 40$, which in turn can be written as $5y - 35 = 40$. Neither one of these transformations has produced a trivial equation, because neither transformation was what was necessary to strip away the camouflage surrounding y in the original equation.

3.2.2 Stripping Away Mathematical Camouflage

Equations like $x + 4 = 6$, $y - 7 = 8$, $3p + 7 = 25$, $6(4 - z) + 5(z + 3) = 7z - 13$, and $\frac{8}{n+4} - \frac{5}{n-8} = \frac{6}{2n+3}$ differ from trivial equations such as $x = 2$ and $y = 15$ in ways other than the obvious one of length. In each case, the naked variable is camouflaged by being combined with numbers and/or occurring more than once. In transforming these equations, one must select the transformation(s) that remove this camouflage.

The insight into the choice of transformations is *opposite operations*. To undo addition, subtract; to undo division, multiply, etc. In $x + 4 = 6$, the x is camouflaged by having 4 added to it. To undo addition, subtract: we subtract 4 from (both sides of) the equation to obtain $x + 4 - 4 = 6 - 4$ or $x = 2$, since $4 - 4 = 0$ and $6 - 4 = 2$. After performing the arithmetic, we are left with a trivial equation and it is indeed $x = 2$. In Sec. 3.2, we observed that 2 satisfies the equation $x + 4 = 6$ (that is, 2 could replace x and produce an arithmetic fact); now we see that no other number satisfies the equation obtained by a transformation of $x + 4 = 6$. Since transformations are designed to preserve the equality, any equation arrived at by a legal transformation from another one has the same solution as the original one, and therefore 2 is the only solution to $x + 4 = 6$. (There are some transformations that can slightly change the numbers that satisfy an equation. These will be mentioned as they crop up and must be watched carefully.) In $y - 7 = 8$, y is camouflaged by subtracting 7, so we add 7 to both sides of the equation to obtain $y = 15$ $(8 + 7)$. In $3p + 7 = 25$, there are two layers of camouflage; first p is multiplied by 3, then 7 is added to the result. Stripping off the top layer first, we subtract 7 to obtain $3p = 18$ and then divide by 3 to get $p = 6$. Since $3 \times 6 + 7 = 25$ is a true statement, 6 does indeed satisfy the original equation. This step is vital, especially in more complicated equations where the possibility of a mistake or one of those slightly changing transformations is greater.

The next equation at the top of this page is still more complicated because it contains parentheses and multiple occurrences of the variable z. The parentheses must be dealt with first, by the distributive law: $6(4 - z) = 6 \times 4 - 6z$ or $24 - 6z$, and $5(z + 3) = 5z - 15$. The equation is now $24 - 6z + 5z + 15 = 7z - 13$. We must now combine the various terms on the left side to as great an extent as possible. The 24 and 15 can be added to get 39, and as for the $-6z$ and $5z$, they can be combined by using the distributive rule going backwards: $-6z + 5z = (-6 + 5)z = -1z$ or simply $-z$. We now have $39 - z = 7z - 13$. The presence of two z's represents an element of camouflage. We get rid of the z with the smaller numerical factor (here the $-z$ on the left) by adding its opposite (that is, $+z$) to both sides of the equation. We now have $39 - z + z = 7z - 13 + z$. $7z + z$ is $7z + 1z$ or $(7 + 1)z$, since $7 + 1 = 8$, the equation becomes $39 = 8z - 13$ in which z only appears once. The z is multiplied by 8 and has 13 subtracted from the product, so we remove this camouflage by adding 13 and then dividing by 8 to obtain $\frac{52}{8} = z$ or $z = 6\frac{1}{2}$. Inserting this value in the original equation results in $6\left(4 - 6\frac{1}{2}\right) + 5\left(6\frac{1}{2} + 3\right) = 7 \times 6\frac{1}{2} - 13$ or $-15 + 47\frac{1}{2} = 32\frac{1}{2}$. Since $47\frac{1}{2} - 15 = 32\frac{1}{2}$, the solution checks.

Equations containing fractions are dealt with in the same way as non-fraction equations; the first step is to multiply both sides of the equation by all the denominators of the fractions. This will enable the denominators to be canceled, reducing the equation to a fractionless one. A simple example is $\frac{x}{2} + 3 = \frac{2x}{3}$. The product of the denominators is six; multiplication by that number results in $\frac{6x}{2} + 18 = \frac{12x}{3}$. Since $6 \div 2 = 3$ and $12 \div 3 = 4$, this can be rewritten as $3x + 18 = 4x$. We remove one of the two x's by subtracting $3x$, leaving us with $18 = x$. If $x = 18$, then $\frac{x}{2} = 9$ and $\frac{2x}{3} = 12$; since $9 + 3 = 12$, this solution checks as well.

When variables occur in denominators, extra care must be used. Whenever an equation is multiplied by an expression containing a variable, the possibility of an additional solution (called an *extraneous* solution) is created. (Example: $3x = 6$ is solved only by $x = 2$, but $3x^2 = 6x$ is satisfied by both zero and 2.) In addition, we must remember that no denominator is allowed to be zero. In the equation $\frac{6}{s-2} = 4$, the expression $s - 2$ is the denominator. Hence, $s - 2$ cannot be zero, and therefore s cannot be 2. If the equation is transformed into $s = 2$, which has 2 as its only solution, we must conclude that the original equation has no solutions at all! Fortunately, this equation does not present such a difficulty; it is transformed successively into $6 = 4(s - 2)$, $6 = 4s - 8$, $14 = 4s$, and, finally, $s = 3\frac{1}{2}$. This solution checks because $\frac{6}{3\frac{1}{2}-2} = \frac{6}{1\frac{1}{2}} = 6 \div \frac{3}{2} = 6 \times \frac{2}{3} = \frac{12}{3} = 4$.

There are equations out there with no solutions. One of the most obvious is $x + 3 = x$. Certainly adding 3 to x will change it, and the left side will not be equal to the right side. We can make this even more obvious by applying algebra: subtract x from both sides, and the "equation" becomes $3 = 0$. This is false, of course, so we conclude that the original equation is a contradiction in terms; that is, it has no solution.

There are also more subtle equations that lack solutions. One of them is $\frac{2x}{x-3} = 7 + \frac{6}{x-3}$. When we look at this equation we note immediately that 3 is not a possible solution for it, since it makes the denominator $x - 3$ equal to zero. With this in mind, we transform the equation as follows:

$2x = 7(x - 3) + 6$	(multiplying both sides by $(x - 3)$; the equation now contains no fractions)
$2x = 7x - 21 + 6$	(expanding the parentheses)
$2x = 7x - 15$	(since $(-21) + 6 = -15$)
$0 = 5x - 15$	(subtracting $2x$ from both sides)
$5x = 15$	(adding 15 to both sides, and switching left and right)
$x = 3$	(dividing both sides by 5)

We have transformed an equation that cannot have 3 as a solution into one which has 3 as its only solution. The only troublesome step was the first one, in which we multiplied the equation by a variable-containing expression. (Indeed, the value 3, when substituted in the first transformed equation, produces the arithmetic fact $2 \times 3 = 7 \times (3 - 3) + 6$, or $6 = 6$.) Whenever an equation is multiplied by an expression containing a variable, the value of that variable which makes the expression zero is introduced as a possible additional solution to the transformed equation. In this example, the value 3 for x is introduced as an additional solution. Since this is the only solution to the transformed equation, we must conclude that the original equation had no solutions at all.

3.2.3 Word (Story) Problems

Most mathematical mysteries are set forth in English rather than the language of algebra. These are called word or story problems. Word problems are frequently the bane of high-school algebra students, and a cartoon by Gary Larson shows "Hell's Library" filled with books of story problems. But they are not all that difficult. The toughest part of a story problem is translating from English to algebra. Here is how this is done:

English	Algebra
The value of d dimes	$10d$ cents
Mary's age three years ago, if she is m years old today	$m - 3$ years
The cost of g gallons of gas, if one gallon is \$2.09	$2.09g$ dollars or $209g$ cents
The amount of alcohol in x quarts of a 40% (80 proof) solution	$0.4x$ quarts
The distance a train travels in h hours, if it is traveling y miles per hour	yh miles

As you can see, most of these situations involve multiplying two quantities, one of which is generally known and the other unknown. The situation described in the problem will tell you which quantities are multiplied and which are to be added or subtracted. Usually, two products are added and the result is to be made equal to some constant. But every problem is different, so no one method can be used to solve all of them.

In "mixture" problems such as "How many pounds of coffee worth \$3.50 a pound need to be mixed with two pounds of \$5 a pound coffee to form a mixture worth \$4 a pound?," there are usually three quantities, two of which are blended together, that is, added, to produce a third. Here, if c is the number of pounds of \$3.50 coffee, then that coffee is worth $3.5c$ dollars. The two pounds of \$5 coffee are worth $2 \times 5 = 10$ dollars. The mixture then has $c + 2$ pounds of coffee, which is to be valued at $4(c + 2)$ dollars. Therefore, $3.5c + 10 = 4(c + 2)$. This equation can be transformed successively into $3.5c + 10 = 4c + 8$, $10 = 0.5c + 8$, $2 = 0.5c$, and, finally, $c = 4$. Therefore, 4 pounds of coffee worth \$3.50 a pound can be blended with two pounds of \$5 coffee to form a mixture worth \$4 a pound. This solution is checked in the original problem, not the equation derived from it: 4 pounds of $3.50 coffee are worth $14 ($3.50 x 4), two pounds of $5 coffee are worth $10, and $4 + 2 = 6$ pounds of the mixture are worth $14 + 10 = 24$ dollars. Since $24 = 6$ x 4, the mixture is worth $4 a pound, so the solution checks.

Now we consider age problems. Example: John is twice as old as Mary. Four years ago, he was three times as old. How old are they now? The first step is to select letters to stand for their ages. Let j be John's age now and m Mary's age. The first sentence tells us that $j = 2m$. Now, four years ago John was $j - 4$ years old and Mary $m - 4$, so the second sentence means $j - 4 = 3(m - 4)$. We now have two letters to deal with rather than one, but we can get rid of j because the first equation says $j = 2m$. Therefore, we substitute $2m$ for j in the second equation to get $2m - 4 = 3(m - 4)$. This means that $2m - 4 = 3m - 12$, $-4 = m - 12$, and $8 = m$. Since $j = 2m$, we have $j = 16$. Therefore, John is 16 years old and Mary 8. We check this solution by noting that four years ago, John was 12 and Mary 4, and $12 = 3$ x 4. A different sort of age problem has Bill and Jane's ages adding to 26. Four years ago, Jane was twice as old as Bill. How old are they now? The first sentence tells us that $b + j = 26$, and the second says $2(b - 4) = j - 4$. To solve this pair of simultaneous equations, we use the first one to write j in terms of b: $b + j = 26$ means $j = 26 - b$. Therefore, $2(b - 4) = 26 - b - 4$, or $2b - 8 = 22 - b$. This equation is solved by $b = 10$, so $j = 26 - b = 16$ and Bill is 10 years old and Jane 16. Four years ago, Jane was then 12 and Bill 6, and $12 = 2$ x 6, so this solution checks.

A similar situation comes up in problems involving coin collections. A collection of nickels and quarters contains 10 coins and is worth $1.70. How many nickels and how many quarters are in the collection? We let n be the number of nickels in the collection, and q be the number of quarters. The first fact tells us that $n + q = 10$, and the second tells us that $5n$ (the value in cents of n nickels) $+ 25q = 170$. We use the first equation to give us n in terms of q: $n = 10 - q$. Substituting $10 - q$ for n in the second equation gives $5(10 - q) + 25q = 170$, or $50 - 5q + 25q = 170$, or $50 + 20q = 170$, so $20q = 120$ and $q = 6$. Hence $n = 10 - q = 4$, and the collection has 4 nickels and 6 quarters in it. Are 4 nickels and 6 quarters worth $1.70? 4 x $5 + 6$ x $25 = 20 + 150 = 170$, so the solution checks

Next we consider problems involving distances. The formula to remember is Distance = Velocity x Time. Suppose there are two cars at opposite ends of a 100-mile-long track. One car drives at 50 mph east, the other at 70 mph west. Both cars start at the same time. When and where will they pass one another? We let t stand for the number of hours that it will take the cars to reach the same point. In this time, the slower car travels $50t$ miles and the faster one $70t$. Since they are traveling in opposite directions, the sum of these two distances must equal the original separation between the cars or 100 miles. Therefore, $50t + 70t = 100$ and $t = 5/6$. So, the two cars will pass each other in 50 minutes. To answer the second question (where) we use the answer to the first one. In 50 minutes, the eastbound car will travel $5/6$ x $50 = 41\frac{2}{3}$ miles, so the two cars pass each other $41\frac{2}{3}$ miles east of the west end of the road. We check this by noting that $5/6$ x $70 = 58\frac{1}{3}$, and $41\frac{2}{3} + 58\frac{1}{3} = 100$. Now suppose we again have two cars at opposite ends of a 100-mile road. This time, one car starts east at 50 mph, and, somewhat later, the other car starts west at 70 mph. The two cars meet exactly halfway between the ends of the road. How many minutes did the first car have as a head start? We let m be the number of minutes of head start. The faster car traveled 50 miles in $5/7$ of an hour because it was going 70 mph, and the slower car traveled 50 miles in $5/7 + m/60$ hours. Since it travels at 50 mph, $(5/7 + m/60)$ x $50 = 50$, so $5/7 + m/60 = 1$. To clear the fractions, we multiply by 7 x $60 = 420$ to get $300 + 7m = 420$, or $m = 17\frac{1}{7}$ minutes. We check this solution by first calculating how far the slower car will travel in those $17\frac{1}{7}$ minutes: $17\frac{1}{7} \times 50 \div 60 = 14\frac{2}{7}$ miles. It therefore has $50 - 14\frac{2}{7} = 35\frac{5}{7}$ miles to travel to get to the midpoint of the road and the fast car still has 50 miles. Will they take the same time to travel those distances? $35\frac{5}{7} \div 50 = \frac{5}{7}$ and $50 \div 70 = \frac{5}{7}$, so both cars take the same amount of time, and the solution checks.

Lastly, we consider problems involving consecutive integers. The key here is that if the smallest of consecutive integers is n, the next is $n + 1$, then $n + 2$ and so on. For example, four consecutive integers add up to 110. What are they? We have $n + (n + 1) + (n + 2) + (n + 3) = 110$ since four integers are being added. This simplifies to $4n + 6 = 110$ or $n = 26$. Hence, the integers are 26, 27, 28, and 29. This answer is checked simply by adding these four numbers; they do indeed add up to 110. Sometimes, a problem mentions consecutive odd or even integers. The gap between consecutive odd (or even) integers is 2, as with 11 and 13. Therefore, if the smallest is n, the next is $n + 2$, then $n + 4$ and so on. Suppose we now have four consecutive odd integers that add up to 112. Representing the first by x, the others are $x + 2$, $x + 4$, and $x + 6$. Therefore, $4x + 12 = 112$ and $x = 25$. Hence, the odd integers are 25, 27, 29, and 31. These four numbers do add up to 112, so the solution checks.

Occasionally, you encounter problems involving the product, not the sum, of consecutive integers. For example, the product of two consecutive integers is 182. What are they? This situation is represented by the equation

$x(x + 1) = 182$. This is a quadratic equation which cannot be solved by the techniques you have read thus far. Turn to Section 3.6 if you wish to find out how to solve this equation. (Hint: the answer to this problem is either 13 and 14 or -14 and -13.)

3.3 Polynomials

We now abandon the subject of equations for a while, and take up the question of manipulating algebraic expressions, specifically the kind known as *polynomials*. An expression consisting of a power of a variable multiplied by a constant is called a *monomial*; the sum of one or more monomials is called a polynomial. (In Greek, "poly" means "many," as in polymer and polygamy.) The constant can be either a numeral, as in $-5x^2$, or a letter as in ay^3, in which case it is called a *literal* constant. Letters from the front of the alphabet are used to denote literal constants, another idea from Descartes. A constant by itself is also considered a monomial.

When a polynomial consists of two or more terms, the same variable is usually used in all the terms that contain variables, although this rule is not hard and fast. Examples of polynomials include $x - 3$, $ax + b$ (x is the variable), $ax + by + cz$ (x, y, and z are the variables), $5x^2 - 7x + 4$, $\frac{y^3}{4} - \frac{3y^2}{5} + 6$, and $ax^2 + 2bxy + cy^2$ (x and y are the variables). Note that the last example contains the product of two variables; this is legal. The following, in which u, v, and w are the variables, are not polynomials: $3(u + 5)$ because of the parentheses, $6v^2 + 3v - 8 + \frac{4}{v}$ because a variable is in a denominator, $\frac{6v^2 + 3v - 8}{4}$ because the entire expression is just one term (the numerator is a polynomial, however), and $(w + 4)^3 - 4(w + 4)^2 + 5(w + 4) - 8$ because the expression includes powers of $(w + 4)$ instead of just w. This expression is sometimes referred to as a polynomial in $(w + 4)$, and if $w + 4$ is considered as a single object, it can be so called. But it is not a polynomial in w.

The *degree* of a monomial is found by adding all the exponents on all the variables that appear in it. The degree of $2xy^3$ is 4 if both x and y are variables, but only three if x is regarded as a literal constant. (Recall that a "missing" exponent is a one.) The degree of a polynomial is the highest of the degrees of the monomials that make up the polynomial. The degree of an isolated constant is zero. Hence, the degree of $5x - 4$ is one, and of $6x^2y + 5xy + 4y^3 - 7x$ is three if both x and y are variables.

Polynomials are also classified by their length. A monomial has just one term, a *binomial* two, and a *trinomial* three. Like "poly," "mono," "bi," and "tri" are Greek in origin. Separate classifications are generally not used with polynomials of more than three terms.

3.3.1 Addition and Subtraction

Adding two polynomials is fairly easy: Just concatenate the terms, and then collect similar terms and combine them through the distributive property. Two terms are similar if they contain the same variables raised to the same powers. Thus $4x^2$ and bx^2 are similar if b is a literal constant, and their sum is $(4 + b)x^2$. If y and x are both variables, $4x^2$ and yx^2 are not similar. These terms can still be added to get $(4 + y)x^2$, but this is not a polynomial because y is being considered a variable. To subtract polynomial B from polynomial A, change the signs of all the terms in B, including the first one, and then add the changed B to A. Some examples: $(3x^2 + 7x + 4) + (8x^2 - 5x - 9) = 11x^2 + 2x - 5$. Here, the $3x^2$ and $8x^2$ are similar, as are $7x$ and $-5x$ as well as 4 and -9. $(-3x^2 - 6xy + 7y^2) + (5x + 3y) = -3x^2 - 6xy + 7y^2 + 5x + 3y$; there are no similar terms. $(9w^3 + 6wz^2 - 4z) - (6wz + 5z^2 - 4z) = (9w^3 + 6wz^2 - 4z) + (-6wz - 5z^2 + 4z) = 9w^3 + 6wz^2 - 6wz - 5z^2$. The only pair of similar terms is $-4z$ and $+4z$, which sum to $(-4 + 4)z$ or $0z$.

3.3.2 Multiplication

To multiply polynomials, we must start by multiplying monomials. This is done by arithmetically multiplying the numerical constants, and concatenating the literal constants and variables. Letters occurring in both monomials have their exponents added (First Law of Exponents, Sec. 3.1.3). Thus, $(2xy)(3ay^2) = 6axy^3$ and $(-4u^2v^2)(3bu^3vw) = -12bu^5v^3w$. Note that a variable without an exponent is treated as if it had an exponent of one. This is common sense in light of the definition of exponents as numbers that count how many times a variable or expression is multiplied together with itself: $x^3 = (x)(x)(x)$ and $x^2 = (x)(x)$, so it is logical that x^1 is just (x). It should be obvious that the product of two monomials is another monomial, and fairly obvious that the degree of the product is the sum of the degrees of the factors.

The product of two polynomials is the sum of the monomials obtained by multiplying each term of one by every term of the other. Thus, $(a + b + c)(x + y + z) = ax + ay + az + bx + by + bz + cx + cy + cz$. Some of the terms in the product may be similar, as in $(x + 5)(x + 7) = xx + 5x + x7 + (5)(7) = x^2 + (5 + 7)x + 35$ or $x^2 + 12x + 35$. Occasionally, similar terms may cancel each other out, as in $(x - y)(x^2 + xy + y^2)$. The products of the individual terms are x^3, x^2y, xy^2, $-x^2y$, $-xy^2$, and $-y^3$. The x^2y and $-x^2y$ cancel each other out, as do xy^2 and $-xy^2$, leaving just $x^3 - y^3$ as the final product. Products involving similar terms and canceling out are likely to be a part of the "special products and factors" sections of algebra textbooks. This book is not meant to be a textbook, but it does have a special products and factors section.

3.4 Special Products

The following results may be verified by applying the above rule.

[3.1] $(a + b)(a - b) = a^2 - b^2$.
[3.2a] $(a + b)^2 = a^2 + 2ab + b^2$.
[3.2b] $(a - b)^2 = a^2 - 2ab + b^2$.
[3.3a] $(a + b)^3 = a^3 + 3a^2b + 3ab^2 + b^3$.
[3.3b] $(a - b)^3 = a^3 - 3a^2b + 3ab^2 - b^3$.
[3.4] $(x + a)(x + b) = x^2 + (a + b)x + ab$.
[3.5] $(ax + b)(cx + d) = acx^2 + (ad + bc)x + bd$.
[3.6] $(a + b + c)^2 = a^2 + b^2 + c^2 + 2ab + 2ac + 2bc$.

Many of these formulas lead directly to multiplication shortcuts of the type described in Chapter 1. For example, the product of 78 and 82 is calculated by writing 78 as $80 - 2$ and 82 as $80 + 2$ and then applying [3.1], with 80 replacing a and 2 being substituted for b. The square of 106 is calculated by writing it as $100 + 6$ and applying [3.2a].

It should be noted that the product of two polynomials is another polynomial, and the degree of the product polynomial is the sum of the degrees of the factors, just as for monomials. (In [3.1] to [3.3b], a and b are the variables, in [3.4] and [3.5], x is the only variable and a through d are constants, and in [3.6] a, b, and c are all variables.)The most important formula is [3.5], which can be considered the general formula for multiplying two binomials. The formula can be memorized by writing the two binomials one above the other, and connecting each term of the top one to each term of the bottom, like this: (See Fig. 3.1.) The terms of the binomials are connected by four straight lines which form a bow-tie-like shape, so this can be called the bow-tie method of binomial multiplication. For Tolkien buffs, the lines form a rune in the Angerthas Moria (dwarvish alphabet) that represents the sound of English d.

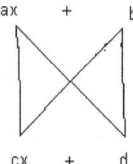

Fig. 3-1. The bow-tie method
of polynomial multiplication.

Some of these formulas can be illustrated graphically, for example [3.4]. The area of a rectangle is the product of its base and height, so the large rectangle has area $(x + a)(x + b)$ and is split into four smaller ones having areas x^2, ax, bx, and ab. (See Fig. 3.2.)

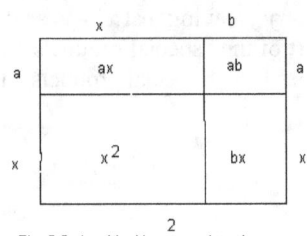

Fig. 3-2. $(x+a)(x+b)=x^2 +ax+bx+ab$.

The illustration of [3.1] is a little more complicated. (See Fig. 3.3.) The area shaded with \\\\ or XXXX is $(a + b)(a - b)$. It includes all of a^2 except the unshaded strip at the top of area ab. It also includes the strip at right (shaded XXXX) which falls short of ab by the amount b^2 (shaded ////) at the upper right. Therefore, $(a + b)(a - b) = a^2 - ab + (ab - b^2) = a^2 - b^2$.

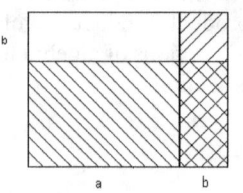

Fig. 3-3. $(a+b)(a-b)=a^2-b^2$.

In addition to the formulas [3.1] – [3.6], a few other important results should be saved for future use. They are:

[3.7a] $(a - b)(a^2 + ab + b^2) = a^3 - b^3$.

[3.7b] $(a - b)(a^{n-1} + a^{n-2}b + ... + ab^{n-2} + b^{n-1}) = a^n - b^n$.

[3.8a] $(a + b)(a^2 - ab + b^2) = a^3 + b^3$.

[3.8b] $(a + b)(a^{n-1} - a^{n-2}b + a^{n-3}b^2 - ... + a^2b^{n-3} - ab^{n-2} + b^{n-1}) = a^n + b^n$, if n is odd.

[3.9] $(a^2 + 2ab + 2b^2)(a^2 - 2ab + 2b^2) = a^4 + 4b^4$.

The dots in the middle of [3.7b] and [3.8b] mean that the exponent on a decreases by 1, and that on b increases by 1, as we proceed from each term to the next, and in [3.8b] that the signs of the terms alternate as well.

Each of these formulas has arithmetic implications. If, for example, we let $a = 1$ and $b = 10$ in [3.9], we get 221 x 181 = 40,001. The reverse substitution $a = 10$, $b = 1$ in [3.8a] results in 11 x 91 = 1,001; since 91 = 13 x 7 (by [3.1] with $a = 10$ and $b = 3$), the magic-number relationship 7 x 11 x 13 = 1,001 is derived. The number 1,729 can be written as the sum of two cubes in two different ways: $1,729 = 1,000 + 729 = 10^3 + 9^3$, and $1,729 = 1,728 + 1 = 12^3 + 1^3$. Application of [3.8a] results in 1,729 = 19 x 91 (palindromic), and 1,729 = 13 x 133, where the first factor is contained in the second. Applying [3.1] to $133 = 169 - 36 = 13^2 - 6^2$ produces 133 = 19 x 7, and we have already seen that 91 = 7 x 13, so both factorings reduce to 1,729 = 7 x 13 x 19. Therefore, when $a = 6$, $(a + 1)(2a + 1)(3a + 1) = (2a)^3 + 1$. This is also true when $a = 0$ or $a = -\frac{1}{2}$.

3.5 Factoring Polynomials

It is frequently desirable to reverse the multiplication process and break a polynomial up into its factors. Just as it is sometimes useful to be able to split 35 into 5 and 7, it is also useful to split $x^2 + 7x + 12$ into $(x + 3)(x + 4)$, for instance, in solving the equation $\frac{2}{x+3} = \frac{10}{x^2+7x+12}$. (This equation is solved by $x = 1$, which makes both sides equal to $\frac{1}{2}$.)

The first step in factoring a polynomial is to remove the largest monomial factor. This will be a monomial that divides all the terms in the given polynomial. In $12ax^2 + 16x$, the factor $4x$ divides both terms, so the polynomial can be factored as $4x(3ax + 4)$. In $25a^2b^3 - 35a^2b^2 + 20ab^3$, the number 5 divides the numerical coefficients 25, -35, and 20, and a and b^2 or higher powers of a and b appear in each term, so the monomial $5ab^2$ is the common factor, and the polynomial can be written as $5ab^2(5ab - 7a + 4b)$.

Most factoring problems involve quadratics, that is, polynomials of degree two. Reversing the formulas [3.1], [3.2a], [3.2b], [3.4] and [3.5] provides the basic framework, with one little problem: We are generally given, not $a^2 - b^2$ but $9x^2 - 16$ to factor; likewise $x^2 + 6xy + 9y^2$ instead of $x^2 + 2xy + y^2$, and $x^2 - 2x - 15$ instead of $x^2 + (a + b)x + ab$. Our job is to match the polynomial we are to factor with the appropriate right side of [3.1], [3.2], [3.4], or [3.5] by finding values of a, b, (c, and d) that produce the proper expression. Thus, $9x^2 - 16$ is matched up with [3.1] by matching $9x^2$ to a^2 or $3x$ to a and 16 to b^2 or 4 to b. The left side of [3.1] now tells us that $9x^2 - 16 = (3x + 4)(3x - 4)$.

The match of $x^2 - 2x - 15$ to $x^2 + (a + b)x + ab$ requires $a + b = -2$ and $ab = -15$. It is not immediately obvious what choices of a and b will satisfy both conditions. Fortunately, a few guesses will establish that 3 and -5 are the pair, so, by [3.4], $x^2 - 2x - 15 = (x + 3)(x - 5)$. For larger numerical coefficients, as in $x^2 + 44x + 315$, a lot of guessing may have to be tried, and there is no guarantee that any choice will work. Shortly, a method for determining the proper pair for a and b, if a pair exists, will be demonstrated, but before it can be used, a digression on factoring by grouping is needed.

3.5.1 Factoring by Grouping

On the surface, the polynomial $x^4 + x^2 + 1$ seems impossible to factor; it cannot be made to match with any of [3.1] – [3.9]. But if we are clever and write it as $x^4 + 2x^2 + 1 - x^2$, the polynomial suddenly becomes factorable. The first three terms now match [3.2a], with $a = x^2$ and $b = 1$, and the last term is a square, so the polynomial can be written as $(x^2 + 1)^2 - x^2$, which now matches [3.1] and thus factors as $(x^2 + x + 1)(x^2 - x + 1)$.

Similar inspired adding and subtracting produced [3.9] itself, in which $4a^2b^2 = (2ab)^2$ was added and subtracted to produce $(a^2 + 2b^2)^2 - (2ab)^2$ to which [3.1] can be applied. There are many other situations in which factoring part of a polynomial leads to factoring all of it. In $9xy + 6x + 3y + 2$, for example, the first two terms share the monomial factor $3x$. Removing this factor from these terms results in $3x(3y + 2) + (3y + 2)$. Since both parenthesized expressions are the same, the distributive law can be applied again, resulting in $9xy + 6x + 3y + 2 = (3x + 1)(3y + 2)$. If the two middle terms are reversed to give $9xy + 3y + 6x + 2$, we can factor $3y$ out of the first two terms and 2 out of the last two to produce $3y(3x + 1) + 2(3x + 1)$. Again, the two parenthesized expressions are equal, and the final factoring is $(3y + 2)(3x + 1)$. All the changing of the order of the middle terms has accomplished is the changing of the order of the factors, and this will be true in general.

Factoring by grouping can also produce answers for quadratic trinomials such as $x^2 - 2x - 15$, which was factored above. After a and b are found to be 3 and -5, we may split the middle term $-2x$ into $3x$ and $-5x$ rather than applying [3.4] directly, and then factor by grouping the resulting $x^2 + 3x - 5x - 15$. The first two terms share an x, the last two a -5, producing $x(x + 3) - 5(x + 3)$ or $(x - 5)(x + 3)$. Note here that $-5x - 15 = -5(x + 3)$, not $-5(x - 3)$. It is important to observe the sign laws in all instances where negatives appear.

3.5.2 The Royal Road to Factoring $ax^2 + bx + c$

Now the systematic method for determining numbers, given their sum and product, can be presented, and applied to factoring. The method is based on Formula [3.1], and was known to Babylonian mathematicians in 1500 B. C. Strangely enough, I have not seen it in any algebra textbook.

Given s, the sum, and p, the product, we can double the pair of numbers with that sum and product to get another pair whose sum is $2s$ and whose product is $4p$. (This is done to avoid fractions.) Now, if one of the numbers is $s + k$, then the other must be $s - k$ in order for their sum to be $2s$. But $(s + k)(s - k) = s^2 - k^2$. We wish to make this equal to $4p$ by choosing the correct value for k, so the equation to be satisfied by k is $s^2 - k^2 = 4p$, or $k^2 = s^2 - 4p$. If $s^2 - 4p$ is a perfect square, we find its square root, which is k, and then calculate $s + k$ and $s - k$. These numbers now sum to $2s$ and multiply to $4p$ because of the condition imposed on k, and we divide the pair by two to get the solution to our original problem with sum s and product p. If $s^2 - 4p$ is not a perfect square, or is negative, the unavoidable conclusion is that *no* integers exist that satisfy the original problem. If $s^2 = 4p$, then $k = 0$ and the solution to the original problem is that both numbers are equal to $\frac{s}{2}$. A few examples will illustrate. Given $s = 7$, $p = 12$: $s^2 - 4p = 1$, so $k = 1$, $s + k = 8$, $s - k = 6$, and half of these are 4 and 3. It is readily seen that $4 + 3 = 7$ and $4 \times 3 = 12$. Given $s = 44$, $p = 315$ as in Sec. 3.5: $s^2 - 4p = 1,936 - 1,260 = 676$, so $k = 26$, $s + k = 70$, and $s - k = 18$; dividing by two gives 35 and 9 as the desired numbers. Given $s = 12$, $p = 9$, $s^2 - 4p = 108$, which is not a perfect square, so no integers exist with this sum and product. Mathematicians, remember, hate problems with no answers, and another type of number has been invented to give this problem an answer. See Sec. 3.9, "Irrational Numbers," at the end of this chapter. Given $s = 17$, $p = 75$, $s^2 - 4p = -11$, a negative number. Another type of number has been invented for this type of problem. See Sec. 3.10, "Imaginary and Complex Numbers."

Now, the application of this method to factoring. Given $ax^2 + bx + c$, determine two numbers whose sum is b and whose product is ac, the product of the other numerical coefficients. Split the bx term into two according to the numbers found in the previous step, and apply factoring by grouping to the result. Thus, given $21x^2 + 44x + 15$, certainly a daunting task in appearance, we find $ac = 315$. The second example just above results in 35 and 9, so $44x$ is split into $35x$ and $9x$. Now, the first two terms share $7x$ as a factor and the last two share 3, so the polynomial factors as $(7x + 3)(3x + 5)$. Given $12x^2 + 23x + 10$, the method results in 15 and 8, and $12x^2 + 15x + 8x + 10 = 3x(4x + 5) + 2(4x + 5) = (3x + 2)(4x + 5)$. Note that when ac is negative, the number $s^2 - 4p$ will be larger than s^2. Thus, given $12x^2 + 7xy$

$-12y^2$, $ac = -144$, and $s^2 - 4p = 49 - (-576) = 49 + 576 = 625$. The rest of the method gives 16 and -9, so $12x^2 + 7xy - 12y^2$ becomes $12x^2 + 16xy - 9xy - 12y^2 = 4x(3x + 4y) - 3y(3x + 4y) = (4x - 3y)(3x + 4y)$. If both a and c are negative, as in $6x - 4 - 3x^2$, it is recommended that a -1 be factored out before proceeding. Thus, $6x - 4 - 3x^2 = -1(3x^2 - 6x + 4)$. Unfortunately, the value of $s^2 - 4p$ proves to be negative for this polynomial, and the conclusion is that it cannot be factored. If we had $7x - 4 - 3x^2$, we could factor it into $-1(3x^2 - 7x + 4) = -1(3x - 4)(x - 1) = (3x - 4)(1 - x)$. This could be written, equally correctly, as $(4 - 3x)(x - 1)$; the choice of which factor to combine with the -1 is immaterial. Other monomial factors, however, should be left out in the open, so you would write $3(x + 2)(2x - 5)$ rather than $(3x + 6)(2x - 5)$.

3.6 Quadratic Equations

Many practical situations lead to quadratic equations, that is, equations containing x^2. For example, a certain rectangle has a width 3 feet greater than its length and an area of 28 square feet. To find its dimensions, let x be the length in feet. Then the width is $x + 3$ feet, so the area, the product of length and width, is $x(x + 3)$ or $x^2 + 3x$ square feet. Since we are told that the area is 28 square feet, $x^2 + 3x = 28$, a typical quadratic equation. I will tell you in advance that this equation is solved by $x = 4$ and $x = -7$, but -7 feet is not a possible length for a real rectangle, so the length must be 4 feet and the width $4 + 3$ or 7 feet. The dimensions duly multiply to 28 ft^2.

3.6.1 Solutions

A moment ago, I said that $x^2 + 3x = 28$ is solved by both 4 and -7. This double answer should come as no surprise; quadratic equations in general have two solutions. The fundamental reason for this is that every number has two square roots; both +4 and -4 have 16 as square, and thus the equation $x^2 = 16$ has both 4 and -4 as solutions. Sometimes, the solutions coalesce, as in $(x - 3)^2 = 0$ which has only 3 as a solution. When we are confronted with a quadratic equation, we must determine both solutions, if they are distinct.

Not every quadratic has a solution; $x^2 = -1$ lacks one because the square of every real number is positive, but -1 is negative. (See Sec. 3.10, "Imaginary and Complex Numbers," for a discussion of how mathematicians got around this difficulty.) So quadratics fall into three classes: those with no (real) solutions, those with two distinct real solutions, and those with two coincident solutions, or just one solution.

3.6.2 The Zero-Product Property

It is obvious that if either a or b is zero, then their product ab is zero as well. What is not obvious is that the reverse is true as well: If ab is zero, then either a or b (or both) must be zero. To prove this, start with the equation $ab = 0$. Now a represents some number, which is either zero or not zero. If it is zero, we have our "either a or b" already, and if it is not, we may divide the equation by a to obtain $b = 0$. So if a is not zero, b is, and the zero-product property is proved.

To use this property to solve quadratic equations, four steps must be gone through: 1. Transform the equation so that one of the sides is zero. 2. Factor the other side, which, in general, will be of the form $ax^2 + bx + c$. (This step is one of the reasons why factoring is important.) 3. Apply the zero-product property by setting each of the factors found in step 2 equal to zero. 4. Solve the resulting linear (first-degree) equations. Each solution to an equation involving a factor is also a solution to the original quadratic. To take an easy example, consider $x^2 - 3x = 0$. This factors as $x(x - 3) = 0$, and use of the zero-product property allows us to split the equation into $x = 0$ and $x - 3 = 0$. Adding 3 to both sides of the latter equation gives the second solution, $x = 3$.

For a somewhat more advanced problem, consider the equation that started this section, $x^2 + 3x = 28$. I pulled the solutions, 4 and -7, out of the hat there; now I will show how they are found. We apply the zero-product property as follows: Subtract 28 from both sides of the equation to put a zero on the right side: $x^2 + 3x - 28 = 0$. We now find numbers whose sum is 3 and product -28 (since the product is negative, one but not both of the numbers will be negative); the method in Sec. 3.5.2 produces 7 and -4. These numbers are substituted as a and b in [3.4] to factor $x^2 + 3x - 28$ into $(x + 7)(x - 4)$. With the factoring accomplished, the zero-product property is applied, producing $x + 7 = 0$ and $x - 4 = 0$ which become $x = -7$ and $x = 4$ respectively once the last bit of camouflage is stripped away. The quadratic polynomial $12x^2 + 23x + 10$ was factored in Sec. 3.5.2 as $(3x + 2)(4x + 5)$, so the quadratic equation $12x^2 + 23x + 10 = 0$ or $(3x + 2)(4x + 5) = 0$ is satisfied by $-\frac{2}{3}$ and $-1\frac{1}{4}$, the solutions to the equations $3x + 2 = 0$ and $4x + 5 = 0$ involving the factors. One last example illustrates the one-solution type of equation: $x^2 + 9 = 6x$. This is transformed into $x^2 - 6x + 9 = 0$ to apply the zero-product method. Applying the technique in Sec. 3.5.2 with $s = -6$ and

$p = 9$ results in $s^2 - 4p = 0$, and the left side factors as $(x - 3)(x - 3) = 0$ where the two factors are identical. The two equations that application of the zero-product property produce thus both yield the same solution, $x = 3$, which is therefore the only solution to this quadratic.

3.6.3 Completing the Square

Not every quadratic polynomial can be factored. (Try $x^2 + 3x + 1$.) Since the zero-product method can only be applied if the polynomial can be factored, another algorithm that can always be used is needed. The method known as completing the square is such a process. It is based on formula [3.2a]: $(a + b)^2 = a^2 + 2ab + b^2$. Replacing a by the unknown variable x results in $x^2 + 2bx + b^2 = (x + b)^2$. Of note in this formula is the fact that the constant term on the left, b^2, is the square of half the number ($2b$) that multiplies x. Thus, given $x^2 + 2bx$, the addition of b^2 changes the expression into a perfect square, or "completes the square." It is easy to determine the value of b needed to complete the square, as it is half of the constant multiplying x.

To apply the method of completing the square, the following steps must be taken: 1. Transform the equation to the form $x^2 + 2bx = c$, where b and c are constants. In particular, no terms involving the variable may appear on the other side of the equation from the x^2 term. 2. Determine b and add its square to both sides of the equation. This transforms the quadratic side into $(x + b)^2$. 3. Take the square root of both sides of the equation, remembering as you do that every number has two square roots, one positive and one negative. 4. Solve the resulting equation, which should involve x only in the expression $x + b$.

We apply the method to $x^2 + 4x + 1 = 0$ which does not readily factor. Step 1 changes the equation to $x^2 + 4x = -1$. Now, $2b = 4$, so $b = 2$ and $b^2 = 4$, so Step 2 is implemented by adding 4 to both sides, giving $(x + 2)^2 = 3$. Extracting the square root of both sides produces $x + 2 = \pm\sqrt{3}$. The strange anchor-like symbol $\pm$ means "plus or minus," and is used to denote both square roots of 3 at once. As to the numerical value of $\sqrt{3}$, there is no integer or fraction which can be squared to produce 3, but mathematicians have not allowed themselves to be stymied by that awkward fact. (See Sec. 3.9, "Irrational Numbers," at the end of this chapter.) For now, it will suffice to say that a number, whose value is close to 1.732, has been deemed to exist and have the property that its square is 3. Now step 4 can be carried out, resulting in the solutions $x = \sqrt{3} - 2$ and $x = -\sqrt{3} - 2$, or approximately -0.268 and -3,732 in numerical values.

To acquaint the reader further with the concept of $\sqrt{3}$, we will substitute the first of these solutions into the original equation (and show that it checks as well). The square of $\sqrt{3} - 2$ is calculated by applying [3.2b], with $a = \sqrt{3}$ and $b = 2$. The formula yields $(\sqrt{3} - 2)^2 = (\sqrt{3})^2 - 2 \cdot \sqrt{3} \cdot 2 + 2^2$. The defining property of $\sqrt{3}$ is that its square is 3, and the commutative property of multiplication holds for $\sqrt{3}$ as it does for any other number, so the expression simplifies to $7 - 4\sqrt{3}$. If $x = \sqrt{3} - 2$, then $4x = 4\sqrt{3} - 8$, and $x^2 + 4x + 1 = (7 - 4\sqrt{3}) + (4\sqrt{3} - 8) + 1$; the terms involving $\sqrt{3}$ cancel, leaving us with $7 - 8 + 1$, which is zero as required. The point to be emphasized is that $\sqrt{3}$ is just another number which obeys the same rules as "ordinary" numbers like 3.

The method of completing the square can lead to complicated arithmetic, especially if the x^2 term contains a numerical factor other than 1. Since the coefficient of x^2 must be 1 before the method can be applied, equations containing ax^2 with $a \neq 1$ must be divided by a before proceeding further. As an example, consider $12x^2 + 23x + 10 = 0$, solved by factoring in Sec. 3.6.2:

$$x^2 + \frac{23}{12}x + \frac{10}{12} = 0 \quad \text{(removing the coefficient from } 12x^2\text{)}$$

$$x^2 + \frac{23}{12}x = -\frac{10}{12} \quad \text{(the } \frac{10}{12} \text{ is subtracted to make the left side match } x^2 + 2bx\text{)}$$

$$x^2 + \frac{23}{12}x + \frac{529}{576} = \frac{529}{576} - \frac{10}{12} \quad (2b = \frac{23}{12}, \text{ so } b = \frac{23}{24}, 23^2 = 529, \text{ and } 24^2 = 576)$$

$$(x + \frac{23}{24})^2 = \frac{49}{576} \quad \text{(completing the square on the left transforms it into } (x + b)^2, \text{ with } b = \frac{23}{24}; \text{ and}$$
$\frac{10}{12} = \frac{480}{576}, \quad 529 - 480 = 49)$

$$x + \frac{23}{24} = \pm\frac{7}{24} \quad \text{(taking square roots)}$$

$$x = \frac{7-23}{24} = -\frac{2}{3} \text{ or } \frac{-7-23}{24} = -1\frac{1}{4} \quad \text{(solving the remaining equations)}$$

The answers, of course, are the same as in the preceding section. Even some of the arithmetic is the same, as obtaining 15 and 8 from their sum 23 and product 120 involved the calculation $529 - 480$. But the factoring solutions kept fractions out of the picture until the very end, while completing the square involved fractions from the start. No one method of solution is the best for all quadratics. Immediately below, the "quadratic formula" and three derivations will be presented; the very fact that three methods of arriving at the same formula are available illustrates the variety of methods of solution for specific quadratic equations.

3.6.4 The Quadratic Formula

We will now apply the methods of solving specific quadratics to the general quadratic, $ax^2 + bx + c = 0$. The traditional presentation uses the method of completing the square and will be presented first.

Step 1(a): $x^2 + \frac{b}{a}x + \frac{c}{a} = 0$ (removing the a from ax^2)

Step 1(b): $x^2 + \frac{b}{a}x = -\frac{c}{a}$ (subtracting $\frac{c}{a}$ to match the left side with $x^2 + 2bx$)

Step 2(a): $x^2 + \frac{b}{a}x + \frac{b^2}{4a^2} = \frac{b^2}{4a^2} - \frac{c}{a}$ (adding the square of half of $\frac{b}{a}$; completing the square)

Step 2(b): $(x + \frac{b}{2a})^2 = \frac{b^2 - 4ac}{4a^2}$ (performing symbolic arithmetic)

Step 3: $x + \frac{b}{2a} = \pm \frac{\sqrt{b^2 - 4ac}}{2a}$ (taking square roots; note the $\pm$ sign and the fact that the square root of a fraction is found by taking the square roots of the numerator and denominator)

Step 4: $x = \frac{-b \pm \sqrt{b^2 - 4ac}}{2a}$ (cleaning up the last bit of camouflage by subtracting $\frac{b}{2a}$)

So there it is, the formula that has intimidated (or inspired) thousands of tenth-graders each year since schools had tenth grades. I, of course, was inspired; the formula's appearance seemed to me wonderfully balanced, with the beginning of the square-root sign located right in the middle of the fraction, appearing directly on top of the denominator $2a$.

Now, we will see how the same formula emerges from an attempt to use the zero-product method on $ax^2 + bx + c = 0$.

First, remove the largest monomial "factor": $x^2 + \frac{b}{a}x + \frac{c}{a} = 0$.

Now, we need two numbers which add to $\frac{b}{a}$ and multiply to $\frac{c}{a}$. We let the numbers be $\frac{b}{2a} + k$ and $\frac{b}{2a} - k$, and impose the condition on k that the product of the two numbers be $\frac{c}{a}$. From [3.1], $\left(\frac{b}{2a} + k\right)\left(\frac{b}{2a} - k\right) = \frac{b^2}{4a^2} - k^2$, so $\frac{b^2}{4a^2} - k^2 = \frac{c}{a}$, or $k^2 = \frac{b^2}{4a^2} - \frac{c}{a} = \frac{b^2 - 4ac}{4a^2}$. (Look familiar yet?) Thus, $k = \frac{\sqrt{b^2 - 4ac}}{2a}$, and our two numbers are $\frac{b + \sqrt{b^2 - 4ac}}{2a}$ and $\frac{b - \sqrt{b^2 - 4ac}}{2a}$. (Still not recognizable?) Note that the $\pm$ sign was not used when we found k from k^2; this is permissible because $+k$ and $-k$ are used interchangeably in finding the numbers with the required sum and product.

Having found our numbers, we use them to "factor" $x^2 + \frac{b}{a}x + \frac{c}{a}$ into $(x + \frac{b + \sqrt{b^2 - 4ac}}{2a})$ times $(x + \frac{b - \sqrt{b^2 - 4ac}}{2a})$. The zero-product property enables us to reduce this quadratic to two linear equations: $x + \frac{b + \sqrt{b^2 - 4ac}}{2a} = 0$ and $x + \frac{b - \sqrt{b^2 - 4ac}}{2a} = 0$, which have solutions $\frac{-b - \sqrt{b^2 - 4ac}}{2a}$ and $\frac{-b + \sqrt{b^2 - 4ac}}{2a}$ respectively. We combine both of these into one expression with the $\pm$ sign: $x = \frac{-b \pm \sqrt{b^2 - 4ac}}{2a}$. Presto! The quadratic formula has reappeared!

For the third method of deriving the quadratic formula, we take a cue from the last half of the process of completing the square. After taking the square root of both sides, we have an equation of the form $x - p = \pm\sqrt{q}$, with solutions of the form $p \pm \sqrt{q}$. We substitute these "solutions" into the original equation, $ax^2 + bx + c = 0$, and determine the values of p and q which actually satisfy that equation. First, applying [3.2a] and [3.2b], we find $(p + \sqrt{q})^2 = p^2 + q + 2p\sqrt{q}$ and $(p - \sqrt{q})^2 = p^2 + q - 2p\sqrt{q}$, so $a(p \pm \sqrt{q})^2 = ap^2 + aq \pm 2ap\sqrt{q}$. Then, $b(p \pm \sqrt{q}) = bp \pm b\sqrt{q}$, so $ax^2 + bx + c$ becomes $ap^2 + aq + bp \pm 2ap\sqrt{q} \pm b\sqrt{q} + c$, where both $\pm$ signs take the same value, either both plus or both minus. For this expression to equal zero, the $\sqrt{q}$ terms must cancel out, so $2ap + b = 0$ or $p = -\frac{b}{2a}$. Substituting this value of p into the previous expression results in $\frac{b^2}{4a} + aq - \frac{b^2}{2a} + c = 0$ or $aq = \frac{b^2}{2a} - \frac{b^2}{4a} - c = \frac{b^2-4ac}{4a}$. This means $q = \frac{b^2-4ac}{4a^2}$ and the solutions $p \pm \sqrt{q}$ are $\frac{-b \pm \sqrt{b^2-4ac}}{2a}$. The quadratic formula strikes again!

Of the three derivations, I like the third one the best because of its unusual approach to the problem and the rather unexpected way in which the quadratic formula appears.

To apply the quadratic formula, you must first reduce the equation given to the standard form $ax^2 + bx + c = 0$. Then read off the values of a, b, and c from your equation, substitute them in the appropriate places in the formula, and perform the arithmetic to arrive at the answer. Thus, for $x^2 + 4x + 1 = 0$, $a = 1$, $b = 4$, and $c = 1$, so the solutions are $\frac{-4 \pm \sqrt{4^2 - 4 \cdot 1 \cdot 1}}{2} = \frac{-4 \pm \sqrt{12}}{2}$. Now, $\frac{\sqrt{12}}{2} = \frac{\sqrt{12}}{\sqrt{4}} = \sqrt{\frac{12}{4}} = \sqrt{3}$, so the solutions are $-2 + \sqrt{3}$ and $-2 - \sqrt{3}$ as in Sec. 3.6.3. Given $8x - 5 = 3x^2$, we first subtract $3x^2$ to get the equation into standard form: $-3x^2 + 8x - 5 = 0$, from which we read off $a = -3$, $b = 8$, and $c = -5$. The formula produces $\frac{-8 \pm \sqrt{64 - 4(-3)(-5)}}{-6} = \frac{-8 \pm 2}{-6}$, giving solutions of $+1$ and $+1\frac{2}{3}$. Checking, we find that letting $x = 1$ produces $8 - 5 = 3$ (rather obviously true), and substituting $1\frac{2}{3} = \frac{5}{3}$ produces $\frac{40}{3} - 5 = 3 \times \frac{25}{9}$ or $\frac{40}{3} - \frac{15}{3} = \frac{25}{3}$. The original equation in this section, $x(x + 3) = 28$, is transformed into $x^2 + 3x - 28 = 0$ ($a = 1$, $b = 3$, $c = -28$), which produces $\frac{-3 \pm \sqrt{9 + 4 \cdot 28}}{2} = \frac{-3 \pm 11}{2} = 4$ or -7 as stated in Sec. 3.6.1.

3.7 Radicals

Now it is time to pay more attention to how numbers behave under square-root signs (also known as radicals). By definition, $\sqrt{a}$ is the number that, when multiplied by itself, produces a. It is a convention to define $\sqrt{a}$ as referring to the positive number only; if we want the negative one, we must write $-\sqrt{a}$, while $\pm\sqrt{a}$ denotes both roots at once. However, the phrase "the square root of a" refers to both values; to specify only one, the sign, positive or negative, must be used.

In general, sums and differences of radicals cannot be simplified. But products and quotients can be worked with. The product of $\sqrt{a}$ and $\sqrt{b}$ is $\sqrt{ab}$, as will be demonstrated: $(\sqrt{a} \cdot \sqrt{b})^2 = \sqrt{a} \cdot \sqrt{b} \cdot \sqrt{a} \cdot \sqrt{b} = \sqrt{a} \cdot \sqrt{a} \cdot \sqrt{b} \cdot \sqrt{b} = a \cdot b$. Thus, the square of $\sqrt{a} \cdot \sqrt{b}$ is ab, meaning that $\sqrt{a} \cdot \sqrt{b} = \sqrt{ab}$. In particular, $\sqrt{a^2 b} = \sqrt{a^2} \cdot \sqrt{b} = a\sqrt{b}$, since it is obvious that $\sqrt{a^2} = a$. (After all, a <u>can</u> be multiplied by itself to give a^2!)

This last equation leads to what is known as *simplification of radicals*: the removal of the largest possible square factor of the number under the square-root sign. To simplify $\sqrt{72}$, we divide 72 by the square 36, producing 2. Therefore, $\sqrt{72} = \sqrt{2 \cdot 36} = \sqrt{2} \cdot \sqrt{36} = 6\sqrt{2}$. Given $\sqrt{275}$, we find that 25 divides 275, so $\sqrt{275} = 5\sqrt{11}$. And from $\sqrt{2,048}$ we get $32\sqrt{2}$ since $2,048 = 2 \times 32^2$. These simplifications enable us to add or subtract certain combinations of radicals. Thus, $\sqrt{99} + \sqrt{275} = 3\sqrt{11} + 5\sqrt{11} = 8\sqrt{11}$, and $\sqrt{128} + \sqrt{2,048} = 8\sqrt{2} + 32\sqrt{2} = 40\sqrt{2}$. These two sums can be written as the single radicals $\sqrt{704}$ and $\sqrt{3,200}$, which serves to reinforce the denial of the plausible but incorrect formula $\sqrt{a} + \sqrt{b} = \sqrt{a + b}$; $99 + 275$ falls far short of 704. Radicals with the same non-square factors such as $\sqrt{99}$ and $\sqrt{275}$ are called similar because they can be added like similar terms such as $4x^2$ and $3x^2$.

As for division of radicals, a process called *rationalizing the denominator* is employed as follows: $\frac{\sqrt{x}}{\sqrt{y}} = \frac{\sqrt{x}}{\sqrt{y}} \cdot \frac{\sqrt{y}}{\sqrt{y}} = \frac{\sqrt{xy}}{y}$ in which the square-root sign has disappeared from the denominator. The same process can be applied if the numerator is free of radicals, producing the result $\frac{x}{\sqrt{y}} = \frac{x\sqrt{y}}{y}$, and, if $x = 1$, $\frac{1}{\sqrt{y}} = \frac{\sqrt{y}}{y}$.

Division by combinations of radicals, such as $\frac{5}{\sqrt{2}+\sqrt{3}}$, would seem impossible, but the notion of *conjugates* makes some simplification possible. Numbers of the form $a \pm b$ are called conjugates, especially if one or both of a and b are radicals. The importance of conjugates stems from [3.1], which states that the product of conjugates is the difference of two squares. This offers a way to get rid of radicals in unwanted positions like denominators. (The obvious way of getting rid of radicals, squaring, will not work in this case: the square of $\sqrt{a} + \sqrt{b}$ is $a + b + 2\sqrt{ab}$, which still contains a radical.) To remove the square-root signs from the denominator of $\frac{5}{\sqrt{2}+\sqrt{3}}$, we multiply both numerator and denominator by the conjugate of $\sqrt{2} + \sqrt{3}$, that is, $\sqrt{2} - \sqrt{3}$. Then, $\frac{5}{\sqrt{2}+\sqrt{3}} = \frac{5(\sqrt{2}-\sqrt{3})}{(\sqrt{2}+\sqrt{3})(\sqrt{2}-\sqrt{3})} = \frac{5\sqrt{2}-5\sqrt{3}}{2-3} = 5\sqrt{3} - 5\sqrt{2}$. If we are interested in obtaining a numerical value, we can "unsimplify" the last expression to $\sqrt{75} - \sqrt{50}$, which requires just a square-root table and a subtraction to calculate.

3.8 More About Exponents

At the risk of moving beyond "basic" algebra, the notion of exponent will be broadened to cover negative and fractional cases. We wish to define negative and fractional exponents in such a way that the three Laws of Exponents (Sec. 3.1.3) will still hold.

First, we will consider the meaning of a zero exponent. Any number divided by itself equals one, as in $27 \div 27 = 1$. Now 27 is 3^3. If Law 3 is to hold for the zero exponent, then $3^3 \div 3^3 = 3^{3-3} = 3^0$ must equal one. In general, $a^0 = a^{n-n} = a^n \div a^n = 1$ (provided that a is not zero, so that the division by a^n can be done). If $a = 0$, the expression 0^0 is what mathematicians call an "indeterminate quantity," to which no value can be assigned. For the rest of this section, the number a, called the *base*, will be restricted to positive values only.

With a^0 defined, we can now define a^{-n}, where n is a positive integer and therefore $-n$ is a negative integer. By Law 1, $a^{-n} \cdot a^n = a^{-n+n} = a^0 = 1$, as just defined. Hence, a^{-n} is the number that can be multiplied by a^n to produce 1, and is therefore equal to $\frac{1}{a^n}$. For example, $3^{-4} = \frac{1}{3^4} = \frac{1}{81}$ and $10^{-3} = \frac{1}{10^3} = \frac{1}{1,000}$ or .001.

To define a^n when n is a fraction, we resort to Law 2. The nth root of a, written $\sqrt[n]{a}$, is the number that can be raised to the nth power to give a, that is, $(\sqrt[n]{a})^n = a$. Since $a = a^1$, $(\sqrt[n]{a})^n = a^1$. Denoting the exponential equivalent to $\sqrt[n]{a}$ by a^p, we see that $(a^p)^n = a^{pn} = a^1$, so that $pn = 1$ and $p = \frac{1}{n}$; that is, $a^{\frac{1}{n}} = \sqrt[n]{a}$. Now, we replace a by b^m and consider $\sqrt[n]{b^m}$. This number is by definition one that can be raised to the nth power to produce b^m, that is, $(\sqrt[n]{b^m})^n = b^m$. As above, we denote $\sqrt[n]{b^m}$ by b^q; this means $(b^q)^n = b^m$ so that $qn = m$ and $q = \frac{m}{n}$; that is, $b^{m/n} = \sqrt[n]{b^m}$. The zeroth root of a number is not defined, since it is the equivalent of an exponent involving division by zero. If $n = 2$, it is customary to omit it in $\sqrt[n]{a}$, because the square (second) root is by far the most often used.

In case $\frac{m}{n}$ is a negative fraction, we note that the derivation of a^{-n} does not depend on n being an integer, since $-n + n = 0$ holds for all numbers, not just integers. So these three definitions extend the concept of exponent to all rational numbers, positive, negative, or zero, while retaining the three Laws of Exponents:

[3.10a] $a^0 = 1$

[3.10b] $a^{p/q} = \sqrt[q]{a^p}$

[3.10c] $a^{-n} = \frac{1}{a^n}$

In calculus (Chapter 7), the notion of exponent will be further extended to all numbers, rational, irrational, and complex, but before we can do that, we must define irrational and complex numbers themselves. That is the subject of the last two sections of this chapter.

3.9 Irrational Numbers

The quadratic formula, which we will assign an equation number to: [3.11] $x = \frac{-b \pm \sqrt{b^2 - 4ac}}{2a}$, contains a square-root sign. If the number under that radical is not a perfect square like 9, 64, or 1,521, it can be shown that no rational number (quotient of integers) exists that can be squared to produce the required number. As was mentioned in the explanation of $\sqrt{3}$ in Sec. 3.6.3, mathematicians simply postulate the existence of such numbers to get around this difficulty. They recognize that these numbers are not rational by calling them "irrational," but that does not mean that they are insane or much more difficult to deal with than integers or rational numbers. In fact, the techniques demonstrated in Sec. 3.7 are methods of working with irrational numbers. (Of a certain type; there are many more kinds of irrational numbers than square roots!)

But there is still the question of the actual value of numbers like $\sqrt{3}$. The square root table shows $\sqrt{3} = 1.732$; if it is bigger, it may say $\sqrt{3} = 1.732051$. But multiplying either of these numbers by itself will produce a number different from 3. (Naturally, the longer decimal will produce a result closer to 3.) The reason should be obvious: a terminating decimal is a rational number (Sec. 2.5) and thus cannot provide the exact value of the irrational $\sqrt{3}$.

Although we cannot compute the exact value of an irrational square root, there is nothing preventing the calculation of an approximation as close as desired. One method starts by pinning down an unknown square root between two consecutive integers. If a and b are both positive, then $(a + b)^2 = a^2 + 2ab + b^2$ is greater than a^2 by the amount $b(2a + b)$. Therefore, if x lies between n^2 and $(n + 1)^2$, $\sqrt{x}$ must lie between n and $n + 1$. To calculate $\sqrt{20}$, we see that 20 is between $16 = 4^2$ and $25 = 5^2$, so $\sqrt{20}$ must be between 4 and 5. We now let $a = 4$ and $(a + b) = \sqrt{20}$, so $(a + b)^2 = 20$ and $b(2a + b) = 4 (20 - 16)$. Since b is much less than a, we approximate $2a + b$ by $2a$ and then calculate the approximate value of b as $\frac{4}{2a} = \frac{1}{2}$, resulting in an approximation for $\sqrt{20}$ of $4\frac{1}{2}$. We now let $a = 4\frac{1}{2}$, $a^2 = 20\frac{1}{4}$, and $b = \sqrt{20} - a$, so $b(2a + b) = -1/4$ and b is approximately $\frac{-1/4}{2a} = -\frac{1}{36}$, resulting in a new approximation for $\sqrt{20}$ of $4\frac{17}{36}$ whose square is $20\frac{1}{1296}$. Further repetitions of this process result in increasingly large numerators and denominators, and increasingly close approximations to $\sqrt{20}$. Only patience (and your supply of paper and pencils) limit the closeness of the approximation that can be obtained.

A second approximation method is based on the fact that if $|x| < 1, |x^2| < |x|$. (The symbol $|x|$ means the absolute value of x, as defined in Sec. 3.1.4.) When applied to square roots, this means that if $\frac{a}{b}$ is a good approximation to $\sqrt{n}$, then $\left|\sqrt{n} - \frac{a}{b}\right|$ will be small, and $\left|n + \frac{a^2}{b^2} - \frac{2a\sqrt{n}}{b}\right| = \left|\frac{nb^2 + a^2 - 2ab\sqrt{n}}{b^2}\right|$ will be even smaller. This implies that $\frac{nb^2 + a^2}{2ab}$ is a better approximation to $\sqrt{n}$ than $\frac{a}{b}$ is. Applying this to $\sqrt{20}$ with $\frac{a}{b} = \frac{4}{1}$, we get $\frac{nb^2 + a^2}{2ab} = \frac{20 + 16}{8} = \frac{9}{2}$, repeating, we get $\frac{80 + 81}{36} = \frac{161}{36}$ as the next approximation.

Still a third method begins with the idea that if $\sqrt{n} \cdot \sqrt{n} = n$, then $\frac{n}{\sqrt{n}} = \sqrt{n}$, and that a good approximation to $\sqrt{n}$ will almost satisfy this last equation. To approximate $\sqrt{20}$, we start with 4, calculate $\frac{20}{4} = 5$, and average 4 and 5 to get $4\frac{1}{2}$. We then divide 20 by $4\frac{1}{2}$, getting $4\frac{4}{9}$, and average $4\frac{1}{2}$ and $4\frac{4}{9}$ to obtain $4\frac{17}{36}$ as the next approximation.

The astute reader will notice that all three methods produced the same series of approximations to $\sqrt{20}$, starting with 4. The reason is simple: all three of these methods are in fact identical! The first and third are merely disguises of the second, most direct method, repeatedly replacing $\frac{a}{b}$ with $\frac{nb^2 + a^2}{2ab}$.

Provided that the first approximation in the sequence is an integer within one of the actual square root, any approximation $\frac{a}{b}$ in the series will be within $\frac{1}{b}$ of the exact value. Since the denominator at least doubles (more stringently, the number of digits in the denominator approximately doubles) with each repetition, it will eventually exceed any number that can be chosen before the computation begins, which means that the difference between the approximation and the exact value will eventually be smaller than any positive number that is chosen before the computation begins. This situation arises often enough that mathematicians have given it a special name: The sequence of approximations has as its *limit* the exact value. (The concept of limit is a key notion in calculus; see Chapter 7.) Mathematicians have agreed to call any number that is the limit of a sequence of rational numbers a *real* number. In practice, everyone, including computers, has only a finite amount of patience (let alone paper and pencils), so there is a practical ceiling on the accuracy of our square root approximations, as well as those to other irrational numbers, like π.

The mention of "real" numbers implies that there are "fictional" or "imaginary" numbers. Such do exist, and are the subject of the next section.

3.10 Imaginary and Complex Numbers

In the quadratic formula ([3.11]), we have so far met only examples in which b^2 was larger than $4ac$ so b^2 –

$4ac$ was positive. Of course, there is no law saying that all equations $ax^2 + bx + c = 0$ must satisfy $b^2 > 4ac$, so we must face the problem of how to deal with $\sqrt{n}$ when n is negative.

As mentioned in Sec. 3.6.1, there are no real numbers whose square is negative because of the law of signs for multiplication (Sec. 3.1). As with the need to create numbers whose squares were 2, 3, 5, etc., mathematicians postulated the existence of numbers whose squares were -1, -2, -3, ..., -½, $-\sqrt{8}$, etc. (This took a lot of courage, and was not done until the 17th century.) These numbers are stated in terms of a basic unit, the square root of -1. This quantity is denoted by i, the first letter in the word "imaginary." It is possible to define imaginary numbers so as to satisfy all the laws of arithmetic that hold for real numbers except two: it is not possible to say that one complex or imaginary number is "greater" or "less" than another, or that an imaginary or complex number is "positive" or "negative."

The square roots of other negative numbers are defined by the equation $\sqrt{-a} = i\sqrt{a}$, where a is positive. Thus, $\sqrt{-49} = i\sqrt{49} = 7i$, and $\sqrt{-256} = i\sqrt{256} = 16i$. With this definition, we can use the quadratic formula to solve $ax^2 + bx + c = 0$ even when $4ac > b^2$. Thus, given $x^2 - 4x + 8 = 0$, we calculate $x = \frac{4 \pm \sqrt{16-32}}{2} = 2 \pm 2i$. (It is customary to write the i after a numerical factor unless that factor is a radical.) Real and imaginary numbers are incommensurable; $2 + 2i$ cannot be further simplified. The sum of a real number and an imaginary one is called a *complex* number; the real and imaginary numbers can be considered subsets of the complex numbers.

Formally, a complex number can be written in the form $a + bi$, where a and b are real numbers. It can also be considered as an ordered pair (a, b) of real numbers. Since imaginary and complex numbers are to be numbers, admitted on a par with real numbers, arithmetic with them must be defined as was done with fractions in Chapter 2. From the $a + bi$ point of view, i is treated as if it were just another variable – except when i^2 crops up; then it is replaced with -1 in keeping with its definition.

With this provision and the polynomial multiplication techniques in Sec. 3.3.2, the following formulas can be established.

[3.12a] $(a + bi) + (c + di) = (a + c) + (b + d)i$.
(In the ordered pair notation: $(a, b) + (c, d) = (a + c, b + d)$.)
[3.12b] $(a + bi) - (c + di) = (a - c) + (b - d)i$, or $(a, b) - (c, d) = (a - c, b - d)$.
[3.13a] $(a + bi) \cdot (c + di) = (ac - bd) + (ad + bc)i$, or $(a, b)(c, d) = (ac - bd, ad + bc)$.

As for division, the notion of conjugates comes to the rescue again. (Conjugates seem to come in handy in dealing with difficult square roots!) Applying [3.13a] to conjugates, $(a + bi)(a - bi) = a^2 + b^2$, or $(a, b) \cdot (a, -b) = (a^2 + b^2, 0)$. Hence, we can "real-ize" the denominator in a complex fraction by multiplying by the conjugate of that denominator. This produces $\frac{a+bi}{c+di} = \frac{(a+bi)(c-di)}{(c+di)(c-di)} = \frac{(ac+bd)+(bc-ad)i}{c^2+d^2}$ or $\frac{(a,b)}{(c,d)} = (\frac{ac+bd}{c^2+d^2}, \frac{bc-ad}{c^2+d^2})$, which we call equation [3.13b].

With complex arithmetic defined, we can now check the solution of our quadratic equation, $x^2 - 4x + 8 = 0$, that produced complex answers. The square of $2 + 2i = (2 + 2i)(2 + 2i) = (4 - 4) + (4 + 4)i = 8i$, and the square of $2 - 2i$ is $-8i$. If $x = 2 + 2i$, then $-4x = -8 - 8i$, and $x^2 - 4x + 8 = 8i - 8 - 8i + 8 = 0$, while substituting $2 - 2i$ gives $-8i - 8 + 8i + 8$ which is also zero. So both complex numbers satisfy the equation.

The reader may now be saying to himself, "It is O.K. to create the concept of i to represent the square root of -1, but what about the square root of i itself? If i is to be a number, it must have a square root too. Perhaps we shall call it j?" These questions are answered by stating that $\sqrt{i}$ already exists as a complex number! To prove this, let $a + bi$ be that number, and set conditions on the real numbers a and b so that $(a + bi)^2 = a^2 - b^2 + 2abi = 0 + 1i$. Equating real and imaginary parts separately, we see that we must have $a^2 = b^2$ and at the same time $2ab = 1$ or $ab = ½$. The first equation means that a is either b or $-b$. If $a = -b$ their product will be negative and cannot be equal to ½. If $a = b$ then $ab = a^2$ and $a^2 = ½$, thus $a = \pm\sqrt{\frac{1}{2}} = \pm\frac{1}{\sqrt{2}} = \pm\frac{\sqrt{2}}{2}$. Since $a = b$ and our complex number was $a + bi$, we have $\sqrt{i} = \pm\frac{\sqrt{2}}{2} \pm \frac{\sqrt{2}}{2}i$. Both $\pm$ signs are the same, so this can be changed to $\pm(\frac{\sqrt{2}}{2} + \frac{\sqrt{2}}{2}i)$. So we find that i, like real numbers, has two square roots, either of which is the negative of the other.

Other complex numbers share this property too; the square root of $x + yi$ is $\pm(a + bi)$, where $a = \sqrt{\frac{x+\sqrt{x^2+y^2}}{2}}$ and $b = \sqrt{\frac{-x+\sqrt{x^2+y^2}}{2}}$. The signs of a and b are the same if y is positive, and opposite if y is negative.

We can go further than square roots. Back in Sec. 1.4 I said that the cube root of 216 is 6 because $6^3 = 216$.

However, there are also two complex numbers, namely $-3 + (3\sqrt{3})i$ and $-3 - (3\sqrt{3})i$ which, when cubed, produce 216. I will prove this startling accusation in the case of the second of these numbers by two applications of [3.13a]: $[-3 - (3\sqrt{3})i]^2 = (9 - 9 \cdot 3) + (2 \cdot 3 \cdot 3\sqrt{3})i = -18 + (18\sqrt{3})i$, and $[-3 - (3\sqrt{3})i]^3 = [-18 + (18\sqrt{3})i][-3 - (3\sqrt{3})i] = +54 + 54 \cdot 3 - (54\sqrt{3})i + (54\sqrt{3})i = 54 + 162 + 0i = 216$. Not only does 216 have "extra" cube roots, but so do all real and complex numbers. (The formula for the cube root of a complex number goes beyond the scope of this chapter.) In general, any number, real or complex, has precisely n nth roots, of which at most two are real. Roots of real numbers always appear as pairs of complex conjugates, like $-3 + (3\sqrt{3})i$ and $-3 - (3\sqrt{3})i$, or are real themselves, like 6.

More generally still, we have what is called the Fundamental Theorem of Algebra: the polynomial $px^n + qx^{n-1} + rx^{n-2} + \cdots + sx + t$ factors, uniquely except for order, as $p(x - a_1)(x - a_2) \ldots (x - a_n)$, where p, q, r, $\ldots$, s, t and a_1, a_2, $\ldots$, a_n are complex numbers. If p, q, r, $\ldots$, s, and t are all real, then the a's are either real or are pairs of complex conjugates. (The proof of this goes well beyond the scope of this book, and was first achieved by Karl Friedrich Gauss, the greatest mathematician of his time and one of the greatest of all time.)

Because of this theorem, there is no practical reason to create any more types of numbers. But mathematics is at its heart a creative discipline, and the theorem has not deterred mathematicians from going on and creating more number-like objects, some of which will be encountered later on in this book. I prefer to end the concept of number with the complex numbers, probably because of the Fundamental Theorem of Algebra; all further elaborations such as quaternions, vectors, and matrices (the last two of which will appear) are merely number-like.

Chapter 4   Number Sequences and Patterns

Probably no branch of mathematics has more improbable and bewildering (hence enjoyable) facts as the study of number sequences and patterns. Who would guess, for example, that each number that is simultaneously square and triangular is exactly 34 times the one before it, minus the one before that, plus two? Or that square triangular numbers are related to the problem of finding integer solutions to $a^2 + b^2 = c^2$ which also satisfy the condition $b = a + 1$? (One instance of how they are related is $119^2 + 120^2 = 169^2$, $119 + 169 = 288$, and $1 + 2 + \ldots + 288 = 41{,}616 = 204^2$.)

Although these two results are fairly advanced examples, interesting facts can be discovered at a more elementary level as well, by studying the simplest number sequences: progressions.

4.1 Arithmetic, Geometric, and Harmonic Progressions

An arithmetic progression is a number sequence in which the difference between any two consecutive numbers is the same. This common difference is denoted by d. If the first term in the sequence is called a_1 (read "a-one"), the second a_2 ("a-two"), ..., the nth a_n ("a-en"), then $a_2 = a_1 + d$, $a_3 = a_2 + d = (a_1 + d) + d = a_1 + 2d$, $a_4 = a_1 + 3d$, ..., $a_n = a_1 + (n - 1)d$. Going backwards from a_n, $a_{n-1} = a_n - d$, $a_{n-2} = a_n - 2d$, etc. Examples of arithmetic progressions include 1, 3, 5, 7, ..., the set of odd numbers; 5, 9, 13, 17, ...; 18, 13, 8, 3, -2, -7, -12, ...; and any row in the addition or multiplication table.

When studying progressions, it is often desired to find the sum of an arbitrary number of consecutive terms. For arithmetic progressions, the following trick does this for us. Let S stand for the sum:
$$S = a_1 + a_2 + a_3 + \cdots + a_{n-2} + a_{n-1} + a_n.$$
Then, since addition is commutative:
$$S = a_n + a_{n-1} + a_{n-2} + \cdots + a_3 + a_2 + a_1.$$
We now write the terms in the first line using a_1 and d:
$$S = a_1 + (a_1 + d) + (a_1 + 2d) + \cdots + (a_1 + (n - 2)d) + (a_1 + (n - 1)d).$$
Then the second line using a_n and d:
$$S = a_n + (a_n - d) + (a_n - 2d) + \cdots + (a_n - (n - 2)d) + (a_n - (n - 1)d).$$
Now, add the two equations together:
$$S + S = (a_1 + a_n) + (a_1 + a_n) + \cdots + (a_1 + a_n),$$
where there are n addends of $(a_1 + a_n)$. Therefore, $S = \frac{n}{2}(a_1 + a_n)$. This method was used by Gauss, when he was six (!), to add the numbers from 1 to 100. Here, $a_1 = 1$, $a_n = 100$, and $n = 100$, so $S = 5{,}050$. If we are given d instead of a_n, we can write $S = \frac{n}{2}(a_1 + (a_1 + (n - 1)d)) = na_1 + \frac{n(n-1)}{2}d$. If $d = a_1$, so the progression is a multiple of 1, 2, 3, 4, ..., then $S = \frac{n(n+1)}{2}d$. If, further, $d = 1$, we have the important result:
$$[4.1] \quad 1 + 2 + 3 + 4 + \cdots + n = \frac{n(n+1)}{2}.$$

For each type of progression, we can define a *mean* between two numbers a and b as the number x that makes a, x, b a three-term progression. For a, x, b to be an arithmetic progression, we need $a - x = x - b$ or $x = \frac{a+b}{2}$. Therefore, the arithmetic mean between a and b is defined to be $\frac{a+b}{2}$. The formula for the sum of an arithmetic progression can be rewritten as $S = n\frac{a_1 + a_n}{2}$, meaning that the sum of such a progression is the number of terms times the arithmetic mean between the first and last terms.

In a geometric progression, two consecutive terms have a common ratio instead of a common difference. Letting r stand for the ratio and a for the first term, the successive terms are ar, ar^2, ar^3, etc. We can find the sum of an n-term geometric progression in two ways. The first one resembles the method used above for finding the sum of an arithmetic progression. Again, we symbolize the sum by S:
$$S = a + ar + ar^2 + \cdots + ar^{n-1}.$$
Then,
$$rS = \quad ar + ar^2 + \cdots + ar^{n-1} + ar^n,$$
And $rS - S$ or $(r - 1)S = ar^n - a = a(r^n - 1)$, so that $S = a\frac{r^n - 1}{r - 1}$.

The second method starts off the same way, by letting $S = a + ar + ar^2 + \cdots + ar^{n-1}$, but now, instead of calculating rS, we factor out the a from all the terms: $S = a(1 + r + r^2 + \cdots + r^{n-1})$. Now, from [3.7b] with $a = 1$ and $b = r$, this last parenthesis is found to be $\frac{1 - r^n}{1 - r} = \frac{r^n - 1}{r - 1}$, so that $S = a\frac{r^n - 1}{r - 1}$ again. The equivalence of the two results constitutes a proof of [3.7b].

If $|r| < 1$, we can define the sum of an *infinite* geometric progression. As n gets large, r^n gets very small, and the difference between $1 - r^n$ and 1 approaches zero. Eventually r^n is small enough that we can replace $1 - r^n$ with 1 without changing its value by more than any positive constant, however small, and the difference between $a\frac{1-r^n}{1-r}$ and $\frac{a}{1-r}$ becomes less than any pre-selected positive number. As in Sec. 3.9, this means that the limit of $a\frac{1-r^n}{1-r}$ as n increases indefinitely is $\frac{a}{1-r}$, and therefore we can say that the sum of an infinite geometric progression with $|r| < 1$ is $\frac{a}{1-r}$ where a is the first term. Therefore, $1 + \frac{1}{2} + \frac{1}{4} + \frac{1}{8} + \frac{1}{16} + \cdots = 2$ since $a = 1$ and $r = \frac{1}{2}$, and $1 - \frac{1}{2} + \frac{1}{4} - \frac{1}{8} + \frac{1}{16} - \cdots = \frac{2}{3}$ since $a = 1$ and $r = -\frac{1}{2}$.

I must be quick to point out that this theorem only applies when $|r| < 1$. Otherwise, we can produce ludicrous results like $1 + 2 + 4 + 8 + \ldots = -1$. Obviously, the sum of positive numbers cannot be negative! However, we can still apply the formula for a *finite* sequence to the sum $1 + 2 + 4 + 8 + \ldots + 2^{n-1}$ and arrive at a result first demonstrated by Euclid. Now Euclid is famous for unifying all of Greek geometry into one book, the *Elements*. But the *Elements* contained things other than geometry, some of which will be mentioned in this chapter. Euclid proved that $1 + 2 + 4 + 8 + \ldots + 2^{n-1} = 2^n - 1$ by methods which need not concern us; the same result can be found from the geometric progression sum formula with $a = 1$ and $r = 2$.

Obviously, the formula $S = a\frac{r^n - 1}{r - 1}$ cannot be used if $r = 1$. However, if $r = 1$, all the terms in the progression are equal to a, the first term, and the sum of n terms is therefore equal to na. Note that such a "progression" can be considered as an arithmetic progression with $d = 0$; the formula $S = na + \frac{n(n-1)}{2}d$ then becomes $S = na$, showing that this formula is consistent.

As with the arithmetic mean, a geometric mean can be defined between a and b as the number x that makes a, x, b a geometric progression. This requires $\frac{a}{x} = \frac{x}{b}$ or $x = \sqrt{ab}$. Those who know something about logarithms will note that $\log x = \frac{\log a + \log b}{2}$; that is, $\log x$ is the arithmetic mean of $\log a$ and $\log b$. (Logarithms will be defined rigorously in Chapter 7; for now it suffices to note that a logarithm has the property that, for all a and b, $\log ab = \log a + \log b$.) If a and b are both negative, their geometric mean is defined to be the negative square root; if they are of opposite sign, the geometric mean is not defined (unless you are willing to accept an imaginary number as one).

A harmonic progression is defined as a number sequence whose reciprocals form an arithmetic progression; that is, a sequence of the form $\frac{1}{a}, \frac{1}{a+d}, \frac{1}{a+2d}, \frac{1}{a+3d}, \ldots$ Thus, 60, 30, 20, 15, 12, 10 is a harmonic progression because it can be described as $\frac{60}{1}, \frac{60}{2}, \frac{60}{3}, \frac{60}{4}, \frac{60}{5}, \frac{60}{6}$. There is no convenient formula for the sum of a harmonic progression. A harmonic mean between two numbers can still be defined, however; by the definition of a harmonic progression, the harmonic mean between $\frac{1}{a}$ and $\frac{1}{b}$ is $\frac{1}{\frac{a+b}{2}}$ or $\frac{2}{a+b}$. Now, interchanging $\frac{1}{a}$ and a, and $\frac{1}{b}$ and b, the harmonic mean between a and b is $\frac{2}{\frac{1}{a}+\frac{1}{b}}$ or $\frac{2ab}{a+b}$. (The interchange is legal because if $\frac{1}{a} = x$, then $\frac{1}{x} = a$.)

Now comes the surprise: Denoting the arithmetic, geometric, and harmonic means between a and b by $A(a,b)$, $G(a,b)$, and $H(a,b)$ respectively, we find that $A(a,b) \cdot H(a,b) = ab$, or, equivalently, $G(a,b) = G(A(a,b), H(a,b))$. This result is interesting in itself, but what is more provocative, it leads to a method of estimating square roots. Since A, G, and H are all between a and b and $G(a,b) = \sqrt{ab}$, we can define two series $a_1, a_2, a_3, \ldots$ and $h_1, h_2, h_3, \ldots$ by $a_1 = 1$, $h_1 = n$, $a_2 = A(a_1, h_1)$, $h_2 = H(a_1, h_1)$, $a_3 = A(a_2, h_2)$, $h_3 = H(a_2, h_2)$, etc. Both the sequences $a_1, a_2, \ldots$ and $h_1, h_2, \ldots$ must converge to $\sqrt{n}$, and both consist of rational numbers exclusively. The sequence $a_1, a_2, \ldots$ is of special interest; it is exactly the sequence produced by any one of the three square-root approximation techniques in Sec. 3.9 when they are started with 1 as first guess! So we are now up to four apparently totally different methods of calculating square roots that are in fact the same procedure in various disguises. Mathematics has a few other surprises in store for its intrepid explorers, and discovering them is part of the feeling meant to be conveyed by the word "Zahlvergnugen."

4.2 Recurrence Relationships

Progressions are the simplest kinds of sequences that are defined by *recurrence relationships*. In these, each term of the sequence, except for at most a few initial ones, is given by an expression involving one or more of the preceding terms of the sequence, so that once the first n terms have been determined, the $(n + 1)$st can be calculated. Then, since the first $n + 1$ are known, the $(n + 2)$nd can be computed, and since the first $n + 2$ are now known, the $(n + 3)$rd can be calculated, and so on, forever. So the first few terms plus the recurrence relationship determine all the terms in the endless sequence. Arithmetic progressions are characterized by $s_1 = a$ and $s_{k+1} = s_k + d$ for $k > 0$. Geometric progressions have $s_1 = a$ and $s_{k+1} = rs_k$; and harmonic progressions are determined by the more complex $s_1 = a$, $s_{k+1} = \frac{s_k}{1+ds_k}$. These recurrence relationships are considered simple because they do not involve k explicitly, and only the term immediately preceding s_{k+1} enters into the calculation.

More complicated recurrence relationships, which violate one or both of the above restrictions, give rise to more interesting sequences. Among these are the factorials ($s_0 = 1$, $s_{k+1} = (k + 1)s_k$); the triangular numbers ($s_1 = 1$, $s_{k+1} = s_k + (k + 1)$); the Fibonacci sequence ($s_1 = s_2 = 1$, $s_{k+1} = s_k + s_{k-1}$); and generalizations of the above. The Fibonacci sequence 1, 1, 2, 3, 5, 8, 13, 21, 34, ... has so many interesting properties that it will have an entire section all to itself in this chapter.

Common manipulations of sequences defined by recurrence relationships include obtaining a formula for the terms in the sequence that does not involve any of the preceding terms, so that a term may be calculated without having to determine all the ones before it; determining a limiting value for the sequence, or a limiting ratio between two consecutive terms, if one exists; and finding any interesting properties of the sequence.

4.2.1 Fibonacci Numbers

The most important example of a sequence generated by a recurrence relationship is the Fibonacci sequence: $x_1 = x_2 = 1$, $x_{n+1} = x_n + x_{n-1}$. It is important both for its many interesting properties that its terms have, and because terms (called Fibonacci numbers) show up all throughout nature. As an example, the arrangement of the florets in composite flowers such as the daisy is controlled by Fibonacci numbers, usually 21 and 34. The Fibonacci sequence, hereafter denoted with an F, so that F_n represents the nth Fibonacci number, also has an intimate connection with the aesthetically pleasing Golden Ratio, which is defined as the positive number x satisfying $x^2 = x + 1$.

This quadratic equation can be solved by use of [3.11] to get $x = \frac{1}{2}(\sqrt{5} + 1) \approx 1.618$. A rectangle with sides in this ratio can be cut into a square and another rectangle the same shape as the original, which is apparent from the ratio's defining equation and Fig. 4-1. The Golden Ratio is usually denoted by φ (Greek letter phi = English ph or f). From the defining equation it is readily apparent that successive powers of φ satisfy the Fibonacci recurrence relationship: $\varphi^{n+1} = \varphi^n + \varphi^{n-1}$. If successive Fibonacci numbers are divided larger by smaller, the resulting quotients approach φ as a limit: $5/3 = 1.667$, $8/5 = 1.600$, $13/8 = 1.625$, $21/13 = 1.615$, $34/21 = 1.619$.

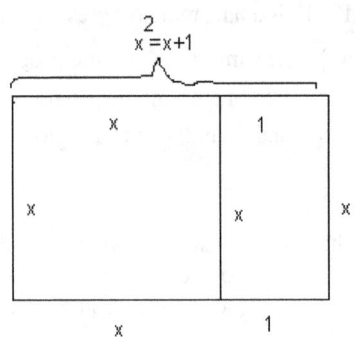

It should come as no surprise that the formula for F_n, the nth Fibonacci number, involves φ. It also contains the negative solution to $x^2 = x + 1$, denoted by φ': $\varphi' = \frac{1}{2}(1 - \sqrt{5}) \approx -0.618 = 1 - \varphi = \varphi - \sqrt{5} = -\frac{1}{\varphi}$. In terms of φ and φ', the nth Fibonacci number, $F_n = \frac{1}{\sqrt{5}}(\varphi^n - (\varphi')^n)$. Since both φ and φ' satisfy $x^{n+1} = x^n + x^{n-1}$, the Fibonacci numbers do indeed satisfy their recurrence relationship. As for the initial values,

Fig. 4-1. A rectangle with sides in the Golden Ratio can be cut into a square and a rectangle similar to the original.

$$\varphi - \varphi' = \frac{1}{2}(1 + \sqrt{5}) - \frac{1}{2}(1 - \sqrt{5}) = \frac{1}{2}(2\sqrt{5}) \quad \text{and} \quad F_1 = $$
$$\frac{1}{\sqrt{5}}(\varphi - \varphi') = 1, \text{ while } F_2 = \frac{1}{\sqrt{5}}(\varphi^2 - (\varphi')^2) = \frac{1}{\sqrt{5}}\left(\frac{1}{4}(6 + 2\sqrt{5}) - \frac{1}{4}(6 - 2\sqrt{5})\right) = \frac{1}{\sqrt{5}}\left(\frac{1}{4}(4\sqrt{5})\right) = 1 \text{ as well.}$$

One immediate consequence of the Fibonacci recurrence relationship is that every positive integer can be written as the sum of Fibonacci numbers without using two consecutive Fibonacci numbers at any point in the summation. For example, $100 = 89 + 8 + 3$, $101 = 89 + 8 + 3 + 1$, $102 = 89 + 13$, and $103 = 89 + 13 + 1$. It turns out that there is only one way to so subdivide each integer because $F_1 + F_3 + F_5 + \cdots + F_{2n-1} = F_{2n}$ and $F_2 + F_4 + F_6 + \cdots + F_{2n} = F_{2n+1} - 1$. To prove these, write $F_1 = F_2$, $F_3 = F_4 - F_2$, $F_5 = F_6 - F_4$, etc., and add these equations so that everything cancels except F_{2n}; for the second, $F_2 = F_3 - F_1$, $F_4 = F_5 - F_3$, $F_6 = F_7 - F_5$, etc.

Note that we have $F_{2n} = F_{2n-1} + F_{2n-3} + \cdots + F_5 + F_3 + F_1$. At first glance, this does not seem to use two consecutive Fibonacci numbers, but remember that F_1 and F_2 are both 1, so $F_3 = 2$ and F_1 are, in a way, consecutive.

As another example of manipulating the recurrence relationship, let us begin with $F_n = F_{n-1} + F_{n-2}$, and repeatedly split the higher-indexed term on the right in two: $F_n = 2F_{n-2} + F_{n-3} = 3F_{n-3} + 2F_{n-4} = 5F_{n-4} + 3F_{n-5} = \cdots$. As can be seen, the coefficients appearing in the successively generated expressions are themselves Fibonacci numbers. The general formula is $F_n = F_{k+1}F_{n-k} + F_k F_{n-(k+1)}$. From the individual equations above, we can see that every third Fibonacci number is even, every fourth is a multiple of 3, and every fifth a multiple of 5. The general formula tells us that if a is a multiple of b, then F_a is a multiple of F_b. (Example: $F_8 = 21$ is a multiple of $F_4 = 3$.) This result can be turned around to show that F_n can be prime only if n is prime, with the exception of F_4, since it must be a multiple only of $F_2 = 1$. The smallest prime n for which F_n is composite is 19.

A more advanced (and therefore more intriguing) result can be discovered by focusing on any one F_n. If we square this number, and multiply the two Fibonacci numbers around it together, the two answers will always differ by 1. (Examples: $13^2 = 169$ and $8 \times 21 = 168$; $21^2 = 441$, $13 \times 34 = 442$.) Symbolically, $F_n^2 = F_{n-1}F_{n+1} \pm 1$. The square is sometimes greater, sometimes less; the reader may conduct "numerical experiments" to see precisely when each case holds, or (s)he may read the discussion below.

By widening the focus further away from F_n^2, we find that the product of the Fibonacci numbers two away from F_n is also one away from F_n^2 (in the other direction, of course); the product of the Fibonacci numbers three away from F_n differs from F_n^2 by four $(= F_3^2)$ in the same direction as the innermost pair, and so on. Symbolically, $F_{n+k}F_{n-k} = F_n^2 \pm F_k^2$. We prove this formula and find the rule determining the choice of the $\pm$ sign from the formula for F_n, remembering that $\varphi\varphi' = -1$ and therefore $\varphi^a (\varphi')^b = (-1)^b \varphi^{a-b}$ if $a > b$, $(-1)^a (\varphi')^{b-a}$ if $a < b$, and finally $(-1)^a$ if $a = b$. Now, $F_{n+k}F_{n-k} = \frac{1}{5}(\varphi^{2n} + (\varphi')^{2n} - (-1)^{n-k}\varphi^{2k} - (-1)^{n-k}(\varphi')^{2k})$, $F_n^2 = \frac{1}{5}(\varphi^{2n} + (\varphi')^{2n} - 2(-1)^n)$, and $F_k^2 = \frac{1}{5}(\varphi^{2k} + (\varphi')^{2k} - 2(-1)^k)$. Therefore, $F_{n+k}F_{n-k} - F_n^2 = \frac{1}{5}(2(-1)^n - (-1)^{n-k}\varphi^{2k} - (-1)^{n-k}(\varphi')^{2k})$. Then, $(-1)^{n-k}F_k^2 = \frac{1}{5}(-2(-1)^n + (-1)^{n-k}\varphi^{2k} + (-1)^{n-k}(\varphi')^{2k}) = -(F_{n+k}F_{n-k} - F_n^2)$. This means $F_{n+k}F_{n-k} = F_n^2 - (-1)^{n-k}F_k^2$, and the $\pm$ sign is $-$ if $n - k$ is even and $+$ if $n - k$ is odd.

A further useful result occurs when we let $k = n - 2$ or $n - 1$ in the last formula, so that F_{n-k} is 1. We thus derive $F_{2n-1} = F_n^2 + F_{n-1}^2$ and $F_{2n-2} = F_n^2 - F_{2n-2}^2$. Examples: $F_{13} = 233 = F_7^2 + F_6^2 = 169 + 64$ and $F_{14} = 377 = F_8^2 - F_6^2 = 441 - 64$.

Now we manipulate the recurrence relationship again. Instead of splitting just one term on the right, we decompose all of them.

$$\begin{aligned} F_n &= F_{n-1} + F_{n-2} \\ &= F_{n-2} + 2F_{n-3} + F_{n-4} \\ &= F_{n-3} + 3F_{n-4} + 3F_{n-5} + F_{n-6} \\ &= F_{n-4} + 4F_{n-5} + 6F_{n-6} + 4F_{n-7} + F_{n-8} \\ &= F_{n-5} + 5F_{n-6} + 10F_{n-7} + 10F_{n-8} + 5F_{n-9} + F_{n-10}, \text{ etc.} \end{aligned}$$

Not only have we created a "Christmas tree" of equations (we can place F_n on top to serve as a "star"), but we have also created a new pattern of numbers in the coefficients. Each coefficient is the sum of the two above it (remember that a "missing" coefficient is a one). These numbers are called *binomial coefficients* because of their connection with powers of the binomial $(x + y)$; the reader may recognize the patterns in [3.2a] and [3.3a] in the last chapter as appearing in the second and third rows above. The coefficient appearing in the nth row and $(r + 1)$st position in that row is denoted by $\binom{n}{r}$; thus $\binom{3}{1} = 3, \binom{2}{0} = 1, \binom{5}{3} = 10$, and $\binom{4}{4} = 1$. Binomial coefficients will be discussed further later in this chapter and in Chapter 9.

To find out what this means for Fibonacci numbers, we let $n = 2k$ be an even number and continue the Christmas tree until the bottom right term is F_0, which is defined to be zero. The last row now reads $F_{2k} = \binom{k}{0}F_k + \binom{k}{1}F_{k-1} + \binom{k}{2}F_{k-2} + \cdots + \binom{k}{k}F_0$. (Example: $F_{10} = 55 = 1 \cdot 5 + 5 \cdot 3 + 10 \cdot 2 + 10 \cdot 1 + 5 \cdot 1 + 1 \cdot 0 = 5 + 15 + 20 + 10 + 5$.) If n is odd, we represent it by $2k + 1$ and stop when the bottom right term is F_1; the last row reads $F_{2k+1} = \binom{k}{0}F_{k+1} + \binom{k}{1}F_k + \binom{k}{2}F_{k-1} + \cdots + \binom{k}{k}F_1$. (Example: $F_{11} = 89 = 1 \cdot 8 + 5 \cdot 5 + 10 \cdot 3 + 10 \cdot 2 + 5 \cdot 1 + 1 \cdot 1 = 8 + 25 + 30 + 20 + 5 + 1$.)

We now switch from analyzing the entire sequence to studying its individual terms. It should be obvious that no integer larger than 1 can divide both of two consecutive Fibonacci numbers. For, if two consecutive Fibonacci numbers were both multiples of k, then their difference would be also. But that difference is the previous Fibonacci number. So if both F_n and F_{n+1} are multiples of k, then F_{n-1} must be as well. Then, since F_n and F_{n-1} are multiples of k, F_{n-2} must be one. And, now that k divides both of F_{n-1} and F_{n-2}, it must divide F_{n-3}. And so we continue backwards in the sequence until we reach F_2. But $F_2 = 1$, and the only number that 1 is a multiple of is 1 itself. This proves that if k is a number that divides both F_n and F_{n+1}, then $k = 1$, which is the same thing as proving that no two consecutive Fibonacci numbers can have a common factor greater than 1.

Next, we note that every integer is a factor of some Fibonacci number; in particular, there is a Fibonacci number that ends in a million zeroes. (Don't ask for it, however; it probably has about $10^{1,000,000}$ digits, which is far too many to write down – even on a sheet of paper the size of the Milky Way, with a pen smaller than the smallest virus.) To show this, let n denote the prospective factor, and write down the Fibonacci numbers modulo n; that is, whenever the process of adding the last two to get the next takes you past n, simply subtract n from the total. The Fibonacci sequence modulo 20 begins 1, 1, 2, 3, 5, 8, 13, 1, 14, 15, 9, 4, 13, 17, 10, 7, 17, 4, 1, 5, 6, 11, 17, 8, ... Once we have a pair of consecutive Fibonacci numbers modulo n, the rest of the sequence modulo n can be determined, as can the sequence preceding that pair. (Example: the number before the pair 15, 9 in the modulo-20 sequence must be $9 - 15$ modulo $20 = 29 - 15 = 14$.) Only numbers from zero to $n - 1$ can appear in the sequence modulo n. Since there are only a finite number of pairs possible, one of them must eventually appear twice. The first repeated pair must be 1, 1 (since we can work the sequence backwards, it is this first pair which must be the first repeated one), and the term before the second occurrence must be a zero – meaning that the actual Fibonacci number corresponding to that zero must be the desired multiple of n. (The smallest Fibonacci number divisible by 20 is F_{30}.) This proof not only guarantees the existence of a Fibonacci multiple of n but places a limit on how far we have to go to find one as well. Since there are only n^2 possible pairs of numbers between 0 and $n - 1$, the first Fibonacci multiple of n can be no further out than F_{n^2}. Closer analysis can push this limit in considerably: For $n > 2$, we need go no farther than $F_{n\varphi(n)}$, where $\varphi(n)$ is Euler's *totient* function, the number of positive integers smaller than n that have no common factor with n. Usually, even this limit is generous.

If a sequence is started with two arbitrary integers, but follows the Fibonacci recurrence rule, it will have properties similar to the original Fibonacci sequence; in particular, the limiting ratio will still be φ. This type of sequence is called a "generalized Fibonacci sequence."

4.3 Figurate Numbers

We now discuss series of numbers called *figurate* because dots in these numbers can be arranged into regular geometric figures. For example, 15 is considered a *triangular* number because 15 dots can be arranged into an equilateral triangle. (See Fig. 4-2(a).), and 16 is a square number because of the square formation shown in Fig. 4-2(b).

Because the 15-dot triangle has 5 dots on each side, 15 is the fifth triangular number. The nth triangular number will be denoted by $\Delta(n)$ in this book.

(a)   (b)

By visualizing the 15-dot array with one of its sides removed, it is apparent that $\Delta(n) = \Delta(n - 1) + n$, and by removing further rows until there are no dots left, the relation $\Delta(n) = 1 + 2 + 3 + \cdots + n$ can be demonstrated. Equation [4.1] gives the formula for this sum, so $\Delta(n) = \frac{n(n+1)}{2}$.

Fig. 4-2. (a) 15 is the fifth triangular number.
(b) 16 is the fourth square number.

It is often convenient to reverse the process and find out whether a given number is triangular, and, if it is, the side of the triangle corresponding to it. With analogy to square roots, this reverse number is called the triangular root of the original number, symbolized in this book by $\sqrt[\Delta]{n}$. Hence, $n = \frac{\sqrt[\Delta]{n}(\sqrt[\Delta]{n}+1)}{2}$, which is quadratic in $\sqrt[\Delta]{n}$. Reducing this equation to $ax^2 + bx + c = 0$ produces $(\sqrt[\Delta]{n})^2 + \sqrt[\Delta]{n} - 2n = 0$, which, by [3.11], has solutions $\frac{-1\pm\sqrt{8n+1}}{2}$. Since only positive numbers can make sense for sides of triangles, and $\sqrt[\Delta]{n}$ is a side of a triangle, only the + sign need be used in the solution, which can be rewritten $\sqrt[\Delta]{n} = \frac{\sqrt{8n+1}-1}{2}$. Now, n is a triangular number if and only if this formula produces an integer, which happens only if $\sqrt{8n + 1}$ is an integer; therefore, n is triangular if and only if $8n + 1$ is square. We test this by substituting $15 = \Delta(5)$ for n and find $\sqrt[\Delta]{15} = \frac{\sqrt{8\times15+1}-1}{2} = \frac{\sqrt{121}-1}{2} = \frac{10}{2} = 5$.

One more example of plane figurate numbers is the *hexagonal* series, H_n. $H_3 = 19$ as shown in Fig. 4-2(c). It is obvious that successive members of the hexagonal series differ by multiples of 6: $H_1 = 1$, $H_2 = H_1 + 6$, $H_3 = H_2 + 12$, ..., $H_{n+1} = H_n + 6n$. This means that $H_n = 6\Delta(n-1) + 1 = 3n(n-1) + 1$, thus, $H_3 = 3$ x 3 x 2 + 1 = 19.

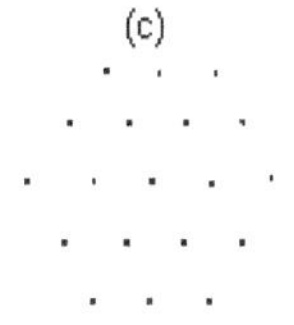

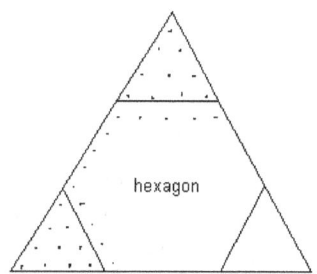
Fig. 4-2(c). 19 is the third hexagonal number.

More interesting, however, is a relation based on cutting a hexagon out of a larger triangle, as shown in Fig. 4-3. When these shapes are filled with dots, the relation $H_n = \Delta(3n - 2) - 3\Delta(n-1)$ can be read off. For H_3, we have $\Delta(7) - 3\Delta(2) = 28 - 3 \cdot 3 = 19$. Equating the two formulae for H_n yields $6\Delta(n-1) + 1 = \Delta(3n-2) - 3\Delta(n-1)$ or $9\Delta(n-1) + 1 = \Delta(3n-2)$. Replacing $n-1$ by n in this equation results in $9\Delta(n) + 1 = \Delta(3n+1)$, meaning that if x is triangular, so is $9x + 1$.

Fig. 4-3. A large triangle cut into three small triangles and a hexagon.

Now, we look at numbers that are simultaneously square and triangular. Since y is triangular if $8y + 1$ is square, y^2 is simultaneously square and triangular if $8y^2 + 1$ is square; that is, if there exists an integer x such that $x^2 = 8y^2 + 1$ or $x^2 - 8y^2 = 1$. This is a special case of the *Pell equation* $x^2 - ny^2 = 1$, where n is not a square. Any solution to a Pell equation, other than the trivial one $x = 1$ and $y = 0$, gives rise to an infinite sequence of such solutions, as will be demonstrated.

If $x^2 - ny^2 = 1$, then $(x^2 - ny^2)^2$ is also 1, and this can be expanded to $[x^4 + n^2y^4] - n[2x^2y^2] = 1$. Adding $2nx^2y^2$ to the first bracket and $2x^2y^2$ to the second (which is equivalent to subtracting $2nx^2y^2$, since the second bracket is multiplied by $-n$) produces $[x^4 + 2nx^2y^2 + n^2y^4] - n[4x^2y^2] = 1$. Now both bracketed terms are squares and we have $[x^2 + ny^2]^2 - n[2xy]^2 = 1$. This means that if x and y satisfy $x^2 - ny^2 = 1$, then so do $x^2 + ny^2$ and $2xy$. Since $x^2 - ny^2 = 1$, $x^2 + ny^2 = 2x^2 - 1$, so the solution can be rewritten as $2x^2 - 1$ and $2xy$, which is $2x(x) - 1$ and $2x(y) - 0$. In other words, the next solution to $x^2 - ny^2 = 1$ is obtained by multiplying the previous one by $2x$ and subtracting the one smaller than that − if the "smaller than that" is the trivial solution. We can use a technique called *mathematical induction* to extend this two-step ladder into an infinite one.

To this end, let $s_1 = (x_1, y_1)$ be the ordered pair containing the smallest non-trivial solution to $x^2 - ny^2 = 1$, $s_0 = (1, 0)$ be the ordered pair corresponding to the trivial solution, and $s_k = (x_k, y_k)$ be the kth solution. Our job is to show that $s_{k+1} = 2x_1s_k - s_{k-1}$ is also a solution, assuming that $s_k = 2x_1s_{k-1} - s_{k-2}$, and so on down to $s_2 = 2x_1s_1 - s_0$, which was demonstrated above.

We assume that $x_{k+1}^2 - ny_{k+1}^2 = 1$ and expand s_{k+1} as $2x_1s_k - s_{k-1}$. This gives us $(2x_1x_k - x_{k-1})^2 - n(2x_1y_k - y_{k-1})^2 = 1$. Expanding this and simplifying, remembering that s_k and s_{k-1} both satisfy the Pell equation, the condition to be satisfied if s_{k+1} is to be a solution is $x_1 = x_kx_{k-1} - ny_ky_{k-1}$. Substituting $s_k = 2x_1s_{k-1} - s_{k-2}$ and again using the fact that s_{k-1} satisfies the Pell equation, the condition is reduced to $x_1 = x_{k-1}x_{k-2} - ny_{k-1}y_{k-2}$. Clearly, this can go on until the subscripts on the x's and y's are 1 and 0: $x_1 = x_1x_0 - ny_1y_0$. Since $x_0 = 1$ and $y_0 = 0$ for all n, the necessary condition is satisfied. The logic of mathematical induction (which resembles an infinite string of falling dominoes, and will be studied in more detail in Chapter 6) then proves that all solutions (x_k, y_k) to $x^2 - ny^2 = 1$ can be produced from the first non-trivial one (x_1, y_1) by the recurrence relationship $s_k = 2x_1s_{k-1} - s_{k-2}$. Finding the first non-trivial solution to $x^2 - ny^2 = 1$ for certain values of n can be very difficult; when $n = 61$, $x_1 = 1,766,319,049$ and $y_1 = 226,153,980$. (And if you think that was fun, try $n = 10,169$. The first solution for that n is in more than 80 digits!)

To apply this to square triangular numbers, recall that y^2 is such a number if $x^2 - 8y^2 = 1$ for some integer x. The first non-trivial solution is not difficult to find; since $8 = 3^2 - 1$, $(3, 1)$ is the first such solution. Hence, 1 is the smallest number that is both square and triangular (no surprise there!). The square root of such a number, of course, is given by y; the triangular root by the formula for $\sqrt[\Delta]{n}$ as $\frac{\sqrt{8y^2+1}-1}{2} = \frac{x-1}{2}$ since (x, y) satisfy $x^2 - 8y^2 = 1$. Further solutions to $x^2 - 8y^2 = 1$ are found by the rule just proven:

$$s_2 = 2x_1s_1 - s_0 = 6(3,1) - (1,0) = (17,6)$$
$$s_3 = 2x_1s_2 - s_1 = 6(17,6) - (3,1) = (99,35)$$
$$s_4 = 2x_1s_3 - s_2 = 6(99,35) - (17,6) = (577,204), \text{ etc.}$$

From these we calculate:

$$T_2 = y_2^2 = 36; \sqrt{T_2} = 6, \sqrt[\triangle]{T_2} = 8$$
$$T_3 = y_3^2 = 1,225; \sqrt{T_3} = 35, \sqrt[\triangle]{T_3} = 49$$
$$T_4 = y_4^2 = 41,616; \sqrt{T_4} = 204, \sqrt[\triangle]{T_4} = 288, \text{ etc.}$$

From the recurrence relationship $y_n = 6y_{n-1} - y_{n-2}$ and the theory of generating functions (which I would like to include in this book but can't), y_k can be shown to equal $\frac{(1+\sqrt{2})^{2k} - (1-\sqrt{2})^{2k}}{4\sqrt{2}}$, and thus $T_k = y_k^2 = \frac{(1+\sqrt{2})^{4k} + (1-\sqrt{2})^{4k} - 2}{32}$. This, in turn, can be used to prove the first assertion at the beginning of this chapter: $T_k = 34T_{k-1} - T_{k-2} + 2$. As for x_k, it satisfies the same recurrence formula as y_k but has a different starting point since $x_0 = 1$ and $x_1 = 3$. Hence, its formula is slightly different: $x_k = \frac{(1+\sqrt{2})^{2k} + (1-\sqrt{2})^{2k}}{2}$.

The numbers $1 + \sqrt{2}$ and $1 - \sqrt{2}$ are interesting in their own right. Both these numbers satisfy $x^{n+2} - 2x^{n+1} - x^n = 0$, as can be seen by solving the related quadratic equation $x^2 - 2x - 1 = 0$. Hence, the sequence $(1 + \sqrt{2})^n$ satisfies $x_n = 2x_{n-1} + x_{n-2}$, and so must the sequences $\frac{(1+\sqrt{2})^n + (1-\sqrt{2})^n}{2}$ and $\frac{(1+\sqrt{2})^n - (1-\sqrt{2})^n}{2\sqrt{2}}$, which are the integer and $\sqrt{2}$-associated parts of $(1 + \sqrt{2})^n$ respectively. $1 + \sqrt{2}$ and $1 - \sqrt{2}$ are related to each other in ways similar to φ and φ' of the Fibonacci sequence.

These numbers can also be used to prove the second assertion at the opening of this chapter. For reasons that will be seen in Chapter 5, importance is attached to sets of three integers a, b, and c such that $a^2 + b^2 = c^2$. These sets are called *Pythagorean triples* after the theorem of Pythagoras which will appear there. If, in addition, a, b, and c have no common divisor other than one, the triple is called *primitive*. All primitive Pythagorean triples can be generated by the following mechanism: Let m and n be two positive integers with no common divisor except 1 and one of them be odd and the other even. Then the numbers $2mn$, $m^2 - n^2$, and $m^2 + n^2$ are a primitive Pythagorean triple, with $m^2 + n^2$ playing the role of c. In particular, if m and n are consecutive members of the sequence $\frac{(1+\sqrt{2})^k - (1-\sqrt{2})^k}{2\sqrt{2}}$, $2mn$ and $m^2 - n^2$ will differ by exactly one. The sequence begins 1, 2, 5, 12, 29, 70, ..., and the numerical calculations give:

m	n	$2mn$	$m^2 - n^2$	$m^2 + n^2$	Sum of smallest and largest	Half the sum of all three
2	1	4	3	5	8	6
5	2	20	21	29	49	35
12	5	120	119	169	288	204

As can be seen by comparing this table with the numerical results immediately above, the last two columns derived from the Pythagorean triples are the triangular and square roots of simultaneously square and triangular numbers, establishing the second assertion at the opening of this chapter.

There are also several puzzle-type situations related to the sequences based on $1 + \sqrt{2}$, but exploring them would be spending too much time on one topic. The reader may wish to explore further by trying to generalize the equations $1 + 2 + 3 + 4 + 5 = 7 + 8$ and $\frac{20 \times 21}{14 \times 15} = 2$. (Hints: In the first equation, all the numbers from 1 to **8** are involved, except for **6**; in the second, **20** and **21** are in the numerator, while the denominator's factors add up to **29**.)

Figurate numbers can form three-dimensional patterns as well as two-dimensional ones. (Think of tennis balls instead of dots.) There are two main types of pyramidal numbers: square-pyramidal, whose arrays are shaped like Egyptian pyramids; and triangular-pyramidal or tetrahedral, whose arrays are shaped like the "Pyraminx," a Rubik's Cube spinoff with 4 triangular sides instead of 6 square ones. (This shape, known formally as a *tetrahedron*, is the simplest possible three-dimensional shape.)

Square-pyramidal numbers are sums of consecutive squares: 1, 5, 14, 30, 55, ... Denoting them by $Sp(n)$, we have $Sp(n) = Sp(n - 1) + n^2$. The method of finite differences, to be explained later in this chapter, produces the

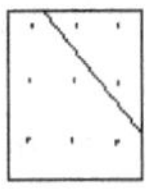

Fig. 4-4. Each square number is the sum of two consecutive triangular numbers.

formula $Sp(n) = \frac{n(n+1)(2n+1)}{6}$. For $n = 5$, $Sp(5) = \frac{1}{6}(5 \times 6 \times 11) = 55$. Tetrahedral numbers are sums of consecutive triangular numbers: 1, 4, 10, 20, 35, ... Any square pyramid of tennis balls can be split into two consecutive tetrahedra in the same way that a square can be split into two consecutive triangles. (See Fig. 4-4.) Denoting tetrahedral numbers by $Tp(n)$, we have $Tp(n) = Tp(n-1) + \Delta(n)$, and, by a method similar to the one for $Sp(n)$, $Tp(n) = \frac{n(n+1)(n+2)}{6}$.

Since there are an infinite number of simultaneously square and triangular numbers, the natural question now is: Are there numbers that are simultaneously square- and triangular-pyramidal? The answer is likely to be no, except for the trivial case of 1, because there is no equivalent to Pythagorean triples in three dimensions; that is, there does not exist a set of three integers a, b, and c such that $a^3 + b^3 = c^3$ (unless a or b is allowed to be zero). There is, however, a number that is simultaneously square and square-pyramidal: $4,900 = 70^2 = Sp(24)$. That 4,900 is the *only* such number is very difficult to prove.

4.4 Number Theory

The branch of mathematics known as number theory is defined as the study of the integers, especially the positive ones. In a broad sense, this entire chapter is number theory, but this section focuses on the part growing out of the concept of divisibility.

4.4.1 Divisibility Rules

In Chapter 1, I mentioned the idea of "casting out nines" and proved that the sum of the digits of any multiple of 9 is also a multiple of 9. This sum-of-digits rule is one of several divisibility rules for integers. All of the rules can be verified by considering the place-value analysis of a general number, as was done for the 9 test in Sec. 1.1.2.

To test a number for divisibility by 2^n, where n is a positive integer, it suffices to divide the last n digits by 2^n. The digits before the last n all have place values of at least 10^n, and, since $10 = 2 \times 5$, $10^n = 2^n \times 5^n$ is divisible by 2^n, as is 10^m for $m > n$. Thus, the number 7,503,841,352 is divisible by 8 ($ = 2^3$) because the last three digits 352 = 8 x 44. Since 5 is also a factor of 10, the same principle applies to its powers; thus 482,907,375 is divisible by $5^3 = 125$ because 375, the last three digits, is 3 x 125.

The sum-of-the-digits rule for 9 works equally well for its factor, 3; a number is divisible by 6 if it passes the tests for 6's factors, 2 and 3. This observation is true in general: a number with factors other than one can be checked by testing for all its factors, or better yet, by all its highest prime-power factors. (A prime number is one with no factors other than itself and one. Primes will be dealt with more fully in the next section.)

This leaves 7 as the smallest number without a divisibility rule so far. It is a prime number, not being divisible by 2 (and therefore 4 and 6), 3, or 5. The best way to test a number for divisibility by 7 runs as follows: Split the number into groups of two digits, working from right to left; 327,498,085 becomes 3|27|49|80|85. We now divide each of these small groups by 7, and write down the remainders, being careful not to disturb their order. For the number above, we have 3, 6, 0, 3, 1. Now, working from left to right, we double the first remainder, add it to the second, and subtract 7's if possible. This gives us a new remainder to be operated on in like fashion. The original number is a multiple of 7 if and only if the addition of the last remainder and subtraction of 7's results in zero. The calculations for the example above go 3 x 2 = 6, + 6 = 12, - 7 = 5, x 2 = 10, + 0 = 10, - 7 = 3, x 2 = 6, + 3 = 9, - 7 = 2, x 2 = 4, + 1 = 5. The final result is not zero, so 327,498,085 is not a multiple of 7.

The rule for 7 is justified by noting that each two-digit group has a place value exactly 100 times greater than the group to its right, and that 100 = (14 x 7) + 2. Thus 53 hundreds = 53 x 100 ones = 53 x ((14 x 7) + 2) ones = (53 x 14 x 7) + (53 x 2) ones. We discard the part that is multiplied by 7 and are left with 53 x 2 ones. In other words, when testing for divisibility by 7, 53 hundreds are equivalent to 53 x 2 ones – the number of hundreds is doubled to convert them into equivalent ones.

The next prime number, 11, has a simpler rule. Simply check off every other digit in your test number, then add together the checked-off digits and, separately, the unchecked ones. Subtract the smaller total from the larger. If the result is zero or consists of two identical digits, your number is a multiple of 11. If it is a non-zero single digit or consists of two different digits, the original number is not a multiple of 11. If your result has three or more digits (this takes a very large number!), repeat the process until you have a number less than 100. Thus, for 7,283,598,608, the underlined digits total 28, as do the non-underlined ones, so this number is a multiple of 11.

The rule for 11 works because $100 = (11 \times 9) + 1$, while $10 = (11 \times 1) - 1$. By reasoning similar to that done for 7 above, this means that hundreds (and ten-thousands, millions, etc.) are equivalent to ones, while tens (thousands, hundred-thousands, etc.) are equivalent to negative ones when testing for divisibility by eleven.

Larger numbers have divisibility rules as well, but in general they are no more efficient than actually dividing by the number in question. We pass now to their results: the presence or absence of factors for the number that is being divided into.

4.4.2 Factors and Primes

A number (positive integer) a is said to have *factors* b and c (also positive integers) if and only if $bc = a$. Since $1 \times a = a$ no matter what a is, all numbers have one and themselves as factors. A number is called *prime* if those are its only factors. (With the exception of 1 itself; because of 1's special role in multiplication, it is referred to as the *unit* or *identity*.) A number is called *composite* if it has factors other than 1 and itself. Thus, 10 is composite because $2 \times 5 = 10$ and neither 2 nor 5 is either 1 or 10. However, 7 is prime since it does not have 2, 3, 4, 5, or 6 as a factor, as was shown in the last section.

We note that factors always come in pairs: with 2 as a factor of 10, must come $10 \div 2 = 5$, and with 14 as a factor of 210 must come $210 \div 14 = 15$. Because multiplication is always increasing (that is, if $a > b$ and $c > d$ then $ac > bd$ if all these numbers are positive), at least one of the factors in each pair must be at least as great as the number's square root, and the other no larger than the square root. Hence, almost all numbers must have an even number of factors. The exceptions are numbers whose square roots are integers, such as 36: when we write $36 = 6 \times 6$, we are demonstrating a factor "pair" both of whose members are the same number. Since 6 can only count once in the list of factors of 36, the latter number must have an odd factor count, and this consideration holds for all perfect squares. (All numbers can have only one positive square root.)

While breaking 210 down into 14×15, you may have noticed that both 14 and 15 have smaller factors of their own: $14 = 2 \times 7$ and $15 = 3 \times 5$. Hence, $210 = 2 \times 7 \times 3 \times 5$, or, rearranging these factors in numerical order, $210 = 2 \times 3 \times 5 \times 7$. All of these numbers are prime, so the process of breakdown cannot go any further. Like a chemist who has broken down sugar into hydrogen, carbon, and oxygen, we have reached the fundamental building blocks. These are called the prime factors of 210 (or any other number we may start with).Note that if we started out splitting 210 into 21×10, we still get $3 \times 7 \times 5 \times 2 = 2 \times 3 \times 5 \times 7$ when we are through. Like the chemist, who will always find 22 hydrogen atoms, 12 carbon atoms, and 11 oxygen atoms in every molecule of table sugar he looks at, we will always get 2, 3, 5, and 7 as the prime factors of 210 no matter how we break down the number. But unlike the chemist, who will get 22 H, 12 C, and 11 O from several other substances (mostly other kinds of sugar), we get 2, 3, 5, and 7 from 210 and 210 only, for the obvious reason that there is only one product of these four numbers.

The two facts just stated about the prime factors of 210 can be generalized into the Fundamental Theorem of Arithmetic: Every positive integer has a unique set of prime factors. (For a number like 36, we must modify the notion of "set" to allow the same number to appear more than once in the set of prime factors; $36 = 2 \times 2 \times 3 \times 3$, so both 2 and 3 are in the set twice, so to speak.) To prove this theorem, we note that any integer N is either divisible by 2 or not; if it is, we write $N = 2N'$, then N' is an integer that is either divisible by 2 or not, and so on until we have an integer M that is odd and we have written N in the form $2^p M$, where p is a non-negative integer (possibly zero). Once M is determined, we work on it and the next prime number, 3, in the same way as we dealt with N and 2. This produces a number L and a non-negative integer q such that L is not divisible by 3 and $N = 2^p 3^q L$. It is important to note that L cannot be divisible by 2 either since it is a factor of M which is not; if L was divisible by 2, then its multiple M would have to be as well. (We have $M = 3^q L$; if L was $2K$, then M would be $2 \times 3^q K$.) We continue in this way, producing non-negative integers r, s, t, etc. such that $N = 2^p \times 3^q \times 5^r \times 7^s \times 11^t$, etc. times a positive integer that eventually gets reduced to 1, and at no time is divisible by any of the prime numbers whose exponents have previously been determined. When the "unresolved" part of N has been reduced to 1, or more frequently a prime number, we have completed the set of N's prime factors. Using the number 84 as an example, we successively derive $84 = 2 \times 42$ (still even) $= 2^2 \times 21$ (odd) $= 2^2 \times 3 \times 7$ (prime, completes the set of prime factors).

It should be fairly obvious that this algorithm can produce only one set of prime factors for any given N. To prove the Fundamental Theorem of Arithmetic, we must show that any other method must produce the same set also. To that end, suppose another algorithm produces a prime factor x of N to which our first algorithm assigned a zero exponent to. Two cases must be distinguished: (1) x is smaller than the largest prime assigned a positive exponent, in which case the above algorithm would have tested for divisibility by x eventually and found it (since N, by hypothesis, is divisible by x), and (2) x is larger than the largest prime assigned a positive exponent, in which case the above algorithm would not have terminated until x was reached. There remains the possibility that two different algorithms might produce the same set of *distinct* prime factors for N, but with different exponents for some of them. In this case, we divide N by a prime raised to the smaller of the two different powers assigned to it, thus reducing the problem to the previous one in which that prime appears in only one set of factors. (This trick of changing a problem into one that has already been solved is another favorite technique of mathematicians.)

4.4.3 The Sieve of Eratosthenes and Other Primality Tests

We have yet to show how to determine whether a given number is prime or composite. By definition, a prime number has no factors other than itself and one, so dividing a number by everything between 2 and one less than itself will weed out the composites. But this method is very inefficient. The observation at the beginning of Sec. 4.4.2 that multiplication is always increasing implies that we can stop dividing once we have reached the number's square root. In addition, once we have divided our number by 2 and found a remainder, there is obviously no need to try 4, 6, 8, etc. as possible factors; in general, it is only necessary to try prime numbers as possible factors. With these improvements, the method becomes fairly efficient and acquires the name of a classical Greek mathematician: it is called the sieve of Eratosthenes.

The sieve is best illustrated by using it on all the numbers less than a moderate upper limit, say, 100. Arranging the numbers as in Fig. 4-5 makes it easier to cross out multiples of 2, 3, and 5, as they fill columns and/or diagonals in the diagram. Multiples of 7 also fall on fairly convenient lines as 3 x 7 = 21 is right next to a multiple of 10. Since the next prime, 11, is greater than $\sqrt{100}$, we are finished. The uncrossed-out numbers are the primes less than 100 and are circled for extra emphasis. There are 25 of them and, except for 2 and 5, all are found in the four columns ending in 1, 3, 7, and 9. The reason should be clear: any other last digit means that a number bearing it is a multiple of 2, 5, or both.

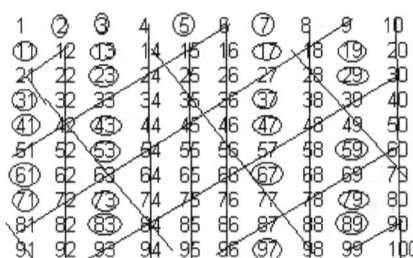

Fig. 4-5. The sieve of Eratosthenes picks out all the prime numbers less than 100.

We may think of 2, 3, 5, and 7 as representing holes in Eratosthenes' sieve. Multiples of those numbers "drain" through the corresponding holes, leaving the primes behind. As larger numbers are tested by the sieve, more holes must be drilled to correspond to the greater number of primes that are less than the square root of the largest number to be tested. Only four holes are needed to test numbers up to 100, but eleven are required for testing the first thousand numbers, and 25 – one for each prime less than 100 – must be drilled to check all numbers less than 10,000. This means two things: the chance of a number being prime goes down as the number itself goes up, and it takes longer to test larger numbers. We will return to the first thing in the next section, and use the second to begin the discussion of other ways to test numbers for primality.

The sieve of Eratosthenes becomes less efficient as the numbers being tested get larger. The main reason for this is that it does much more than merely check a number for primality; it also produces a pair of factors for numbers that are composite. (Example: the 91 in the bottom left of the above array is crossed out by a line belonging to 7, meaning that 7 is a factor of 91: 91 = 7 x 13.) The general agreement among mathematicians is that factoring is much harder than primality testing, so methods of primality testing that do not involve factoring should be more efficient. It is worth noting that the methods in use now for testing large numbers (which may have hundreds of digits!) are all derived from a 350-year-old theorem attributed to the French mathematician Pierre de Fermat (who, among other things, developed the concept of "probability"), and named "Fermat's Little Theorem" in his honor. This theorem states that if p is a prime and n any positive integer, then $n^p - n$ is a multiple of p, or, for purposes of primality testing, that if $n^p - n$ is <u>not</u> a multiple of p, then p is not prime. This theorem can be proved by a variety of combinatorial techniques; see Chapter 9. It is important to note that it can give an absolute proof of non-primeness, but cannot absolutely prove any number is prime (although Fermat thought so), because there are numbers, called pseudoprimes, that satisfy "$n^p - n$ is a multiple of p" for certain choices of n; thus, $2^{341} - 2$ is a multiple of 341 even though 341 = 11 x 31. Fortunately, pseudoprimes are rare, and get scarcer as you go out to larger numbers. This allows variations on the basic Fermat test to guarantee the primality of any number that passes them.

Testing certain numbers for primality is even easier; in particular, any number that is one less than a power of

2 can be tested relatively quickly. The test runs as follows: Define a sequence $x_1, x_2, \ldots$ by $x_1 = 4$ and x_n = the remainder when $(x_{n-1})^2$ is divided by M_p, where $M_p = 2^p - 1$ is the number being tested. Then M_p is prime if and only if x_{p-1} is exactly zero. To see how it works, we use the example of $M_7 = 2^7 - 1 = 127$: $x_1 = 4$, $x_2 = 4^2 - 2 = 14$, x_3 = the remainder when $14^2 - 2 = 194$ is divided by $127 = 67$, x_4 = the remainder when $67^2 - 2 = 4,487$ is divided by $127 = 42$, x_5 = the remainder when $42^2 - 2 = 1,762$ is divided by $127 = 111$, and x_6 = the remainder when $111^2 - 2 = 12,319$ is divided by $127 = 0$. The zero result for x_6 proves that 127 is a prime number. The amount of computation necessary to test a small number like 127 may seem excessive compared to the five operations required by trial division, but consider what happened when I tried, not 127, but $M_{127} = 2^{127} - 1$, which has 34 digits. My mother's computer, with a 66-megahertz processor, took about a minute to prove this number prime; the world's fastest computer would take months if trial division were used. And, emboldened by that success, I then tried the same program on M_{1279}. That took about a day to prove prime, while the world's fastest computer would take much longer that the age of the universe to test with trial division. M_{127} is the largest number proved prime before the electronic computer, in 1914; the calculations involved would probably take a few weeks of 8-hour days to do by hand. With ultrafast machines and the above-described method, much larger numbers can be tested; on March 25, 1992, a new largest prime was announced: $2^{756,839} - 1$, which has 227,832 (!) digits, beginning 1741 and ending 887. Just writing out this number would take dozens of pages of this size and type.

4.4.4 Euclid's Prime Theorem and the Prime Abundance Theorem

The ever-increasing speeds of computers produce ever-larger prime numbers (in 1983, the largest known was a mere $2^{44,497} - 1$), and, at the same time, the sieve of Eratosthenes tells us that the chance that a number is prime gets smaller the larger a number becomes. Will there be a point at which Eratosthenes' sieve has so many holes that it can't hold any water at all? In other words, does the list of prime numbers have an end?

Euclid, the Greek mathematician mentioned earlier, supplied us with the answer to this question, which is, "No, there is no largest prime number." His method of proving this is quite instructive.

Suppose that you think you have found the last prime – call it L. Now list all the primes, up to and including L, and multiply them together. Now, add 1 to the result and call this number N: $N = (2 \times 3 \times 5 \times 7 \times \ldots \times L) + 1$. Certainly N will be greater than L. Since L is supposed to be the last prime, N must have some prime factors other than 1 and itself. But what can those factors be? When N is divided by each of the primes on the list (which, remember, is supposed to include each and every prime number), there will be a remainder of 1 in every division, because of the "+ 1" at the end of N's definition. Therefore, either N is itself prime, or it has smaller prime factors that are not on the list; in either case, a prime not on the list must exist. Therefore, there can be no last prime, and the list of primes must be infinite.

The method of this proof is central to mathematics: to prove statement "A," you assume "not A" and then derive a contradiction. Showing that "not A" cannot be true is equivalent to showing that "A" must be true.

Euclid's prime theorem tells us that Eratosthenes' sieve will never become all holes, but does not tell us how close it comes to being all holes. A much later and more sophisticated piece of mathematics called the prime number theorem (it should be called the prime abundance theorem, because it has information about the relative abundance of primes) does this for us. The prime abundance theorem says that as N gets large, the chance that it is prime gets very close to $\frac{1}{\ln N}$, where "ln" means "natural logarithm of." Logarithms are defined in Chapter 7; it suffices for now to note that $\ln N$ is about 2.3 times the number of digits in N. As N approaches infinity, so does $\ln N$; therefore the chance that N is prime approaches zero – but Euclid's prime theorem tells us it never quite gets there!

4.5 Modular Arithmetic - For Beginners

In many mathematical situations, only remainders after division by a certain number are important. In these situations, we are actually dealing with *modular* arithmetic. This subject, under the guise of "clock arithmetic," has been sneaked into the curriculum of junior-high students recently; they learn that, since five hours after 9:00 is 2:00, 5 + 9 = 2 in clock arithmetic with a 12-hour clock. The true mathematical statement reads 5 + 9 = 2 (mod 12), where "mod" stands for "modulo." In this equation (strictly, *congruence*), 12 is called the *modulus*.

In general, $a = b$ (mod m) if a and b have the same remainder after division by m; that is, if there exists an integer k such that $a = b + km$. The usefulness of modular arithmetic stems from two facts: If $a = b$ (mod m) and $c = d$ (mod m), then $a + c$ (or $a - c$, ac) = $b + d$ ($b - d$, bd) modulo m; and a is a multiple of m if and only if $a = 0$ (mod m). The first fact enables modular arithmetic to check ordinary arithmetic. "Casting out nines" is simply using arithmetic modulo 9 in this manner, as was done in Sec. 1.1.2 to check the calculation 103 x 107 = 11,021. There it was found that

$103 = 4$ (mod 9) and $107 = 8$ (mod 9), so 11,021 must be congruent to 32 (mod 9). Since $11,021 = 5$ (mod 9) and $32 = 5$ (mod 9), the calculation does indeed check.

The second fact above enables us to apply Fermat's Little Theorem (Sec. 4.4.3). To test the number 119 for primality, we calculate $2^{119} - 2$ (mod 119), which must be zero for 119 to be prime. We do not need to calculate 2^{119} itself, but only its remainder modulo 119. To do this we convert 119 into a sum of powers of two: $119 = 64 + 32 + 16 + 4 + 2 + 1$. Then we determine 2^1 mod $119 = 2$, 2^2 mod $119 = 4$, 2^4 mod $119 = (2^2)^2$ mod $119 = 4^2$ mod $119 = 16$, 2^8 mod $119 = 16^2$ mod $119 = 18$ ($256 - 2 \times 119$), 2^{16} mod $119 = 18^2$ mod $119 = 86$, 2^{32} mod $119 = 86^2$ mod $119 = 18$, and 2^{64} mod $119 = 86$. Then we multiply $2 \times 4 \times 16 \times 86 \times 18 \times 86 = 2 \times 4 \times 16 \times 18 \times 86 \times 86 = 2 \times 4 \times 16 \times 18 \times 18$ (mod 119) $= 2 \times 4 \times 16 \times 86$ (mod 119) $= 128 \times 86$ (mod 119) $= 9 \times 86$ (mod 119) $= 774$ (mod 119) $= 60$. Therefore, $2^{119} - 2 = 58$ (mod 119). Since 58 is not zero, 119 cannot be prime. Note that this test gives absolutely no clue as to the factors, 7 and 17, of 119.

Modular arithmetic is most useful when the modulus is a prime number. When m is prime, we can calculate $a \div b$ (mod m) even when a is not a multiple of b; $a \div b = c$ (mod m) if and only if $b \times c = a$ (mod m). If m is prime, this congruence can always be solved for c. The answer is $c = a \times b'$ (mod m), where b' is the reciprocal of b modulo m; that is, the solution to the congruence $b \times b' = 1$ (mod m). Unfortunately, there is no quick way to solve this congruence in general. See Fig. 4-6 and Table 3 for some reciprocal tables.

Using these tables, we divide $6 \div 9$ (mod 13). This must be equal to $6 \times \frac{1}{9}$ (mod 13) $= 6 \times 3$ (mod 13) $= 18$ (mod 13) $= 5$. Sure enough, $9 \times 5 = 6$ (mod 13) because $45 = (3 \times 13) + 6$. The most obvious feature in the tables is their near-total lack of patterns; this is what makes finding reciprocals in modular arithmetic fairly difficult. Two rather expectable patterns do appear: if $\frac{1}{a} = b$ (mod m), then $\frac{1}{b} = a$ (mod m), and $\frac{1}{-a}$ (mod m) $= -(\frac{1}{a})$ (mod m).

m=5	
b	b'
1	1
2	3
3	2
4	4

m=7	
b	b'
1	1
2	4
3	5
4	2
5	3
6	6

m=11	
b	b'
1	1
2	6
3	4
4	3
5	9
6	2
7	8
8	7
9	5
10	10

m=13	
b	b'
1	1
2	7
3	9
4	10
5	8
6	11
7	2
8	5
9	3
10	4
11	6
12	12

Fig. 4-6. Modular reciprocal tables for the first few prime numbers.

4.5.1 The Chinese Remainder Theorem

The Chinese remainder theorem is a way to combine information regarding several different moduli into information about a larger modulus that is the product of the others. Given that $n = 6$ (mod 7) and $= 2$ (mod 13), what is n (mod 91)? The theorem tells us that n (mod 91) can be split into two parts, one of which is a multiple of 13 but has a remainder of 6 when divided by 7, and the other of which is a multiple of 7 but has a remainder of 2 when divided by 13. The sum of these parts will have the required remainders for both moduli. For the first part, we ensure that it will be a multiple of 13 by writing 13 as one of its factors. To achieve the correct result modulo 7, we write 6 as another factor, and make the third be one such that, when multiplied by 13, is congruent to 1 mod 7; that is, the third factor is $\frac{1}{13}$ (mod 7). Since we are only concerned with modulo 7 arithmetic, we substitute 13 (mod 7) for 13. So the first part is $6 \times 13 \times \frac{1}{13 \pmod 7}$ (mod 7), and, similarly, the second part is $2 \times 7 \times \frac{1}{7 \pmod{13}}$ (mod 13). Now, $13 = 6$ (mod 7) and the reciprocal tables show $\frac{1}{6} = 6$ (mod 7), while $\frac{1}{7} = 2$ (mod 13). Thus, $n = 6 \times 13 \times 6 + 2 \times 7 \times 2$ (mod 91) $= 468 + 28$ (mod 91) $= 496 - (5 \times 91)$ (mod 91) $= 41$ (mod 91). Note that $41 = 5 \times 7 + 6$ and $3 \times 13 + 2$.

More generally (and symbolically), if $N = a_1$ (mod m_1), a_2 (mod m_2), ..., a_k (mod m_k) and $M = m_1 \times m_2 \times ... \times m_k$ with $M_i = \frac{M}{m_i}$, then $N = \left[a_1 M_1 \frac{1}{M_1 \bmod m_1} (\text{mod } m_1) + a_2 M_2 \frac{1}{M_2 \bmod m_2} (\text{mod } m_2) + \cdots \right]$ (mod M). This formula only holds if all the m's are prime, or, at worst, no two of them have a common factor greater than 1.

This theorem can be used to multiply large numbers. Keeping track of the remainders when a large number is divided by, say, 101, 103, 107, ..., 197, and 199 (the primes between 100 and 200) is equivalent to keeping track of the number itself because of the Chinese remainder theorem, provided that the number does not exceed the product of those primes, which is about 10^{45}. So a multiplication of two 20-digit numbers can be replaced by 21 multiplications of two- and three-digit ones, which each take about 1% as much time. Even recovering the 40-digit product from the remainders does not involve too much computation, since it requires multiplying 20-digit numbers by two-digit ones, which is still quicker than multiplying two 20-digit numbers.

4.6 Finite Differences

There are number patterns hidden in the sequence of squares, cubes, and other powers of integers. Consider the squares: 1, 4, 9, 16, 25, 36, …. Now write down the difference between each pair of consecutive terms: 3, 5, 7, 9, 11, …. The pattern should be obvious. For cubes, two steps of differences are required: 1, 8, 27, 64, 125, 216, 343, … becomes 7, 19, 37, 61, 91, 127, …, which in turn becomes 12, 18, 24, 30, 36, …, where there is an obvious pattern. In both cases, the pattern is actually a constant difference between successive terms in the last sequence, so two steps of differences produce a constant sequence in squares, and three steps do so in cubes.

This holds true for higher powers as well: If you start with nth powers, you must do n steps of differencing to arrive at a constant, which turns out to be n factorial (written $n!$, and defined in Sec. 4.2 via a recurrence relationship). Factorials get large very quickly, as $10! = 3{,}628{,}800$, while $100!$ has 158 digits of which the last 24 are zeroes. (The number of zeroes at the end of $N!$ is $\frac{N}{5} + \frac{N}{25} + \frac{N}{125} + \cdots$, where each term has the remainder fraction dropped from it before it is added in. Eventually the denominators, which are successive powers of 5, get bigger than N; at this point the sum terminates.)

The fact that all sequences of powers can be differenced in this way may be generalized to all sequences based on polynomials, such as $s_k = 5k^3 - 7k^2 + 4k - 8$. This holds because if two sequences, s_k and t_k, have first differences c_k and d_k respectively, then the sequence $(s_k + t_k)$ has as its first difference $(c_k + d_k)$, $(s_k - t_k)$ has as first difference $(c_k - d_k)$, and (ns_k) has as first difference (nc_k) if n is a constant. What is more, the number of differencing operations required to arrive at a constant equals the degree of the polynomial, and that constant will be the factorial of the degree times the coefficient of the highest-power monomial of the polynomial (called the *leading* coefficient).

Not only can any polynomial-based sequence be differenced to produce a constant, but any sequence that can be differenced to produce a constant is polynomial-based. The technique of finding the polynomial, given the sequence, is called "the method of finite differences." We will see how this technique works by imagining that we have completely differenced a sequence to get a constant called a. Arranging the work upside down, we have:

$$
\begin{array}{cccccccccc}
 & & a & & a & & a & & a & \\
 & b & & b+a & & b+2a & & b+3a & & b+4a \\
 & c & & c+b & & c+2b+a & & c+3b+3a & & c+4b+6a & & c+5b+10a \\
d & & d+c & & d+2c+b & & d+3c+3b+a & & d+4c+6b+4a & & d+5c+10b+10a, \text{ etc.}
\end{array}
$$

If we focus on the coefficients of any one letter, we see that each coefficient is the sum of the one to its upper left and the one to its horizontal left, a pattern similar to the arrangement in Sec. 4.2.1. Therefore, these numbers are binomial coefficients. Using the notation introduced in that example, we can rewrite the arrangement above as follows:

$$
\begin{array}{cccc}
\binom{0}{0}a & \binom{1}{0}a & \binom{2}{0}a & \binom{3}{0}a \\
\binom{0}{0}b & \binom{1}{0}b + \binom{1}{1}a & \binom{2}{0}b + \binom{2}{1}a & \binom{3}{0}b + \binom{3}{1}a & \binom{4}{0}b + \binom{4}{1}a \\
\binom{0}{0}c & \binom{1}{0}c + \binom{1}{1}b & \binom{2}{0}c + \binom{2}{1}b + \binom{2}{2}a & \binom{3}{0}c + \binom{3}{1}b + \binom{3}{2}a & \binom{4}{0}c + \binom{4}{1}b + \binom{4}{2}a \\
\binom{0}{0}d & \binom{1}{0}d + \binom{1}{1}c & \binom{2}{0}d + \binom{2}{1}c + \binom{2}{2}b & \binom{3}{0}d + \binom{3}{1}c + \binom{3}{2}b + \binom{3}{3}a, \text{ etc.}
\end{array}
$$

In general, if it takes n steps of differences to obtain a constant (above n is 3), and we replace $d, c, b, \ldots, a$ with $a_0, a_1, a_2, \ldots, a_n$ and denote the kth term in the bottom row by s_k (counting the far-left term as the zeroth), then we have the equation

$$
s_k = \binom{k}{0}a_0 + \binom{k}{1}a_1 + \binom{k}{2}a_2 + \cdots + \binom{k}{n}a_n.
$$

Now, the binomial coefficients $\binom{k}{n}$ can be written as polynomials of the nth degree in k: $\binom{k}{0} = 1$, $\binom{k}{1} = k$, $\binom{k}{2} = \frac{1}{2}(k^2 - k)$, $\binom{k}{3} = \frac{1}{6}(k^3 - 3k^2 + 2k)$, $\binom{k}{4} = \frac{1}{24}(k^4 - 6k^3 + 11k^2 - 6k)$, etc. In general,

$\binom{k}{r} = \frac{1}{r!}(k)(k-1)(k-2)\ldots(k-r+1)$, where there are r factors involving k. (Note that $\binom{k}{r} = 0$ if r is bigger than k.) Therefore, s_k is the sum of polynomials in k, and is therefore a polynomial itself. We can verify this in the case of third powers ($s_k = k^3$), with a little arithmetic:

$$
\begin{array}{ccccccc}
a_3 = 6 & 6 & 6 & 6 & & & \\
a_2 = 6 & 12 & 18 & 24 & 30 & & \\
a_1 = 1 & 7 & 19 & 37 & 61 & 91 & \\
a_0 = 0 = 0^3 & 1 = 1^3 & 8 = 2^3 & 27 = 3^3 & 64 = 4^3 & 125 = 5^3 & 216 = 6^3
\end{array}
$$

The formula above gives $s_k = \binom{k}{0}0 + \binom{k}{1}1 + \binom{k}{2}6 + \binom{k}{3}6 = 0 + k + 3(k^2 - k) + (k^3 - 3k^2 + 2k)$
$= k + 3k^2 - 3k + k^3 - 3k^2 + 2k = k^3$.

4.7 Patterns in Binomial Coefficients

There are numerous patterns in the array of binomial coefficients. We begin by rewriting the expression at the top of this page: $\binom{n}{r} = \frac{1}{r!}(n)(n-1)(n-2)\ldots(n-r+1)$ as $\binom{n}{r} =$

$$
\frac{n(n-1)(n-2)\ldots(n-r+1)}{r(r-1)(r-2)\ldots(3)(2)(1)}
$$

$$
= \frac{n(n-1)(n-2)\ldots(n-r+1)(n-r)(n-r-1)\ldots(3)(2)(1)}{r(r-1)(r-2)\ldots(3)(2)(1)(n-r)(n-r-1)\ldots(3)(2)(1)}
$$

$$
[4.2] \quad = \frac{n!}{r!(n-r)!}
$$

Since $n - (n - r) = r$, we have $\binom{n}{n-r} = \frac{n!}{(n-r)!(n-(n-r))!} = \frac{n!}{(n-r)!r!} = \binom{n}{r}$, that is, the triangle of binomial coefficients is symmetrical.

The statement that each binomial coefficient is the sum of two others can be written $\binom{n}{r} = \binom{n-1}{r} + \binom{n-1}{r-1}$. This, in turn, can be used to prove the *binomial theorem*, which says

$$
(1+x)^n = \binom{n}{0} + \binom{n}{1}x + \binom{n}{2}x^2 + \cdots + \binom{n}{n}x^n.
$$

To do this, we split each binomial coefficient in half to get

$$
(1+x)^n = \binom{n-1}{-1} + \binom{n-1}{0} + \left[\binom{n-1}{0} + \binom{n-1}{1}\right]x + \left[\binom{n-1}{1} + \binom{n-1}{2}\right]x^2 + \cdots
$$

$+\left[\binom{n-1}{n-1} + \binom{n-1}{n}\right]x^n$. Now, $\binom{n}{r}$ is defined to be zero if $n < r$ or if r is negative, so the two coefficients at either end can be dropped to get

$$
(1+x)^n = \binom{n-1}{0} + \left[\binom{n-1}{0} + \binom{n-1}{1}\right]x + \left[\binom{n-1}{1} + \binom{n-1}{2}\right]x^2 + \cdots + \binom{n-1}{n-1}x^n.
$$

We now collect terms with matching binomial coefficients, instead of matching powers of x, to get

$$
(1+x)^n = \binom{n-1}{0}(1+x) + \binom{n-1}{1}(x+x^2) + \binom{n-1}{2}(x^2+x^3) + \cdots + \binom{n-1}{n-1}(x^{n-1} + x^n).
$$

Each of the x parentheses has $(1+x)$ as a factor, so we can rewrite the right-hand side of this last equation as

$$
(1+x)\left\{\binom{n-1}{0} + \binom{n-1}{1}x + \binom{n-1}{2}x^2 + \cdots + \binom{n-1}{n-1}x^{n-1}\right\}.
$$

The expression inside the braces is like the one we started with, except it has $n-1$ instead of n. Thus, if we let $F(n) = \binom{n}{0} + \binom{n}{1}x + \binom{n}{2}x^2 + \cdots + \binom{n}{n}x^n$, we have just shown that $F(n) = (1+x)F(n-1)$. Also, $F(1) = \binom{1}{0} + \binom{1}{1}x = (1+x)$, and therefore $F(n) = (1+x)^n$, which is the binomial theorem.

Having proved the theorem, we now replace x with 1 in its statement to arrive at the useful result $\binom{n}{0} + \binom{n}{1} + \binom{n}{2} + \cdots + \binom{n}{n} = 2^n$; thus, $1 + 4 + 6 + 4 + 1 = 2^4 = 16$.

Additional relations among binomial coefficients can be established by techniques similar to the ones applied to the Fibonacci sequence in Sec. 4.2.1. We have $\binom{n}{r} = \binom{n-1}{r} + \binom{n-1}{r-1}$; repeatedly splitting the term on the right yields $\binom{n}{r} = \binom{n-1}{r} + \binom{n-2}{r-1} + \binom{n-3}{r-2} + \cdots + \binom{n-r-1}{0}$. For example, $\binom{6}{4} = \binom{5}{4} + \binom{4}{3} + \binom{3}{2} + \binom{2}{1} + \binom{1}{0} = 5 + 4 + 3 + 2 + 1 = 15$. Using $\binom{n}{r} = \binom{n}{n-r}$ and changing variables, we can rewrite this formula as $\binom{n+r+1}{r+1} = \binom{n+r}{r} + \binom{n+r-1}{r} + \binom{n+r-2}{r} + \cdots + \binom{r}{r}$; the $\binom{6}{4}$ example above becomes $\binom{6}{2} = \binom{5}{1} + \binom{4}{1} + \binom{3}{1} + \binom{2}{1} + \binom{1}{1}$. Since $\binom{n}{1} = n$, this means that $\binom{n}{2} = 1 + 2 + 3 + 4 + \cdots + (n-1) = \Delta(n-1)$, showing that triangular numbers are binomial coefficients. As tetrahedral numbers are sums of consecutive triangular numbers, they are binomial coefficients too: $Tp(n) = \Delta(1) + \Delta(2) + \cdots + \Delta(n) = \binom{2}{2} + \binom{3}{2} + \cdots + \binom{n+1}{2} = \binom{n+2}{3}$.

The "Christmas tree" technique done to Fibonacci numbers back in their section can be applied to binomial coefficients as well:

$$\binom{n}{r} = \binom{1}{0}\binom{n-1}{r-1} + \binom{1}{1}\binom{n-1}{r}$$
$$= \binom{2}{0}\binom{n-2}{r-2} + \binom{2}{1}\binom{n-2}{r-1} + \binom{2}{2}\binom{n-2}{r}$$
$$= \binom{3}{0}\binom{n-3}{r-3} + \binom{3}{1}\binom{n-3}{r-2} + \binom{3}{2}\binom{n-3}{r-1} + \binom{3}{3}\binom{n-3}{r}$$
$$\vdots$$
$$= \binom{r}{0}\binom{n-r}{0} + \binom{r}{1}\binom{n-r}{1} + \binom{r}{2}\binom{n-r}{2} + \cdots + \binom{r}{r}\binom{n-r}{r},$$

provided that $n - r$ is at least as big as r. Example: $\binom{8}{3} = \binom{3}{0}\binom{5}{0} + \binom{3}{1}\binom{5}{1} + \binom{3}{2}\binom{5}{2} + \binom{3}{3}\binom{5}{3} = 1 \cdot 1 + 3 \cdot 5 + 3 \cdot 10 + 1 \cdot 10 = 1 + 15 + 30 + 10 = 56$. A particularly useful case arises when $n = 2r$ so that $n - r = r$. We then have $\binom{2n}{n} = \binom{n}{0}^2 + \binom{n}{1}^2 + \binom{n}{2}^2 + \cdots + \binom{n}{n}^2$, so that $\binom{10}{5} = \binom{5}{0}^2 + \binom{5}{1}^2 + \binom{5}{2}^2 + \binom{5}{3}^2 + \binom{5}{4}^2 + \binom{5}{5}^2 = 1 + 25 + 100 + 100 + 25 + 1 = 252$.

The binomial coefficients are most useful in counting arrangements of objects. Chapter 9 on combinatorial theory deals with this aspect of them. For now, however, we will abandon numbers for a while, and concentrate on other parts of mathematics – which, as it turns out, will involve numbers, too, but in a less central way.

Chapter 5   Geometry, Trigonometry, and Analytic Geometry

Mathematics has essentially two main sub-parts: that dealing with numbers, and that dealing with shapes. The two are not entirely separate; in the last chapter we discussed triangular numbers, and triangles are shapes. This chapter will discuss triangles (and other shapes), but numbers will, inevitably, enter the discussion. The part of mathematics dealing with shapes is called *geometry* (which in Greek means earth-measuring; it may have originated with ancient Greek attempts to measure the size and shape of the earth), and its more numerical offshoots include *trigonometry*, which deals specifically with triangles, and *analytic* geometry, which brings in algebra.

5.1.1 Points, Lines, and Angles

We begin the discussion of geometry with the ultimate fundamental shape: the *point*. This is an object with no size, only position. In diagrams, points are represented by dots, usually labeled with letters, but a point is smaller than the smallest dot that can be drawn, smaller than an atom, smaller even than the nucleus of an atom. Every possible position is occupied by one and only one point. Points are undefined objects whose existence must be taken on faith, so to speak; the preceding sentences are designed to give an impression of some of their properties. Most of this chapter will be concerned only with *plane* geometry, in which all points are considered to lie on one flat surface called a plane.

The *line* is another undefined notion. It may have length, but does not have any width or thickness. A critical concept is the *infinite straight line*; this is a line which has no bends or curves in it, and goes on forever with no ends; even though it may be "drawn" on a sheet of paper with edges, it must be thought of as extending off the paper in both directions. The term "line" without further modification shall mean an infinite straight line from now on.

Lines (infinite, straight) have important properties. Through any two points, exactly one line can be drawn, and two lines can have at most one point in common; if they do, they are said to *intersect* in that point. Fig. 5-1 shows two lines intersecting in the point labeled *A*. (But in reality, the lines have no thickness, while the figure shows "lines" with a definite thickness.) Two lines do not have to have a point in common. In plane geometry, lines which do not intersect are called *parallel*.

Fig. 5-2. The line segment XY.

More complicated objects are built up from simple ones, a trick mentioned earlier as being widely used in mathematics. The *line segment* is defined as two points together with that portion of the line they determine that lies between them. A line segment is named by writing the names of its two endpoints together; thus, Fig. 5-2 represents the line segment *XY*. A *ray* is that part of a line which lies entirely to one side of one of its points; to determine which side, we use any other

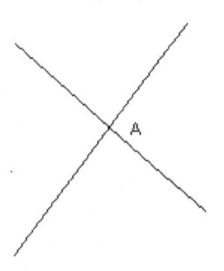

Fig. 5-1. Two lines intersecting in point A.

point of the ray. The dividing point is called the end-point or *origin* of the ray. A ray is named by writing the name of its origin followed with another point of the ray. Thus, Fig. 5-3 represents the ray *XY*. Note that *XW* and *XZ* are names for the same ray. The arrow serves as a reminder that the ray extends indefinitely in one direction (unlike the infinite line, which extends indefinitely in two directions). Two rays with the same origin form an *angle*; they are called *sides* of the angle and their common origin is called the *vertex* of the angle. Points other than those on an angle's sides are called either *interior* or *exterior* depending on whether or not they are in the smaller of the two sections that the angle divides the plane into. If both rays of an angle are part of the same straight line, the angle is called a *straight* angle, which has no interior or exterior. Angles are named with three points, one on each ray with the name of the vertex between them. Fig. 5-4 shows the angle *ABC* (*CBA* is just as good a name) with an interior point *I* and an exterior point *E*. If no confusion can arise, an angle can be named just by its vertex, so the angle in Fig. 5-4 can just be called angle *B*.

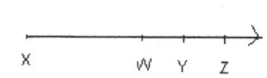

Fig. 5-3. The ray XY (or XW or XZ).

To avoid writing extra words, geometric figures like these are referred to with characteristic notations: the straight line *AB* is denoted by $\overleftrightarrow{AB}$, the line segment *AB* by $\overline{AB}$ (without the arrows) or just *AB*, the ray *AB* by $\overrightarrow{AB}$ with one arrow, and the angle *ABC* is written $\angle ABC$.

To each pair of points is assigned a non-negative number called the *distance* between them and written $d(A, B)$ for the distance between A and B. The distance satisfies three rules called the *metric space axioms*: (1) $d(A, B) \geq 0$, and $d(A, B) = 0$ if and only if A and B are two names for the same point; (2) $d(A, B) = d(B, A)$; and (3) $d(A, B) + d(B, C) \geq d(A, C)$. This last is referred to as the *triangle inequality*. In it, the equality sign holds if and only if B is between A and C. The length of a line segment is defined to be the distance between its endpoints; since a ray has only one endpoint and a line none, they do not have numerical lengths.

Angles are assigned measures, too. Intuitively, the wider the angle's opening is, the larger the measure. This is stated as the *Angle Addition Postulate*: If two angles share a common side, and that side is interior to the angle formed by the other two sides, then the measure of that angle is the sum of the measures assigned to the two smaller angles. In

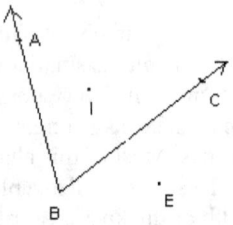

Fig. 5-4. The angle ABC (or CBA) with an interior point I and an exterior point E.

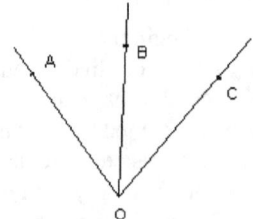

Fig. 5-5. ∠AOC= ∠AOB+∠BOC.

Fig. 5-5, $\angle AOC = \angle AOB + \angle BOC$. There are several systems for assigning numbers to angles. In the degree or *sexagesimal* system, a straight angle is assigned a measure of 180 degrees, written 180°, and in the *radian* system, the same angle is assigned a measure of 3.14159265... radians. (There are other systems in use, but they will not be mentioned here.) The term "sexagesimal" is derived from the Latin word for sixty; a degree is subdivided into 60 *minutes*, each of which is divided into sixty *seconds*, as in time measurement. It should be noted that early Mesopotamian mathematicians based their arithmetic on sixty. As for the number in the radian system, it is one of the most famous numbers in mathematics. Symbolized by π, the Greek letter pi, it is defined as the circumference of a circle divided by its diameter. Pi will be encountered again and again in this and subsequent chapters of this book. It is irrational, but not the root of an algebraic equation with integer coefficients; this is the first example of such a number.

Two segments or two angles are said to be equal if they have the same measure. They are also referred to as *congruent* figures. Congruent figures can be superimposed on one another without stretching or other distorting actions. Much of geometry consists of determining whether or not two figures are congruent. Related to the concept of congruence is that of *similarity*: two figures are similar if one can be superimposed on the other with the help of a regular stretching or shrinking, which must be of the same amount in all directions.

Two lines are called *perpendicular* if all the angles formed at the intersection are equal. In Fig. 5-6, line AB is perpendicular to line CD because $\angle AXC = \angle AXD = \angle BXC = \angle BXD$. The angles formed by perpendicular lines are called *right* angles; the extra symbol ∟ is used to show that an angle is right. Angles larger than a right angle are called *obtuse*; those smaller are called *acute*. Since two right angles add up to one straight angle by the Angle Addition Postulate, the measure of a right angle is 90 degrees or $\frac{\pi}{2} = 1.57079633...$ radians.

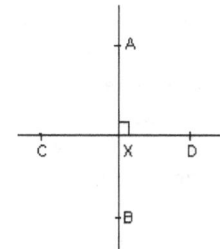

Fig. 5-6. Line AB is perpendicular to line CD; the four angles at X are right angles.

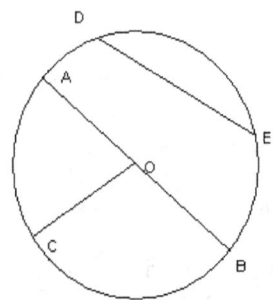

Fig. 5-7. A circle with center O, diameter AB, radius OC, and a chord DE that is not a daimeter.

5.1.2 Circles

A circle is defined in plane geometry as the set of all points that are a given distance, called the *radius*, from a fixed point, called the *center*. Thus, if a circle has center O and radius r, then any point P on the circle satisfies $d(O, P) = r$. The term *radius* is also used to designate a line segment having one endpoint on a circle and the other at its center; it should be obvious that all such line segments for any one circle are equal. A line segment with both endpoints on the same circle is called a *chord*; if a chord passes through the circle's center it is called a *diameter*. This word is also used to designate the length of such a segment; it should be obvious that the diameter of a circle is unique and twice the radius. In Fig. 5-7, O is the center of the circle, OA, OB, and OC are radii (plural of radius), AB is a diameter, and DE is a chord which is not a diameter. Connected parts of a circle are called *arcs*; the measure of an arc is the same as the associated *central* angle, which is an angle whose vertex is at the circle's center. In Fig. 5-7, the measure of arc AC is the same as that of the central angle AOC, written $m(\overarc{AC}) = m(\angle AOC)$. (Since arcs and angles are different figures, we cannot simply write $\overarc{AC} = \angle AOC$.) Although the length of a curve is difficult to define, we assume that it can be measured like that of a line segment and call the length of one trip around the circle the *circumference*. An angle of 1 radian is defined as a central angle which *subtends* (includes in its interior) an arc whose length is equal to the radius. Since π is defined as circumference $\div$ diameter = circumference $\div$ ($2 \times$ radius) = $\frac{1}{2}$ (circumference $\div$ radius), the circumference contains $2\pi = 6.283185307\ldots$ radians. A straight angle contains half the circumference, so the statement above that a straight angle measures π radians follows from the definitions of the radian and of pi.

5.1.3 Polygons

A polygon is defined as a group of line segments that have all of the following characteristics: (1) There are a finite number of them. (2) Each segment shares each of its endpoints with exactly one other segment. (3) No segments intersect except at their endpoints. Since each segment has two endpoints and each endpoint belongs to two segments, the number of endpoints is the same as the number of segments. The endpoints are called *vertices*, and the segments *sides*, of the polygon. Polygons are named triangles, *quadrilaterals*, *pentagons*, *hexagons*, *heptagons*, …, *n*-gons as they have 3, 4, 5, 6, 7, …, n sides. The "-lateral" comes from the Latin word for "side" and the "-gon" from the Greek word for "angle." Mathematicians have drawn freely from both classical languages to describe their concepts. A polygon is called *equilateral* if all its sides have the same length, *equiangular* if the angles defined at each vertex by the associated line segments are equal, and *regular* if it is both equilateral and equiangular. A regular quadrilateral is known as a *square*, but no other special names are given to other regular polygons. Polygons are named by listing their vertices in the order that they are joined, usually in counterclockwise fashion.

5.1.4 Area

A geometric figure is called *closed* if there is a point, not part of the figure, such that every ray originating at that point intersects the figure. The set of all such points is called the *interior* of that figure. Interiors are also called *regions*; the interior of a circle is called a circular region. Sometimes no verbal distinction is made between a figure and its interior.

(Strictly speaking, the definitions of "closed" and "interior" above hold only for *convex* figures, those that satisfy the following: Any segment between two points of the figure does not otherwise intersect the figure, unless both points are on the same line segment belonging to the figure. For other, nonconvex figures, the word "ray" must be replaced with "line, straight or curved, that extends indefinitely far from the point.")

The interiors of closed figures are assigned a positive number called their *area*. All areas of figures can be derived from two statements, plus the notion of limit: (1) The area of an equiangular quadrilateral (*rectangle*) is equal to the product of two adjacent side lengths. (2) The area of a region formed from two or more smaller ones that overlap only along boundaries is the sum of the areas of the smaller figures. In Fig. 5-8, the area of triangle ABC is the sum of the areas of the smaller triangles ABD and ACD that share only the side AD. Congruent figures have the same area (or share the property of lacking area, if they are not closed).

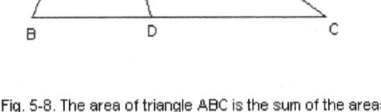

Fig. 5-8. The area of triangle ABC is the sum of the areas of triangles ABD and ACD.

5.2 Axiom Systems and Mathematical Proof

The first several pages of this chapter have consisted almost exclusively of definitions of the many specialized terms used in geometry. (And the list is by no means exhausted; more will come later.) Definitions are a vital necessity, but other facts are needed before deduction can proceed. The Greek geometer Euclid, mentioned in Chapter 4 for his contributions to the theory of numbers, recognized the importance of stating "obvious" facts. In his *Elements* he begins with a list of definitions similar to what has been written above, but in a compressed style, and then he lists "Common Notions," "Axioms," and "Postulates." The first two of these are assumptions of things in general (e. g. if two things are equal to the same (a third) thing, they are equal to each other) and are not specific to geometry. The "Postulates," however, are geometric in nature. The list of five is: (1) A straight line can be drawn through any two points. (2) Any straight line can be extended indefinitely and still be straight. (3) A circle can be drawn with any center and radius. (4) All right angles are equal. (5) If two lines are cut by a third line (called a *transversal*), and the internal angles on one side of the transversal add up to less than a straight angle, then the first two lines will, if extended sufficiently, intersect on that side. In Fig. 5-9, $\overleftrightarrow{AB}$ is the transversal, and if $\angle XAB + \angle YBA < 180°$, then the lines $\overleftrightarrow{AX}$ and $\overleftrightarrow{BY}$ must intersect to the right of $\overleftrightarrow{AB}$.

On these five geometric postulates, as well as the "axioms," "common notions," and definitions, Euclid (and other Greek geometers) built an elaborate structure that modern mathematics calls an *axiom system*. An axiom system can be compared to a game; the axioms (postulates, common notions, assumptions, definitions) are the rules, and the object of the game is to deduce as many facts as possible, using the technique of mathematical proof. Euclid's goal in writing the *Elements* was to create an axiom system for geometry. Although he left out a few assumptions, which were undoubtedly to him so obvious that he did not bother mentioning them, he did remarkably well. Even today, high school geometry is little more than a rearrangement of the first part of the *Elements*.

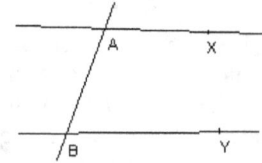

Fig. 5-9. If ∠ XAB+ ∠YBA add up to less than a straight angle, then AX and BY will intersect on the right side of AB

Creating a good axiom system requires the proper choice of the foundation material – the axioms. They should be self-evident, not too numerous, and above all consistent – that is, there can be no way to show that a statement is both true and false. In Euclid's case, his "axioms" and "common notions" are self-evident, and the geometric "postulates" are too – except for the last one, which is longer than the other four combined and not very self-evident. In Fig. 5-9, it is not obvious just by looking that $\overleftrightarrow{AX}$ and $\overleftrightarrow{BY}$ will intersect on the required side of $\overleftrightarrow{AB}$, or even intersect at all. Much mathematics has sprouted up around attempts to prove this postulate as a theorem, and the following seemingly more obvious statements have been shown to be equivalent to it, in the sense that any of them, along with Euclid's other foundation structures, imply the postulate, and the postulate implies all of them:

1. The sum of the angles in any triangle is 180°.
2. The sum of the angles in every triangle is the same.
3. Geometric figures can be similar but not congruent.
4. Two lines parallel to the same line are parallel to each other.
5. Given a line and a point not on it, exactly one other line can be drawn through the point and parallel to the line. (See Fig. 5-10.)

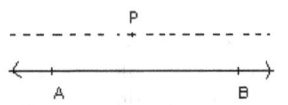

Fig. 5-10. According to the parallel postulate, there is exactly one line through P parallel to AB.

I also have contributed an equivalent to the parallel postulate: Lines can be equidistant at three different spots. (The distance between two lines is measured along a line segment perpendicular to either of them.) In Fig. 5-11, the statement $AD = BE = CF$ is equivalent to the parallel postulate (which is so called because of its equivalence to statements 4 and 5 above involving parallels).

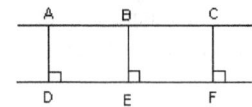

Fig. 5-11. The statement AD=BE=CF is equivalent to the parallel postulate.

The surprising results of the attempts are: (1) The parallel postulate cannot be proved or disproved from the rest of Euclid's foundation, and (2) If Euclid's foundation, including the parallel postulate, is consistent, then so are geometries which include a denial of that postulate. Geometries which are like Euclid's, except for the parallel postulate, are called *non-Euclidean*; there are two of them. In *elliptic* geometry, the phrase "exactly one" in statement 5 above is replaced with the word "no"; that is, there are no parallel lines; and in *hyperbolic* geometry, "exactly one" is replaced with "more than one."

Going back to Euclidean geometry, with the parallel postulate, we must now define a mathematical proof in an axiom system. Such a proof consists of a given situation and a series of statements about it, such that each statement comes from applying a foundation structure (axiom or definition) or a previously proved result to the given situation or earlier statements about it. We will show a few such proofs, both to provide instruction and to begin playing the "Axiom System Game."

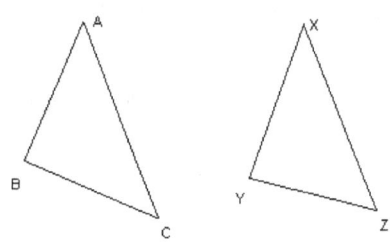

Fig. 5-12. If AB=XY, AC=XZ, and ∠BAC= ∠YXZ, then triangle ABC is congruent to triangle XYZ.

Theorem 5-1 (SAS Congruence Theorem):

If two sides and the angle between them are identical in a pair of triangles, then the triangles are congruent. (See Fig. 5-12.)

We are given two triangles, *ABC* and *XYZ*, such that $AB = XY$, $AC = XZ$, and $\angle BAC = \angle YXZ$. We superimpose point *A* on point *X*, and then line up *AB* with *XY*. Since $AB = XY$, *B* will match up with *Y* (the definition of line segment congruence). Since $\angle BAC = \angle YXZ$ and *AB* is on top of *XY*, *AC* will lie on top of *XZ* (the definition of angle congruence). And, finally, since $AC = XZ$, *C* will match up with *Z* (definition of line segment congruence). Since *B* and *C* match up with *Y* and *Z* respectively, segment *BC* will match segment *YZ*. (There is only one line, and therefore only one line segment, connecting any two points.) We now have all the parts of triangle *ABC* superimposed on triangle *XYZ*, so the two triangles must be congruent (again, the definition of congruence).

Theorem 5-2 (ASA Congruence Theorem):

If, in a pair of triangles, two angles and the side they share are identical, then the triangles are congruent. (See Fig. 5-13.)

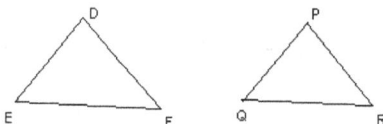

Here, we are given triangles *DEF* and *PQR* such that $\angle DEF = \angle PQR$, $\angle DFE = \angle PRQ$, and $DF = PR$. As before, we superimpose *D* on *P* and line up *DF* with *PR*. Since $DF = PR$, *R* and *F* will match, and because of the angle congruences, the segments *DE* and *FE* will fall along *PQ* and *RQ* respectively. Since two lines can intersect in at most one point, *E* must coincide with *Q*, meaning that all the parts of triangle *DEF* are superimposed on triangle *PQR*, and hence the two triangles are congruent.

Fig. 5-13. If ∠DEF= ∠PQR, ∠DFE= ∠PRQ, and segment EF= segment QR, then triangle DEF is congruent to triangle PQR.

Before we proceed with the next proof, we note that all straight angles are equal since they are all assigned the same measure (180° or π radians).

Theorem 5-3:

When two lines intersect, the angles opposite to each other from the point of intersection (called *vertical angles*) are equal.

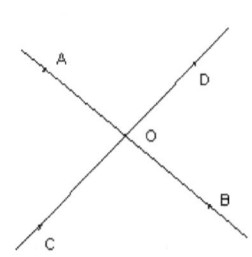

Fig. 5-14. Vertical angles are equal.

In Fig. 5-14, we have lines $\overleftrightarrow{AB}$ and $\overleftrightarrow{CD}$ intersecting at *O*; we must prove that $\angle AOD = \angle BOC$ and $\angle AOC = \angle BOD$.

We begin by noting that $\angle AOD + \angle DOB = \angle AOB$. (Angle Addition Postulate, which in Euclid is included under the "axiom" that the whole is equal to the sum of its parts.)

Then, $\angle COB + \angle DOB = \angle COD$. (A. A. P. again)

$\angle AOB = \angle COD$ since they are both straight angles.

$\angle AOB - \angle DOB = \angle COD - \angle DOB$. ("Common notion," when equals are subtracted from equals, the results remain equal.)

$\angle AOD = \angle COB$. (Combining statements 1, 2, and 4.)

By similar reasoning, we can prove that $\angle AOC = \angle BOD$. (We need merely exchange the letters *C* and *D* in each of the preceding statements.)

The next two theorems illustrate the theme of building on previous results. An *isosceles* triangle is one that has two sides equal; a *bisector* of an angle is a ray that divides it into two equal parts; the *base* of an isosceles triangle is the third, unequal, side; and the base angles of such a triangle are those that include the base as a side.

Theorem 5-4 (Pons Asinorum):
The base angles of an isosceles triangle are equal.

We are given isosceles triangle ABC with $AB = AC$; we must prove that $\angle ABC = \angle ACB$.

First, we draw line AD, which intersects BC in the point D, and is so positioned that $\angle BAD = \angle DAC$. We can do this because we can continuously vary the position of point D along BC, and thus continuously vary the size of $\angle BAD$, so we can choose $\angle BAD$ to be half as big as $\angle BAC$, which, by the A. A. P., will make $\angle DAC$ the same size (and thus AD bisects $\angle BAC$).

Now, in the triangles ABD and ACD, we have (1) $AB = AC$ (given to us), (2) $\angle BAD = \angle CAD$ (by choice of the point D), and (3) $AD = AD$ (anything is equal to itself). Therefore, ABD and ACD are congruent triangles (Theorem 5-1, SAS). Since we can superimpose ABD on ACD with the help of a reflection, we can superimpose $\angle ABC$ on $\angle ACB$, so the two must be equal.

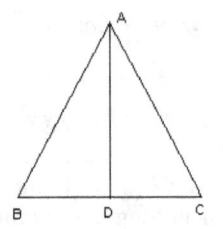

Fig. 5-15. Pons Asinorum; if AB=AC, then angle ABC is congruent to angle ACB.

Note the use of the extra line AD in this proof, as well as the use of the previously proved Theorem 5-1.

Theorem 5-5:
In the circle in Fig. 5-16, $\angle OAB = \angle OBA$. (O is the center of the circle.)

We start by noting that both OA and OB are radii of the circle. This means, in turn:
(1) $OA = OB$ (all radii of a circle are equal);
(2) Triangle AOB is isosceles with base AB (definition of isosceles above); and
(3) $\angle OAB = \angle OBA$ (Theorem 5-4).

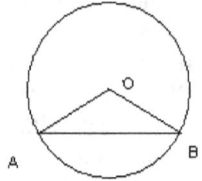

Fig. 5-16. $\angle OAB = \angle OBA$.

This theorem is rather specialized, and is shown primarily as another example of building on a previous result. Th. 5-5 rests on Th. 5-4, which in turn rests on Th. 5-1, like blocks in an Egyptian pyramid. The axiom system of geometry can be built up to unbelievable heights. For the next section, intermediate steps will only be mentioned, if at all, to avoid making that section look too much like a textbook.

5.3 More Basic Results

We will now state, without rigorous proof, other basic geometric theorems. These have many applications to more advanced results.

Th. 5-6. When parallel lines are cut by a transversal:
(1) Corresponding angles are equal.
(2) Alternate interior angles are equal.

In Fig. 5-17, the four angles marked X are equal, as are the four angles marked Y. We note that this theorem, unlike the previous ones, depends on Euclid's fifth postulate, and is therefore not a part of non-Euclidean geometry. Therefore, all theorems that depend on Th. 5-6 (and there are a lot of them) are likewise not found in non-Euclidean

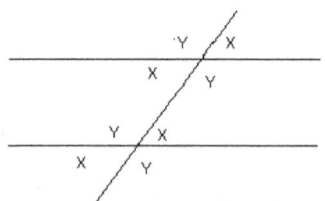

Fig. 5-17. Parallel lines are cut by a transversal: the four angles marked X are equal, as are the four angles marked Y.

geometry.

Th. 5-7. The angles of a triangle sum to 180°.

Th. 5-8. In a *parallelogram* (a quadrilateral in which both pairs of opposite sides are parallel) the opposite sides are equal, the opposite angles are equal, and adjacent angles add up to 180°.

Th. 5-9. Two lines perpendicular to (or parallel with) the same line are parallel to each other.

The *altitude* ("height") of a triangle, parallelogram, or *trapezoid*, which is a quadrilateral with just one pair of opposite sides parallel, is (the length of) a line segment perpendicular to the base of the figure, which can be taken to be any side of a triangle or one of two parallel sides of a trapezoid or parallelogram, and extending to the top of the figure (the opposite vertex of a triangle). In Fig. 5-18, all the vertical line segments are altitudes.

Note that the side taken for the base may have to be extended.

Th. 5-10. (A) The area of a parallelogram is a base times the altitude.

(B) The area of a triangle is half the base times the altitude.

(C) The area of a trapezoid is half the product of the altitude and the sum of the two parallel sides.

Th. 5-11. (A) The angles of a quadrilateral sum to 360°.

(B) All four angles of a rectangle are right angles.

(C) The angles of an *n*-gon sum to $(n-2)$ x 180° or $\pi(n-2)$ radians.

(D) Each angle of a regular *n*-gon is $\frac{n-2}{n}$ 180° or $\frac{\pi(n-2)}{n}$ radians.

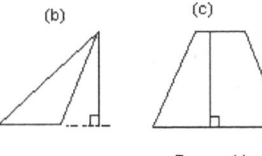

(a)   (b)   (c)

Triangles   · Trapezoid

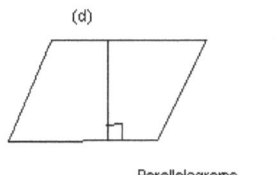   

(d)   (e)

Parallelograms

Fig. 5-18. Altitudes to triangles and quadrilaterals.

We can see why Th. 5-11 is true (once we have Th. 5-7, which depends on Th. 5-6) by slicing up a polygon into triangles. It should be obvious that $n - 2$ triangles can be put together to form the *n*-gon in Fig. 5-19, where *n* is 7. Each angle in these triangles is part of (or all of) an angle in the polygon, so the sum of the angles of all the triangles is the same as the sum of the angles in the polygon.

Th. 5-12 (SSS Congruence Theorem).

If the three sides of one triangle are equal to the sides of another triangle, the two triangles are congruent.

Th. 5-13(A). Linear features of similar figures are in proportion to corresponding sides.

Th. 5-13(B). Areas of similar figures are in proportion to squares of corresponding sides.

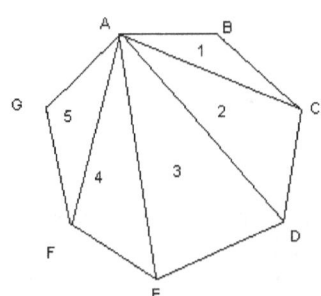

Figure 5-19. An n-gon sliced up into n-2 triangles.

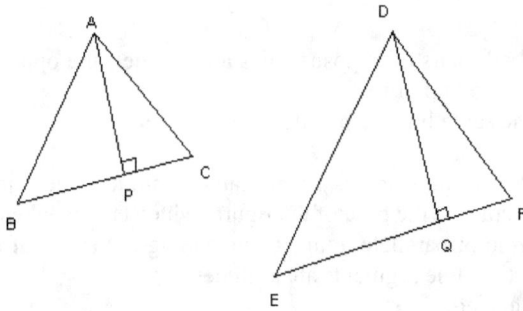

In Fig. 5-20, if triangles *ABC* and *DEF* are similar, then the altitudes *AP* and *DQ* satisfy $\frac{AP}{DQ} = \frac{AB}{DE}$, and the areas A_1 and A_2 satisfy $\frac{A_1}{A_2} = \frac{(AB)^2}{(DE)^2}$. Note that triangles are similar if and only if their angles are equal. It should be noted also that uniform stretching or shrinking preserves the measures of angles.

For the basic results involving circles, we need a few more definitions. A line is *tangent* to a circle if it intersects it in exactly one point. An angle is *inscribed* in a circle if its vertex is on the circle; it is inscribed in a semicircle if, in addition, one point on each of the sides of the angle is at the end of a diameter of the same circle. In Fig.5-21(a), *AB* is tangent at the point *C*. In Fig. 5-21(b), ∠*DEF* is inscribed in the circle with *P* as center, and in Fig. 5-21(c), ∠*XYZ* is inscribed in a semicircle with center *Q*. The arc *DF* is said to be *intercepted* by ∠*DEF*.

Fig. 5-20. If ABC and DEF are similar, then AP/DQ = AB/DE and (area ABC)/(area DEF) = (AB/DE) squared.

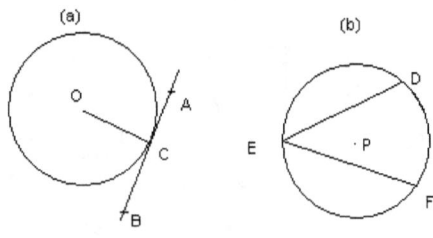

Fig. 5-21 (a) and (b). Ab is tangent to circle O at C, and angle DEF is inscribed in circle P. Since OC is a radius, it is perpendicular to AB.

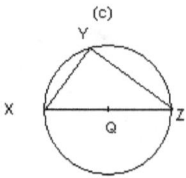

Fig. 5-21(c). ANgle XYZ is inscribed in a semicircle with center Q, and is therefore right.

Th. 5-14. A tangent is perpendicular to its associated radius.

In Fig. 5-21(a), *OC* is a radius and hence it is perpendicular to *AB*.

Th. 5-15. An inscribed angle has a measure half that of its intercepted arc. (See Fig. 5-21(b).)

Th. 5-15A. An angle inscribed in a semicircle is a right angle. (See Fig. 5-21(c).)

Th. 5-16. If a quadrilateral is inscribed in a circle, the opposite angles add up to 180°.

Th. 5-17A. The angle between two chords which intersect inside their circle is equal to half the sum of the intercepted arcs.

Th. 5-17B. The angle between two chords which do not intersect inside their circle is equal to half the difference of the intercepted arcs.

In Fig. 5-22, $\angle BYD = \frac{1}{2}(m(\overset{\frown}{AC}) + m(\overset{\frown}{BD}))$ and $\angle BXD = \frac{1}{2}(m(\overset{\frown}{BD}) - m(\overset{\frown}{AC}))$.

Th. 5-17C. Parallel chords intercept equal arcs.

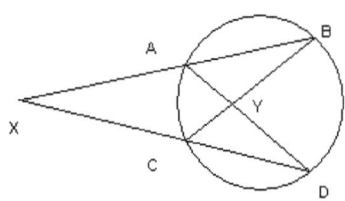

Fig. 5-22. m(∠BYD) = 1/2 m(AC) + m(BD) and m(∠BXD)= 1/2 m(BD) - m(AC).

The *perpendicular bisector* of a line segment is perpendicular to the segment (naturally) and contains the point midway between the ends of the segment. A *median* of a triangle is a segment connecting a vertex and the midpoint of the opposite side. In Fig. 5-23, *EF* is the perpendicular bisector of *BC*, and *AD* is a median of triangle *ABC*, because *BD* = *DC*.

Th. 5-18. The perpendicular bisector of a segment is

unique.

Th. 5-19. All points on the perpendicular bisector of a segment are equidistant from the endpoints of the segment.

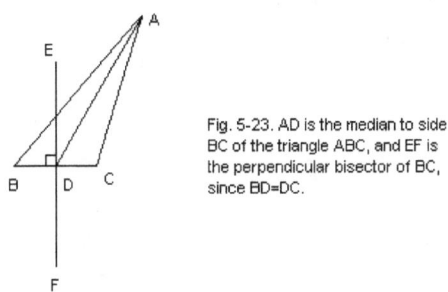

Fig. 5-23. AD is the median to side BC of the triangle ABC, and EF is the perpendicular bisector of BC, since BD=DC.

5.4 Pythagoras' Theorem and Its Many Proofs

We are now in a position to discuss one of the most famous theorems in geometry. It appears in Euclid as the 47th proposition (the 48th and last is its reverse) of the first book of the *Elements*. There are over 300 known proofs, including one found by a President of the U. S. (James Garfield). The first proof presented below is the one found in Euclid.

Th. 5-20 (Pythagoras' Theorem).

In a right-angled triangle, the square on the *hypotenuse* (side opposite the right angle) is the sum, in area, of the squares on the other two sides.

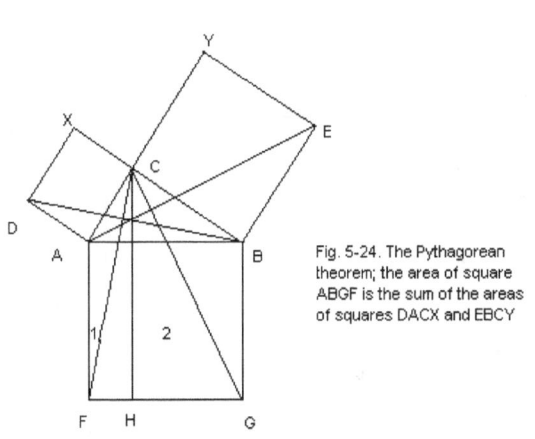

Fig. 5-24. The Pythagorean theorem; the area of square ABGF is the sum of the areas of squares DACX and EBCY

In Fig. 5-24, $\angle ACB$ is right, AB is the hypotenuse, and the area of square $ABGF$ is the sum of the areas of the squares $DACX$ and $EBCY$.

To prove the theorem, we draw the additional lines DB, CF, AE, CG, and CH, making the last parallel to the sides AF and BG of the square $ABGF$. That divides the square into two rectangles, called 1 and 2.

Now, angles DAC and BAF are both right and therefore equal (Postulate 4). Therefore, $\angle DAC + \angle CAB = \angle BAF + \angle CAB$ (If $a = b$, then $a + c = b + c$), and $\angle DAB = \angle CAF$ (Angle Addition Postulate).

Also, $AC = AD$ and $AB = AF$ since each pair of segments are sides of the same square. Therefore, triangle ABD is congruent to triangle AFC (Th. 5-1), so they have the same area.

Now, the point X is on the straight line BC since angles BCA and XCA are both right angles. ($\angle BCA$ is the right angle in triangle ABC, and $\angle XCA$ is part of a square.) Also, AD is parallel to XC (and therefore XB) because they are on opposite sides of a square. Therefore, triangle ABD can be said to have base AD and altitude equal to AC. This means (Th. 5-10B) that the area of ABD is half that of the square $DACX$. Similarly, since CH is parallel to AF, triangle ACF can be said to have base AF and altitude equal to FH, so its area is half of rectangle 1. But the congruent triangles ABD and AFC have the same area, so the area of rectangle 1 must equal that of square $DACX$.

With this proved, identical reasoning involving triangles BEA and BCG, which are congruent like ABD and AFC, and the square $EBCY$ and rectangle 2 shows that the latter two are equal in area. Since the area of square $ABGF$ is the sum of its parts, rectangles 1 and 2, the theorem is proved.

In return for reading through all the gory details of Euclid's proof, I now reward you with this atrocious pun: In an African jungle, there are three large animal skins: an elephant, a hippopotamus, and a rhinoceros. There is a squire standing on each skin, and the one standing on the elephant is married to the one on the rhino. Their son is in between them on the hippo. Therefore (get your laughs ready) the squire on the hippopotamus is the son of the squires on the other two hides! (Don't blame me for this pun – it is not original. I am only responsible for putting it into this book.)

Although the theorem is named for Pythagoras, a Greek philosopher and geometer who lived about two centuries before Euclid, at least one application of enormous practical importance (if you want your buildings to stand

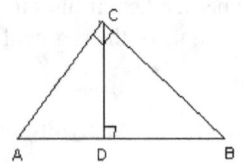

Fig. 5-25. Right traingles ABC, ACD, and CBD are similar to each other.

up straight) was known to Egyptians of pyramid-building ancientry. Since $3^2 + 4^2$ = 5^2, and hence 3, 4, and 5 form a Pythagorean triple as in Chapter 4, a triangle having sides 3, 4, and 5 units in length will contain a right angle. Inside the pyramids are pictures featuring, among other things, a man holding a rope divided into 12 equal parts by knots and shaped into a 3, 4, 5 triangle. This is a handy way of creating a right angle when no sophisticated tools such as a circular compass are available. It should be noted that the bases of the pyramids are squares to within a few inches in as much as 400 feet.

As mentioned above, there are many other proofs of this theorem besides the one found in Euclid. Here are two that use Th. 5-13 on similarity.

Draw the altitude CD to the hypotenuse. (See Fig. 5-25.) Since the acute angles in a right triangle must sum to 90° (Th. 5-7), $\angle ACD = \angle B$ since both are in right triangles that also contain angle A. Similarly, $\angle BCD = \angle A$. Since $\angle ADC = \angle BDC = \angle ACB = 90°$, the three triangles ABC, ACD, CBD are similar to each other, with corresponding parts indicated by the order of the letters. Using only the first half of Th. 5-13, this means that $\frac{AD}{AC} = \frac{AC}{AB}$ or $AD = \frac{(AC)^2}{AB}$, and $\frac{BD}{BC} = \frac{BC}{BA}$ or $BD = \frac{(BC)^2}{BA}$. Since $AD + BD = AB$, $AB = \frac{(AB)^2}{AB} = \frac{(AC)^2 + (BC)^2}{AB}$, so $(AB)^2 = (AC)^2 + (BC)^2$, which is Pythagoras' theorem. Or, we can use the second half of Th. 5-13 to get area ABC : area ACD : area $CBD = (AB)^2 : (AC)^2 : (BC)^2$. Since triangle ABC is just ACD and BCD put together, $(AB)^2 = (AC)^2 + (BC)^2$ again.

There are also proofs of the theorem based on manipulating various area formulas. Fig. 5-26 is said to constitute a proof by itself, but at least some explanation is required. The large square (area = c^2) is cut up into four identical right triangles (area of each = $\frac{1}{2}ab$) and the small square (area = $[a - b]^2 = a^2 - 2ab + b^2$). When the large square's area is equated with the sum of its parts, a little algebra results in $a^2 + b^2 = c^2$, Q. E. D. Similar in spirit is Garfield's proof. In Fig. 5-27, the area of the trapezoid is $\frac{1}{2}(a + b)^2$ (Th. 5-10D), and the areas of the triangles making it up are $\frac{1}{2}ab$, $\frac{1}{2}c^2$, and another $\frac{1}{2}ab$. Again, algebraic manipulation of the equation resulting from equating the whole with the sum of its parts yields $a^2 + b^2 = c^2$.

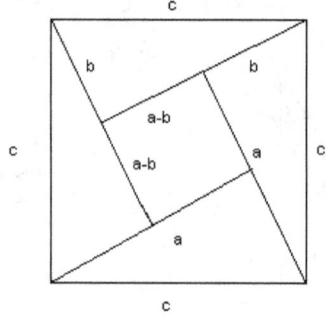

Fig. 5-26. The Pythagorean theorem confirmed with a square cut up into triangles and a smaller square.

One of the most unusual proofs is derived from manipulating, not area formulas for the pieces of a whole, but the pieces themselves. In Fig. 5-28, the larger of the squares on the sides is cut into four pieces by two perpendicular lines through its center, one of which is parallel to the hypotenuse. These pieces and the small square can now be slid, without turning, into positions indicated at upper right in the figure that form the square on the hypotenuse. This is not a rigorous proof as it stands; it can be made rigorous by considering the dotted lines that have been drawn, dividing the already cut-up square into four small squares and interacting with the cut lines to produces four right triangles (one is shaded) that are similar to and exactly half the size of the original triangle.

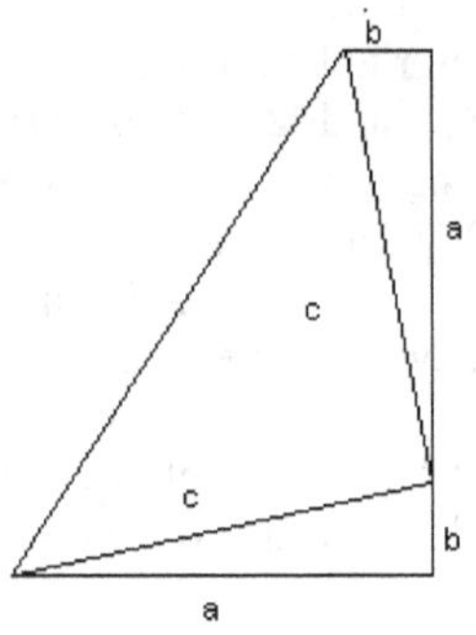

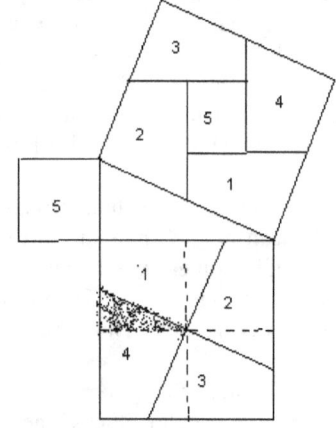

Fig. 5-28. The Pythagorean theorem confirmed with scissors.

Fig. 5-27. James Garfield's proof of the Pythagorean theorem.

5.5 Of Triangles and Their Centers

Squares and circles have centers, but triangles? Yes, they do: at least FIVE. The following theorems give each triangle its five centers.

Th. 5-21. (A) The perpendicular bisectors of the sides of any triangle all intersect in one point, called the *circumcenter* of the triangle.

(B) The bisectors of the angles of a triangle intersect in one point, called the *incenter* of the triangle.

(C) The altitudes of a triangle share a point, the *orthocenter*.

(D) The medians of a triangle share a point, the *centroid*.

Th. 5-22. In any triangle, the following nine points lie on a single circle, whose center is called the triangle's *nine-point* center: the midpoints of the three sides, the feet of the three altitudes, and the midpoints of the segments joining the orthocenter to each vertex.

And if that isn't enough, we also have the following:

Th. 5-23A. The circumcenter, orthocenter, centroid, and nine-point center of any triangle lie on one line, called the *Euler line* after Leonhard Euler, another of the greatest mathematicians of all time.

Th. 5-23B. Denoting the above centers by O, H, G, and N respectively, in any triangle they appear on the Euler line in the order O, G, N, H; and G is one-third of the way, and N halfway, from O to H.

We will prove Th. 5-21(A) and (B) and give non-rigorous proofs of Th. 5-21(D) and Th. 5-22. Before we can prove 5-21(B), we need the following result, which shall be stated and proved as:

Th. 5-24. The bisector of an angle is equidistant from the angle's sides.

In Fig. 5-29, $\overrightarrow{BD}$ bisects $\angle ABC$ and AD and CD are perpendicular to AB and CB respectively; the theorem is proved by showing that $AD = CD$.

We note that in triangles ABD and CBD we have $\angle ABD = \angle CBD$ because BD is a bisector and $\angle BAD = \angle BCD$ since both are right. Therefore, by Th. 5-7, $\angle ADB = \angle CDB$. Since BD belongs to both triangles, we have triangle ABD congruent to triangle CBD (Th. 5-2), and hence $AD = CD$.

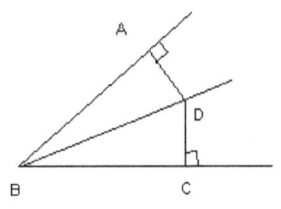

Fig. 5-29. ∠ ABC is bisected by BD, which is equidistant from BA and BC, the sides of the angle (AD=CD).

Now for the proof of Th. 5-21(A). Given the triangle ABC, any point on the perpendicular bisector (p. b.) of AB is equidistant from A and B (Th. 5-19). Also, any point on the p. b. of BC is equidistant from B and C. (See Fig. 5-30.) Therefore, the point of intersection of the two p. b.'s, O, satisfies $OA = OB$ and $OB = OC$, and therefore $OA = OC$. Hence, it is on the p. b. of AC, showing that all three p. b.'s pass through O, which is therefore the triangle's circumcenter. We note that the equality $OA = OB = OC$ guarantees that a circle with center O and radius OA will contain B and C as well. This circle which contains the three vertices of a triangle is said to be *circumscribed* about the triangle, and "circumcenter" stands for "center of the circumscribed circle."

The proof of 5-21(B) is similar. I is the intersection of the bisectors of angles BAC and ABC. (See Fig. 5-31.) By Th. 5-24, $IE = IF$ since I is on the bisector of $\angle BAC$, and $ID = IF$ since I is on the bisector of $\angle ABC$. Therefore, $ID = IE$, and I is on the bisector of $\angle BCA$ as well, showing that all three bisectors pass through I, which is therefore the incenter of the triangle. As with the circumcenter O, the equality $ID = IE = IF$ means that I is the center of a circle containing D, E, and F. Also, since IF is perpendicular to AB, and similarly for ID and IE, the sides of the triangle ABC are tangent to this circle (Th. 5-14). Such

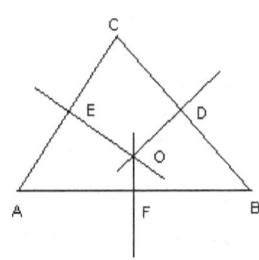

Fig. 5-30. The three perpendicular bisectors of the sides of triangle ABC intersect at O, the circumcenter.

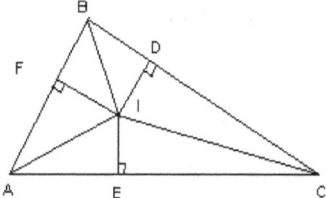

Fig. 5-31. The bisectors of the angles of triangle ABC intersect at I, the incenter. Note the equality ID=IE=IF.

a circle is said to be *inscribed* in the triangle, and "incenter" is short for "center of the inscribed circle."

For the "proof" of Th. 5-21(D), we treat the position of a point as something that can be added to other positions and/or multiplied by numbers. (Underlying this non-rigorous notion is the rigorous concept of *vector analysis*, which will be touched upon in Chapter 10.) The position of the midpoint of a segment is in this way the average of the positions of the endpoints, and the position of a point that is r of the way from A to B (where r is a number between zero and one) is $(1 - r)$ times A's position plus r times B's position. Using these ideas, the position of G in Fig. 5-32, which is two-thirds of the way along the median AD, can be calculated to be $G = \frac{2}{3}D + \frac{1}{3}A = \frac{2}{3}\left(\frac{1}{2}B + \frac{1}{2}C\right) + \frac{1}{3}A = \frac{1}{3}(A + B + C)$. The points two-thirds of the way along the other medians are found to have the same position formula, "proving" Th. 5-21(D) and also showing that the centroid is in a sense the average of the three vertices.

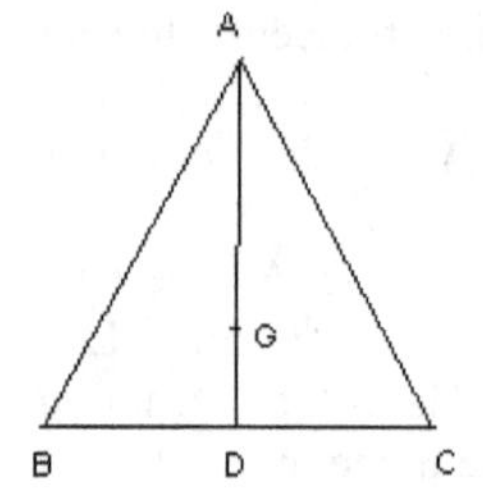

To prove Th. 5-22, we need a fairly large diagram (Fig. 5-33) to show all nine points: the midpoints M, N, P; the feet of the altitudes J, K, L; and the midpoints X, Y, Z of the segments connecting A, B, C to the orthocenter H.

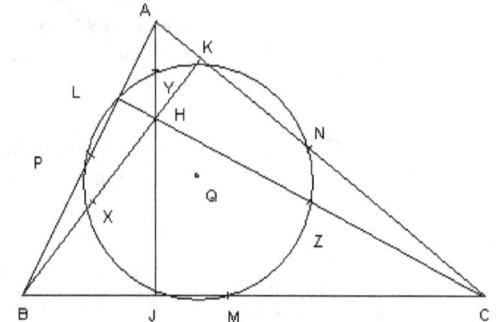

Fig. 5-32. The point G is two-thirds of the way along the median AD; the points two-thirds of the way along the other medians are found to be the same point as G.

We then note that $PNZX$ is a rectangle, and the circle circumscribed around it must also contain K and L; that $PMZY$ is a rectangle, and the circle around it must also contain J and L; and, finally, that $NMXY$ is a rectangle, and its circumscribed circle must also contain J and K. In addition, these three circles, which together contain each of M, N, P, J, K, L, X, Y, Z twice over, share diameters PZ, NX, and YM (each circle contains two of the three), and therefore the "three" circles really are just one circle which contains all nine points. This proof is non-rigorous only because some intermediate steps have been left out; for example, we must prove that $PNZX$ is a rectangle. (This is fairly easy once we know that the line joining the midpoints of two sides of a triangle is parallel to the third side; that in turn follows from the note on similarity after Th. 5-13 and Th. 5-6.)

The even more remarkable Th. 5-23 can be proved once the sub-categories of trigonometry and analytic geometry are brought into the picture, which is the purpose of the next few sections. Before I do that, I should present the (rigorous) proof of Th. 5-6, since it underlies most of the other important theorems, and its method of proof is instructive.

In Fig. 5-34, the parallel lines PQ and RS do not intersect on either side of the transversal XY. If $\angle QAB + \angle SBA$ total less than $180°$, then PQ and RS will intersect to the right of XY by the parallel postulate. At the same time, if

Fig. 5-33. The nine-point circle theorem. The circle with center Q passes through J, K, L, M, N, P, X, Y, and Z.

$\angle PAB + \angle ABR$ total less than $180°$, PQ and RS will intersect to the left of XY. But PQ and RS are parallel, so neither intersection takes place. Therefore, $\angle QAB + \angle SBA \geq 180°$ and $\angle PAB + \angle ABR \geq 180°$. However, the sum of all four angles is equal to the two straight angles PAQ and RBS, which together total $360°$. The only way out is to conclude that $\angle QAB + \angle SBA$ total exactly $180°$, as do $\angle PAB$ and $\angle ABR$. Once this is known, the theorem quickly follows: $\angle QAX + QAB = 180° = \angle SBA + \angle QAB$, so $\angle QAX = \angle SBA$. Similarly, $\angle PAB + \angle PAX = 180° = \angle PAB + \angle ABR$, so $\angle PAX = \angle ABR$. Application of Th. 5-3 to these equalities then produces the other equalities required by the theorem. It is probable that Euclid stated the parallel postulate the way he did, rather than the other, seemingly simpler, ways he could have, in order to more easily apply it to this theorem.

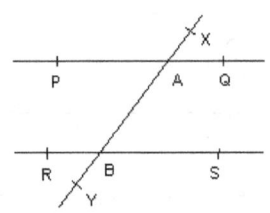

Fig. 5-34. Two parallel lines cut by a transversal, to aid in the proof of Theorem 5-6.

5.6 Not Every Sin Is Deadly (And Not All Tans Are Acquired at the Beach)

Although all triangles display regularities such as Th. 5-23, right triangles have even more properties, e. g., Th. 5-20. For this reason, geometers like to work with them whenever possible. In Fig. 5-35 are several right triangles with different values of the acute angle at bottom left. It is apparent that the shape of the triangle changes in a regular way as this acute angle increases. The shape (as opposed to size) of a right triangle is quantified in terms of ratios of various sides. With 3 sides to choose from, and two sides to a ratio, combinatorial theory (Chapter 9) tells us that there are six ratios. These are called the *trigonometric functions* of the acute angle.

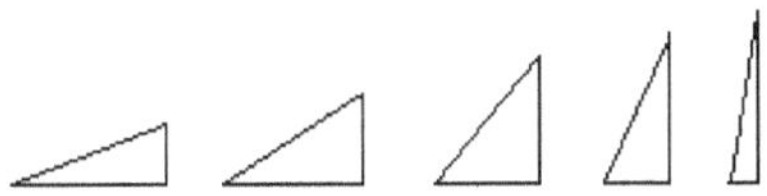

Fig. 5-35. Right triangles with varying sizes of the acute angle at bottom left. Note how the shape of the triangle changes as this angle does.

A right triangle is traditionally named as triangle *ABC*, the point *C* being the vertex of the right angle. The capital letters *A* and *B* refer to the acute angles with these points as vertices, and the lower-case letters *a*, *b*, and *c* refer to the lengths of the sides opposite the vertices named with the corresponding capital letters; therefore, *c* denotes the triangle's hypotenuse. With this notation, the six trigonometric functions are defined as follows:

$\frac{a}{c}$ is the *sine* of angle *A*, written sin *A*.

$\frac{b}{c}$ is the *cosine* of angle *A*, written cos *A*.

$\frac{a}{b}$ is the *tangent* of angle *A*, written tan *A*.

$\frac{b}{a}$ is the *cotangent* of angle *A*, written cot *A*.

$\frac{c}{b}$ is the *secant* of angle *A*, written sec *A*.

And finally, $\frac{c}{a}$ is the *cosecant* of angle *A*, written csc *A*.

Reflecting the basic triangle about the bisector of its right angle exchanges sides *a* and *b* and angles *A* and *B*. Since $A + B = 90°$ and exchanging *a* and *b* in the definitions of the functions either removes or adds "co" from or to the function name, we have trig A = cotrig $(90° - A)$, where "trig" means any of sin, cos, tan, cot, sec, or csc, and "cotrig" is obtained from "trig" by adding a "co" to the name if it is not already there or removing "co" if it is there.

Also, directly from the definitions, we see that $1/\sin = \csc$, $1/\cos = \sec$, and $1/\tan = \cot$, as well as sin/cos = tan, tan/sec = sin, csc × tan = sec, and other quotient and product relationships. By Th. 5-20, $a^2 + b^2 = c^2$, so $\frac{a^2}{c^2} + \frac{b^2}{c^2} = 1$, or $(\frac{a}{c})^2 + (\frac{b}{c})^2 = 1$. Since $\frac{a}{c} = \sin A$ and $\frac{b}{c} = \cos A$, we have, for any acute angle *A*, $(\sin A)^2 + (\cos A)^2 = 1$. (Powers of trigonometric functions are usually written $\sin^2 A$ for $(\sin A)^2$, $\cos^3 B$ for $(\cos B)^3$, etc.) We can divide the Pythagorean result by a^2 or b^2 as well, getting the other Pythagorean relations $1 + \tan^2 A = \sec^2 A$ and $1 + \cot^2 A = \csc^2 A$. We can calculate $\tan^2 A - \sin^2 A$ as follows: $\tan^2 A - \sin^2 A = \frac{a^2}{b^2} - \frac{a^2}{c^2} = \frac{a^2(c^2 - b^2)}{b^2 c^2} = \frac{a^2 a^2}{b^2 c^2}$ (Pythagoras again) = $\tan^2 A \times \sin^2 A$. I could go on and on deriving other relations between trigonometric functions, but perhaps it would save some time if I merely wrote a list of the most important identities.

Reciprocals: $\sin A = \frac{1}{\csc A}$, $\cos A = \frac{1}{\sec A}$, $\tan A = \frac{1}{\cot A}$, $\cot A = \frac{1}{\tan A}$, $\sec A = \frac{1}{\cos A}$, $\csc A = \frac{1}{\sin A}$.

Products: $\sin A = \tan A \times \cos A$, $\cos A = \cot A \times \sin A$, $\tan A = \sin A \times \sec A$, $\cot A = \cos A \times \csc A$, $\sec A = \csc A \times \tan A$, and $\csc A = \sec A \times \cot A$.

Quotients: $\sin A = \frac{\cos A}{\cot A} = \frac{\tan A}{\sec A}$, $\cos A = \frac{\sin A}{\tan A} = \frac{\cot A}{\csc A}$, $\tan A = \frac{\sin A}{\cos A} = \frac{\sec A}{\csc A}$, $\cot A = \frac{\cos A}{\sin A} = \frac{\csc A}{\sec A}$, $\sec A = \frac{\tan A}{\sin A} = \frac{\csc A}{\cot A}$, and $\csc A = \frac{\cot A}{\cos A} = \frac{\sec A}{\tan A}$.

Pythagorean: $\sin^2 A + \cos^2 A = 1$, $1 + \tan^2 A = \sec^2 A$, and $1 + \cot^2 A = \csc^2 A$.

Angle-sum: $\sin (A + B) = \sin A \cos B + \cos A \sin B$, $\cos (A + B) = \cos A \cos B - \sin A \sin B$, and $\tan(A + B) = \frac{\tan A + \tan B}{1 - \tan A \tan B}$.

Double-angle: $\sin 2A = 2\sin A\cos A$, $\cos 2A = \cos^2 A - \sin^2 A$, and $\tan A = \dfrac{2\tan A}{1-(\tan A)^2}$.

Half-angle: $\sin\dfrac{A}{2} = \sqrt{\dfrac{1-\cos A}{2}}$, $\cos\dfrac{A}{2} = \sqrt{\dfrac{1+\cos A}{2}}$, and $\tan\dfrac{A}{2} = \sqrt{\dfrac{1-\cos A}{1+\cos A}}$.

5.6.1 Solving Right Triangles

The main practical importance of trigonometry is in measuring lengths and angles that are inaccessible, using measurements of other lengths and angles that are. For instance, determining the height of a skyscraper. (I know the joke about using a barometer to measure such a height, but here you have access to neither the building's roof nor its superintendent.) Instead, you walk a measured 1,000 feet away from its base, and find that the building's top has an angle of elevation of 35°. Assuming that the building's side is vertical (we should hope so), you now have two elements of a right triangle, as shown in Fig. 5-36. Given two elements, you can always find the others, as long as at least one of the known elements is a side. Labeling the triangle as shown, you are given A and b, and wish to find a. A trigonometric ratio involving a, b, and A is $\tan A = a/b$, or $a = b\tan A$. It so happens that $\tan 35°$ is very close to 0.7, so a here is 1,000 x 0.7, or 700 feet. Since a is the height of the skyscraper, our problem is finished.

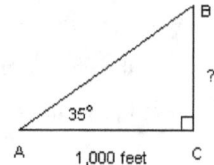

This procedure can be used to solve any right triangle. If you are given one angle and any side, there is always a trigonometric ratio containing the angle, the known side, and an unknown side, while the other angle can be found from $A + B = 90°$. Similarly, if you are given two sides, the theorem of Pythagoras can be used to find the third side, and a trigonometric ratio involving the two known sides can be used to find the sine, cosine, or tangent of an unknown angle. Tables of trigonometric functions have been compiled for this purpose, and any good math book will have a brief table. (This one does too; look in Table 8.)

Fig. 5-36. Using trigonometry to determine the height of a skyscraper.

5.6.2 Solving Oblique Triangles

When a triangle does not contain a right angle, and is therefore called *oblique*, the trigonometric functions cannot be directly applied because there are no right triangles. Fortunately, geometry comes in handy, creating right triangles to which trigonometry can be applied. The result is equations that can be used to solve oblique triangles.

The two most important equations are the law of sines and the law of cosines. To derive the law of sines, we take an arbitrary triangle ABC and draw its circumscribed circle (Th. 5-21(A)) with circumcenter O. We also draw the

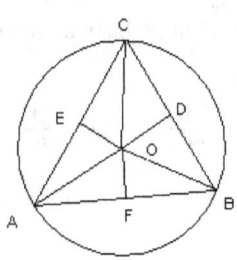

perpendicular bisectors OD, OE, and OF of the three sides a, b, and c respectively, intersecting at O, and the radii OA, OB, and OC. Each of these segments, of course, is equal to the other two; their common value is called the *circumradius*, symbolized by R. Concentrating on the triangles OFA and OFB, we have $OA = OB$, $OF = OF$, and $FA = FB = c/2$ because OF is the perpendicular bisector. Hence, triangles OFA and OFB are congruent (Th. 5-12) and $\angle FOA = \angle FOB$. Now, $\angle ACB = C$ is inscribed in the circle and hence C is half of $m(\overset{\frown}{AB})$, its intercepted arc (Th. 5-15). Since $\angle AOB$ is the *central* angle intercepting $\overset{\frown}{AB}$, it follows that $\angle AOB = 2C$. Now $\angle AOB$ is divided equally into the angles FOA and FOB, so $\angle FOA = C$. Therefore, in the *right* triangle OAF, $\dfrac{FA}{OA} = \sin\angle FOA$, or $\dfrac{c/2}{R} = \sin C$ or $2R = \dfrac{c}{\sin C}$. Repeating this process with the other two sides shows $\dfrac{a}{\sin A} = \dfrac{b}{\sin B} = \dfrac{c}{\sin C} =$

Fig. 5-37. The law of sines derived from a triangle and its circumscribed circle.

$2R$, which is the law of sines.

To derive the law of cosines, we again take an arbitrary triangle ABC, and this time draw the altitude CD as shown in Fig. 5-38. This creates the two right triangles ABD and BCD. In triangle ACD, $CD = b\sin A$ and $AD = b\cos A$, while in triangle BCD, $BD = c - b\cos A$, and therefore $(CB)^2 = a^2 = (b\sin A)^2 + (c - b\cos A)^2 = b^2\sin^2 A + c^2 - 2cb\cos A + b^2\cos^2 A = c^2 + b^2(\sin^2 A + \cos^2 A) - 2bc\cos A$. Since $\sin^2 A + \cos^2 A = 1$ (Pythagorean identity), we have $a^2 = b^2 + c^2 - 2bc\cos A$, which is the law of cosines. By drawing the altitudes to the other

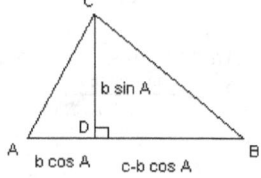

Fig. 5-38. The law of cosines derived by drawing the altitude to one side of a triangle.

sides (or just permuting the letters), we derive its companions, $b^2 = a^2 + c^2 - 2ac \cos B$ and $c^2 = a^2 + b^2 - 2ab \cos C$. Note that if C is a right angle, we have $c^2 = a^2 + b^2$ (Th. 5-20), and therefore $\cos 90°$ must be zero.

To use these laws to solve oblique triangles, our procedure varies with the information we are given. If we have two angles and a side, we find the third angle from Th. 5-7 and then use the law of sines to find the other two sides. With two sides and the angle between them, we use the law of cosines to find the third side and then the law of sines for the other two angles. With three sides, we solve the equations of the law of cosines for the cosines of the unknown angles. If we are given two sides and an angle opposite one of them, we must deal with the possibility of two triangles that satisfy the conditions. In Fig. 5-39, the triangles ABC and ABC' both have the same valuse of AB, BC, and the *non*-included angle A. This figure shows why this is called the *ambiguous case*. Since triangle BCC' is isosceles, it can be proven that $\angle ACB + \angle AC'B = 180°$. Once the sine of C is calculated with the law of sines, we find the acute angle with that sine; if it is larger than $\angle A$, but not $90°$, two solutions exist. With the number of solutions determined, we use Th. 5-7 to determine the value (or values) of B, and then the law of cosines to find the third side (or sides).

Oblique triangles can contain things that right triangles cannot: obtuse angles. The trigonometric functions of an obtuse angle cannot be found by studying right triangles, but they are defined and can be calculated from the following formulas: $\sin(A + 90°) = \cos A$, $\cos(A + 90°) = -\sin A$, $\tan(A + 90°) = -\cot A$, $\cot(A + 90°) = -\tan A$, $\sec(A + 90°) = -\csc A$, and $\csc(A + 90°) = \sec A$. Later on, we will see how trigonometry can be extended to angles greater then $180°$ and even negative angles, even though these cannot be found in triangles.

Geometry can enable us to derive other useful formulas for triangles. Taking, for example, Fig. 5-31, we see that $AE = AF$, $BD = BF$, and $CD = CE$, and the sum of all six segments is the total perimeter of the triangle, and therefore $AF + BD + CE =$ half the perimeter or the *semiperimeter*, written s; $s = \frac{1}{2}(a + b + c)$. In addition, $BD + CD = BC = a$, so $AF = s - a$, and similarly $BD = s - b$ and $CE = s - c$. In the right triangle AIF, IF is the radius of the inscribed circle, symbolized by r, and $\angle IAF = \frac{1}{2}A$ since AI bisects $\angle BAC$. Therefore, $\tan\frac{1}{2}A = \frac{r}{s-a}$, and similarly, $\tan\frac{1}{2}B = \frac{r}{s-b}$ and $\tan\frac{1}{2}C = \frac{r}{s-c}$. These formulas can serve as alternatives to the law of cosines in solving a three-side problem, once we have calculated r. The formula for r is

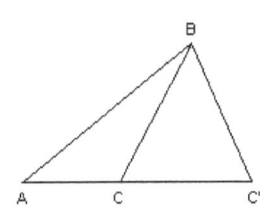

$\sqrt{\frac{(s-a)(s-b)(s-c)}{s}}$, as can be found by comparing various area formulas.

5.6.3 Areas of Triangles

By Th. 5-10(B), the area of a triangle, which is denoted by K, is half a side times the altitude to that side. Fig. 5-38 shows that the altitude to side c, say, is $b \sin A$ or $a \sin B$; therefore, $K = \frac{1}{2}bc \sin A = \frac{1}{2}ca \sin B = \frac{1}{2}ab \sin C$. From the law of sines, $a = \frac{b \sin A}{\sin B}$; substituting this in the previous formulas results in $K = \frac{a^2 \sin B \sin C}{2 \sin A} = \frac{b^2 \sin C \sin A}{2 \sin B} = \frac{c^2 \sin A \sin B}{2 \sin C}$. To get a formula for area in terms of just the three sides, we take $K = \frac{1}{2}bc \sin A$ and expand it as follows:

$$K = \frac{1}{2}bc \sin A$$
$$= \frac{1}{2}bc\sqrt{1 - \cos^2 A} \text{ (Pythagorean identity)}$$
$$= \frac{1}{2}bc\sqrt{1 - \left(\frac{b^2+c^2-a^2}{2bc}\right)^2} \text{ (law of cosines)}$$
$$= \frac{1}{4}\sqrt{(2bc)^2 - (b^2 + c^2 - a^2)^2} \text{ (algebra; dividing outside the radical by } 2bc \text{ and multiplying inside it by}$$
$(2bc)^2$)
$$= \frac{1}{4}\sqrt{(2bc + b^2 + c^2 - a^2)(2bc - b^2 - c^2 + a^2)} \text{ (Equation [3.1] with "}a\text{" replaced with "}2bc\text{" and "}b\text{"}$$
with "$b^2 + c^2 - a^2$")
$$= \frac{1}{4}\sqrt{[(b + c)^2 - a^2][a^2 - (b - c)^2]} \text{ (Equation [3.2a] in the first bracket, [3.2b] in the second)}$$
$$= \frac{1}{4}\sqrt{(b + c + a)(b + c - a)(a - b + c)(a + b - c)} \text{ (Equation [3.1] used twice)}$$

$K = \sqrt{s(s-a)(s-b)(s-c)}$ by the definition of s above.

This formula is called Hero's (sometimes Heron's) formula after the later classical Greek mathematician who discovered it. It is not found in Euclid.

Figure 5-31 yields an area formula as follows: The main triangle is divided into six right triangles, each of which has r, the inscribed circle's radius, for an altitude and whose bases add up to the entire perimeter. Therefore, $K = \frac{1}{2}r(a+b+c) = rs$, since s is the semiperimeter. Combining this with Hero's formula gives the formula for r at the end of the previous section. To find a formula for K in terms of a, b, c, and R, we start off with $K = \frac{1}{2}ab\sin C$ and then plug in the law of sines, $\frac{c}{\sin C} = 2R$ or $\sin C = \frac{c}{2R}$ to get $K = \frac{abc}{4R}$.

5.6.4 Trigonometry in Pictures

As a reminder and memory help the right triangles shown in Fig. 5-40 feature all six trigonometric functions of an acute angle.

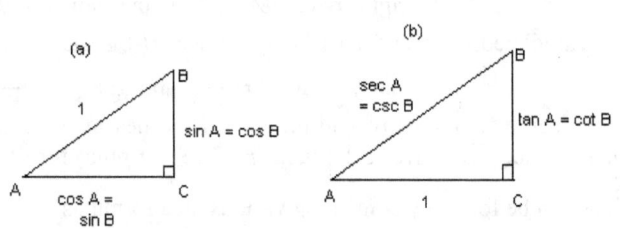
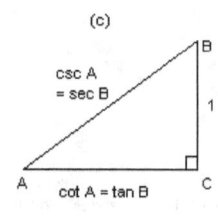

Fig. 5-40. The trigonometric functions of an acute angle displayed.

We can combine the above triangles into one diagram and in addition show directly the trigonometric functions of any angle with the following: Draw a unit circle with center O and radius 1 (inch, foot, megaparsec, what have you). Fix one side of the test angle as the horizontal OA, and also draw the vertical OB. Then draw the tangents (not to be confused with the function!) at A and B and call their point of intersection X. (See Fig. 5-41(a).) Now, the other, variable side of the test angle can be taken as $\overrightarrow{OP}$, which intersects AX in T and BX in V. The ray can be extended backwards through O if necessary, as is the case in Fig. 5-41(b). Now, the sine of the test angle AOP is the length of QP or OR, its cosine is OQ or RP, its tangent is AT, its cotangent is BV, its secant is OT, and its cosecant is OV. The signs of the functions are found like this: The sine is negative if R is below O, the cosine is negative if Q is to the left of O, the tangent is negative if T is below A, the cotangent is negative if V is to the left of B, the secant is negative if $\overrightarrow{OP}$ must be extended backwards to intersect AX, and the cosecant is negative if $\overrightarrow{OP}$ must be extended backwards to intersect BX. Otherwise the functions are positive. If $\overrightarrow{OP}$ is parallel to AX, the tangent and secant functions are undefined, and if $\overrightarrow{OP}$ is parallel to BX, the cotangent and cosecant functions are undefined. The pattern of signs of the functions divides the plane naturally into four *quadrants*, as seen in Fig. 5-42(a). The quadrants are referred to by Roman numerals, for some strange reason, and are applied to the side

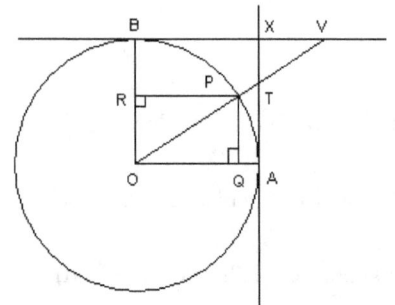

Fig. 5-41(a). The six trigonometric functions of angle AOP in one diagram. The sine of the angle is the length of QP or OR, and so on as given in the text.

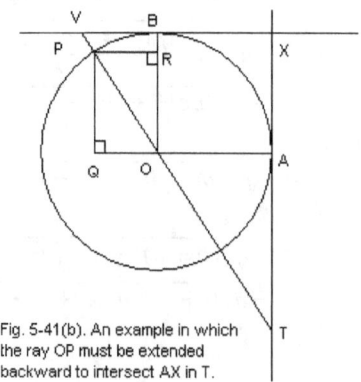

Fig. 5-41(b). An example in which the ray OP must be extended backward to intersect AX in T.

$\overrightarrow{OP}$ of the test angle. If $\overrightarrow{OP}$ does not fall on either of the lines OA, OB, then all six functions are defined, and the angle is in one of the four quadrants, which determines a range for its measure and the signs of the functions, as shown in Fig. 5-42(b). It should be noted that angles greater than 180° are to be thought of in terms of rotation: the fixed side OA rotates through the angle in a counterclockwise direction until it reaches $\overrightarrow{OP}$. A complete rotation describes an angle of 360° or 2π radians. Because rotation through 360° doesn't change anything, the very important *periodic* property of trigonometric functions is apparent: trig(x + 360°) = trig x, where trig means any or all of the six functions. Negative angles are defined as clockwise rotations; thus, the minute hand of a clock sweeps out an angle of - 6° every 60 seconds. From a symmetry argument, the values of the functions of a negative angle can be found as follows: sin (- x) = - sin x, cos (- x) = cos x, tan (- x) = - tan x, cot (- x) = - cot x, sec (- x) = sec x, and csc (- x) = - csc x. The trigonometric functions of negative angles or angles greater than 180° can have no application to triangles, but they are of importance in calculus, where the functions are viewed as being applied to numbers rather than angles – but numbers which are radian measures of corresponding angles.

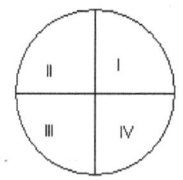

Fig. 5-42(a). The circle divided into the four quadrants by vertical and horizontal lines.

Quadrant	Angle's Measure		Function Signs					
	Degrees	Radians	sin	cos	tan	cot	sec	csc
I	$0 < x < 90$	$0 < x < \pi/2$	+	+	+	+	+	+
II	$90 < x < 180$	$\pi/2 < x < \pi$	+	-	-	-	-	+
III	$180 < x < 270$	$\pi < x < 3\pi/2$	-	-	+	+	-	-
IV	$270 < x < 360$	$3\pi/2 < x < 2\pi$	-	+	-	-	+	-

Fig. 5-42b. The quadrants and the signs of the six trigonometric functions.

5.7 The Cartesian Coordinate System

Rene Descartes, mentioned briefly in the chapter on algebra, was primarily a philosopher, but he made one huge contribution to mathematics: he invented the idea of designating points in the plane by ordered pairs of real numbers. The system he used is called *Cartesian* in his honor.

Numbers are assigned to points based on their positions relative to two perpendicular lines whose intersection is called the *origin*, and an arbitrary unit of length. Once the two basic lines (the *coordinate axes*) have been fixed, the first number assigned to a point is its perpendicular distance from the first axis, and the second number is its perpendicular distance from the second axis, relative to the arbitrary unit. It is conventional to take the two axes as horizontal and vertical, and call the two numbers assigned to a point the x- and y-coordinates of the point. The axes are named the x-axis and y-axis, but, confusingly, the x-coordinate is the distance from the y-axis and the y-coordinate is the distance from the x-axis. It is also a convention to take the x-axis as the horizontal line and the y-axis as vertical, as seen in Fig. 5-43. With these conventions, the coordinates are given signs as follows: the x-coordinate is negative if P is to the left of the y-axis and positive if P is to the right of that line, and the y-coordinate is negative if P is below the x-axis and positive if P is above it. If P is on the x-axis, its y-coordinate is zero, and if P is on the y-axis, its x-coordinate is zero. If P is on both axes, that is, if P is the origin O, both coordinates are zero. We write $P(a, b)$ to show that P's x-coordinate is a and that its y-coordinate is b; we also write that P's coordinates are (a, b). Note that (a, b) and (b, a) are different ordered pairs and refer to different points unless $a = b$. (See Sec. 2.1 on ordered pairs if you are not sure of this.)

5.7.1 Equations and Graphs

An equation in Cartesian coordinates can appear as $y = f(x)$, $x = g(y)$, $f(x) = g(y)$, or, most generally, $F(x, y) = 0$. Here, $f(x)$ represents an algebraic expression involving x (but not y), $g(y)$ is an expression containing y (but not x), and $F(x, y)$ is an expression involving x, y, or (usually) both. Numerical or literal constants may also appear.

The *graph* of an equation consists of all those points, and only those points, whose coordinates satisfy the equation. Thus, the point (2, 1) is on the graph of the equation $2x^2 - 3xy + 5y - 7 = 0$ because $2(2^2) - 3 \cdot 2 \cdot 1 + 5 \cdot 1 - 7 = 0$. In general, the graph of an equation will take the form of a curve

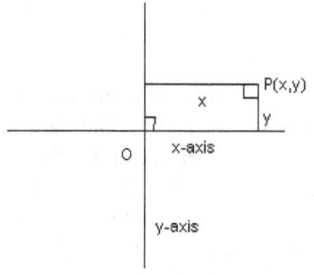

Fig. 5-43. The Cartesian coordinate system.

– the more complicated the equation, the more complicated the curve. The simplest curve is the straight line; equations whose graphs are straight lines are called *linear* for that reason. We now will investigate what types of equations are linear and how to describe lines in terms of coordinate geometry.

5.7.2 Linear Equations, Slopes, and Intercepts

We begin by drawing a line that intersects both axes; that is, one that is neither horizontal nor vertical. This line is determined by any two points on it, so let us take two arbitrary ones: $P_1(x_1, y_1)$ and $P_2(x_2, y_2)$. Between P_1 and P_2 the line has a *run* defined as $x_2 - x_1$ and a *rise* defined as $y_2 - y_1$. Using the point $Q(x_2, y_1)$ in Fig. 5-44, we see that the run is P_1Q and the rise QP_2. Note that the run is measured along a horizontal line and the rise along a vertical one. The line P_1P_2 has a *slope* defined by rise ÷ run, that is, slope of $P_1P_2 = \frac{y_2 - y_1}{x_2 - x_1}$. The slope of a line is denoted by m. (Why m? The letter may begin the German (or Latin, or Greek, or French, etc.) word for slope. Anyway, the letter "s" is used in analytic geometry for other purposes.) It is important to note that the slope of a straight line is independent of the choice of P_1 and P_2 on it. If we pick two other arbitrary points $P_3(x_3, y_3)$ and $P_4(x_4, y_4)$, we define the point $Q'(x_4, y_3)$ and construct the right triangle P_3P_4Q' as we did P_1P_2Q. Since P_3Q', like P_1Q, is horizontal, these two lines are parallel. Similarly, $Q'P_4$ is parallel to QP_2. Therefore, by Theorem 5-6, $\angle Q'P_3P_4 = \angle QP_1P_2$ and $\angle Q'P_4P_3 = \angle QP_2P_1$. Hence, the two triangles P_3P_4Q' and P_1P_2Q are similar, and $\frac{P_4Q'}{Q'P_3} = \frac{P_2Q}{QP_1}$ (Th. 5-13(A)). So, both expressions for the slope yield the same result.

Now, we use this fact to determine the equation of the line through P_1 and P_2. $P(x, y)$ is on this line if and only if the slope of P_1P is the same as that of P_1P_2. Using this in the slope formula above, (x, y) is on P_1P_2 if and only if $\frac{y - y_1}{x - x_1} = \frac{y_2 - y_1}{x_2 - x_1}$. This equation is therefore the equation of the line P_1P_2. It is called the *two-point* form of the equation of a straight line because the coordinates of two points are used to create it. We can obtain other forms of the straight-line equation by various tricks.

Replacing $\frac{y_2 - y_1}{x_2 - x_1}$ with m (both represent the slope) and multiplying by $(x - x_1)$, we obtain the equation $y - y_1 = m(x - x_1)$. Instead of two points, we have one point and the slope m, so this is known as the *point-slope* form of the equation. We can simplify this further by taking P_1 to be the point $I_y(0, b)$ where the line crosses the y-axis. This point, and the value of b, are both called the *y-intercept*. Substituting the coordinates of I_y in the point-slope equation results in $y = mx + b$, which is called the *slope-intercept* form of the equation. The two-point form can be simplified if we take P_1 and P_2

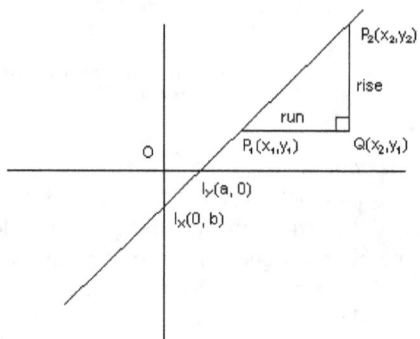

Fig. 5-44. An oblique line in a Cartesian plane; the line's slope is $(y_2\text{-}y_1)/(x_2\text{-}x_1)$.

to be I_y and $I_x(a, 0)$, the latter being the point where the line crosses the x-axis. I_x and a are referred to as x-*intercepts* by analogy to I_y and b. The algebra goes like this: $\frac{y - b}{x - 0} = \frac{0 - b}{a - 0}; \frac{y - b}{x} = -\frac{b}{a}; \frac{a(y - b)}{x} = -b; ay - ab = -bx; ay + bx = ab$. If the line does not pass through the origin, neither a nor b will be zero, so we may divide the last equation by ab to obtain $\frac{x}{a} + \frac{y}{b} = 1$, which is called the *two-intercept* form of the equation. Note here that x and y are divided by the x- and y-intercepts respectively; this is why the division by ab is carried out. Also note that if the line does pass through the origin, there is no two-intercept form of its equation, but that the slope-intercept form still exists, and in fact reduces to $y = mx$.

Note that in all these equations, each side is (or can be changed into) a polynomial of the first degree in x and/or y. This is the hallmark of a linear equation; indeed, first-degree polynomials are called linear because of this. To summarize: <u>Any</u> equation that is linear in both x and y has a straight line as its graph, provided that it is neither an identity (e. g., $4x + 2y + 3 = 4x + 2y + 3$) nor a contradiction (e. g., $4x + 2y + 3 = 4x + 2y + 4$).

So far, our equations have been for a line that intersects both axes. The special forms for lines that are parallel to an axis are as follows: a line parallel to the x-axis (a horizontal line) has as equation $y = b$, and a line parallel to the y-axis (a vertical line) has as equation $x = a$. These follow immediately from the definitions of the coordinates and the fact that parallel lines are everywhere equidistant. The horizontal line equation is a special case of the slope-intercept form $y = mx + b$ with $m = 0$, hence a horizontal line has zero slope. This is consistent with the definition of slope as rise ÷ run; horizontal lines have run but no rise. The equation for a vertical line, however, cannot be matched to any of the forms given for an oblique line. A vertical line has rise but no run, so the calculation of slope as rise ÷ run cannot be carried out. Hence, a vertical line has no slope (which is quite different from having zero slope). Since all the forms of the basic linear equation, except the two-intercept form, have a slope, either explicit as m or implicit as $\frac{y_2-y_1}{x_2-x_1}$, it is not surprising that the equation for a vertical line cannot be put into these forms. (The two-intercept form fails as well, because a vertical line has no y-intercept.)

We bring trigonometry into all of this by looking a little closer at the right triangle P_1P_2Q used to define slope. We have $m =$ rise/run $= \frac{QP_2}{P_1Q} = \tan \angle QP_1P_2$. This angle is called the *angle of inclination* and symbolized by φ, the golden-ratio letter. By Th. 5-6, this angle is the same anywhere on the line, and we write $m = \tan \varphi$ to show that the slope is the tangent of the inclination angle. For a vertical

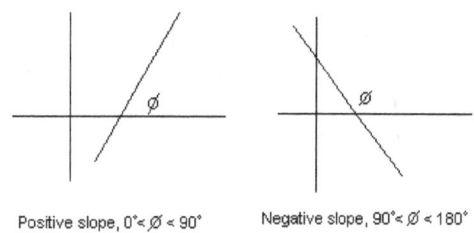

Positive slope, $0° < \varnothing < 90°$ Negative slope, $90° < \varnothing < 180°$

Fig. 5-45. The angle of inclination for lines with positive and negative slopes.

line, $\varphi = 90°$, and an angle of 90° does not have a tangent. The equation $m = \tan \varphi$ shows us that parallel lines have the same slope (or share the property of lacking a slope, if they are vertical). Lines which fall as one moves from left to right have negative slopes. Since φ is defined as a positive angle, this means $90° < \varphi < 180°$. For lines with negative slopes, the definition of φ as $\angle QP_1P_2$ must be replaced with one that involves a fixed horizontal line, such as the x-axis. φ is defined as the smallest positive angle that can be measured from the x-axis to the line, as seen in Fig. 5-45. To complete the definition, we define φ to be zero for a horizontal line; note that $\tan 0° = 0$, so that $m = \tan \varphi$ still holds.

5.7.3 The Distance Formula and Equations

Reference to Fig. 5-44 and the Pythagorean theorem shows that the distance P_1P_2 is the hypotenuse of a right triangle whose sides are the rise and run from P_1 to P_2, and therefore $d(P_1, P_2) = \sqrt{(rise)^2 + (run)^2} = \sqrt{(x_2-x_1)^2 + (y_2-y_1)^2}$. This distance formula is important since many curves that are more complicated than straight lines are defined in terms of distances.

The *circle*, of course, is defined as the set of all points a given distance (called r in this context) from a given point, whose coordinates are taken to be (h, k). Therefore, the equation for the circle is $r = \sqrt{(x-h)^2 + (y-k)^2}$ or $(x-h)^2 + (y-k)^2 = r^2$ or $x^2 + y^2 + 2hx + 2ky + h^2 + k^2 - r^2 = 0$. If we are given an equation $x^2 + y^2 + ax + by + c = 0$, we determine the center and radius of the circle as follows: First, the squares in x and y are completed, resulting in $(x + \frac{a}{2})^2 + (y + \frac{b}{2})^2 + c = \frac{a^2+b^2}{4}$. Then, c is transposed to the other side of the equation, which can now be matched with $(x-h)^2 + (y-k)^2 = r^2$, resulting in $h = -\frac{a}{2}, k = -\frac{b}{2}$, and $r = \frac{1}{2}\sqrt{a^2 + b^2 - 4c}$. Note that if c is too big, the value of r will be imaginary. Only real numbers are used in drawing graphs, so we conclude that if $4c > a^2 + b^2$, the equation has no graph – it is a contradiction (relative to real values of x and y). If $4c = a^2 + b^2$, then the graph consists of the single point $(-\frac{a}{2}, -\frac{b}{2})$, and in all other cases the graph is a circle.

The simplest form of the equation for a circle occurs when $h = k = 0$, that is, when the circle is centered at the origin, and $r = 1$. This reduces the equation to $x^2 + y^2 = 1$. The resulting graph is called the *unit* circle. We can place the rest of Fig. 5-41(a) on that circle, with A being $(1, 0)$, $B(0, 1)$, $X(1, 1)$, the line AX having the equation $x = 1$, and BX having the equation $y = 1$. Denoting the test angle AOP by φ, we have the slope of $\overrightarrow{OP} = m = \tan \varphi$. Since the line passes through the origin its equation is $y = mx = x \tan \varphi$, hence the coordinates of T are $(1, \tan \varphi)$ and of V are $(1/\tan \varphi, 1)$ or $(\cot \varphi, 1)$. The Pythagorean identities in Sec. 5.6 state that $\sin^2 \varphi + \cos^2 \varphi = 1$, and a product identity says $\cos \times \tan = \sin$, so we can take the coordinates of P to be $(\cos \varphi, \sin \varphi)$. Therefore, any point on $\overrightarrow{OP}$ has coordinates $(a \cos \varphi, a \sin \varphi)$, where a is the distance from O to the point.

The circle is a special member of a family of curves called *conic sections*. Now, a cone is a three-dimensional figure and therefore was not mentioned earlier; it consists of a circle, a point not in the circle's plane, and all the lines connecting the point to points on the circle. In a *right circular* cone, the fixed point, called the vertex, is so positioned that the line from it to the circle's center is perpendicular to the circle's plane; that is, it is perpendicular to every line in that plane that intersects it. Conic sections are intersections of a right circular cone with various planes. All of them can be characterized as sets of points whose distance from a fixed point, called the *focus*, is a constant multiple, called the *eccentricity* and denoted by e, of their distance from a fixed line, called the *directrix*. We place this definition into the Cartesian system by taking the focus to be the origin $(0, 0)$ and the directrix to be the vertical line $x = -a$. Since distance from a point to a line is measured perpendicular to the line, the distance between $P(x, y)$ and the directrix is $x + a$; therefore, the equation of the conic section is $x^2 + y^2 = e^2(x + a)^2$ or $(1 - e^2)x^2 + y^2 - 2ae^2x - e^2a^2 = 0$. The coefficient $1 - e^2$ of x^2 suggests a subdivision of the conic sections into three categories by the value of e: $e < 1$, which gives rise to a curve called the *ellipse*; $e = 1$ (which causes the x^2 term to vanish) whose graph is a *parabola*; and $e > 1$, whose graph is a *hyperbola*. The circle is a special case of the ellipse for which $e = 0$ and the directrix is considered to lie infinitely far away.

For a conic that is not a parabola, we can complete the square in x in the equation and transform it into $(1 - e^2)(x - \frac{ae^2}{1-e^2})^2 + y^2 = a^2e^2(1 + \frac{e^2}{1-e^2})$. This form is analogous to the equation of a circle $(x - h)^2 + (y - k)^2 = r^2$, where $h = \frac{ae^2}{1-e^2}$ and $k = 0$. We can always find constants A and B, which depend on a and e, so that the equation can be rewritten as $\frac{(x-h)^2}{A^2} \pm \frac{y^2}{B^2} = 1$, and $h = \frac{ae^2}{1-e^2}$. The point $(\frac{ae^2}{1-e^2}, 0)$ is called the center of the conic. The $\pm$ sign is $+$ if the conic is an ellipse (i. e., $e < 1$), and $-$ if the conic is a hyperbola.

If the conic is a parabola, we can change its equation to $y^2 = 2a(x + \frac{a}{2})$. Note that since y^2 must be at least zero, x must be at least $-\frac{a}{2}$ for (x, y) to be on the graph. The point $(-\frac{a}{2}, 0)$ on the graph is called the vertex of the parabola; it is midway between the focus $(0, 0)$ and the directrix of the parabola.

For any conic, we can exchange x and y in the general equation or any of the more specific ones to get the equation for a curve whose directrix is the line $y = -a$, or replace a with $-a$ to get an equation for a curve whose directrix is $x = +a$ or $y = +a$.

It is convenient when dealing with conic sections to place the origin at the center or vertex of the conic. When this is done, the equations simplify to $\frac{x^2}{A^2} + \frac{y^2}{B^2} = 1$ for the ellipse, $y^2 = 2ax$ for the parabola, and $\frac{x^2}{A^2} - \frac{y^2}{B^2} = 1$ for the hyperbola. The focus of the conic lies at $(\frac{ae^2}{e^2-1}, 0)$ for the ellipse and hyperbola and $(\frac{a}{2}, 0)$ for the parabola.

Since in the equations for the ellipse and hyperbola only even powers of x and y appear, the graphs are symmetric about both axes: if (x, y) is on the graph, then so are $(-x, y)$, $(x, -y)$, and $(-x, -y)$. This implies the existence of a second focus and a second directrix. The ellipse has the property that the sum of the distances from any point on it to the two foci is a constant. The hyperbola has an analogous property with "sum" replaced with "difference." Each conic section has a reflection property as well. For the ellipse, any ray emerging from one focus is reflected off the curve toward the other focus; for this reason, whispering galleries are built in elliptical shapes. The hyperbola's reflection property causes rays coming from one focus to be reflected so as to appear as if they came from the other focus. And the parabola reflects rays from its focus into a bundle of parallel lines, which are all lined up with its axis (the line between its focus and vertex). This phenomenon accounts for the parabolic shape of headlight reflectors, and, acting in reverse, the similar shape of the mirrors in reflecting telescopes and of the network microphones occasionally seen at the sidelines of football games. The actual light or sound sensing device is located at the parabola's focus, to where the incoming light or sound is reflected.

The distance formula can be used to find equations corresponding to other geometric conditions. For example, the set of all points that are twice as far from $(3, 0)$ as from $(0, 0)$ has the equation $\sqrt{(x - 3)^2 + y^2} = 2\sqrt{x^2 + y^2}$, or $x^2 + y^2 - 6x + 9 = 4x^2 + 4y^2$ or $x^2 + y^2 + 2x - 3 = 0$. This can be seen to describe a circle with center $(-1, 0)$ and radius 2. The set of points for which the product of their distances from $(2, 0)$ and $(-2, 0)$ is 4 has as equation $[(x + 2)^2 + y^2][(x - 2)^2 + y^2] = 16$. The left side can be written as a polynomial in

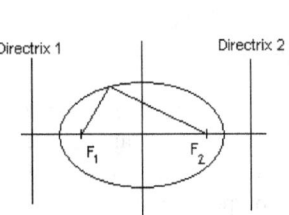

Fig. 5-46(a). An ellipse with its foci and directrices at the center of the Cartesian plane.

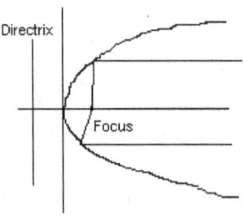

Fig. 5-46(b). An approximate parabola with its focus and directrix.

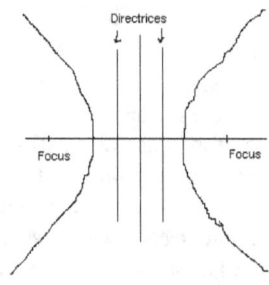

Fig. 5-46(c). An approximate hyperbola and its foci and directrices.

the fourth degree, making it more complicated than the quadratic polynomials that give rise to conic sections. Therefore, the curve is more complex than a conic section; it is called the *lemniscate of Bernoulli* and looks like a two-bladed propeller. (See Fig. 5-47.)

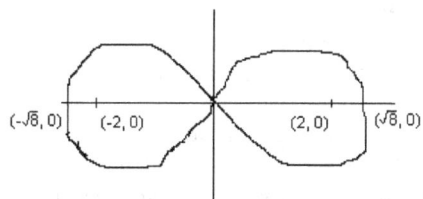

Fig. 5-47. The lemniscate of Bernoulli.

5.7.4 The Proof of the Euler Line Theorem

We are now in position to prove Theorem 5-23. To do so, we place Fig. 5-48, which includes all the relevant points, into the Cartesian coordinate system. A is chosen to be the origin $(0, 0)$ and the side AB is taken to lie on the x-axis. Using the notation introduced in Sec. 5.6, the coordinates of B are found to be $(c, 0)$ and those of C to be $(b \cos A, b \sin A)$ because of the observation of a couple of pages ago about the coordinates of points on a line through the origin. The coordinates of the centroid G are the average of those of the three vertices ("proof" of Th. 5-21(D)), and are therefore $(\frac{c}{3} + \frac{b \cos A}{3}, \frac{b \sin A}{3})$. The coordinates of the side midpoints M, P, and Q are $(\frac{c}{2} + \frac{b \cos A}{2}, \frac{b \sin A}{2})$, $(\frac{b \cos A}{2}, \frac{b \sin A}{2})$, and $(\frac{c}{2}, 0)$ respectively. To find the coordinates of the orthocenter H, we first note that it is on the vertical line CL, and its x-coordinate is therefore that of C, namely, $b \cos A$. Also, H is on AJ, which line has equation $y = mx$ with $m = \tan \angle BAJ$ since A is the origin. The angle BAJ can be found as follows: since AJ is an altitude, $\angle AJB$ is right; therefore, $\angle BAJ = 90° - \angle JBA$, which is the angle B in the triangle. Therefore, $\tan \angle BAJ = \cot B$, and the coordinates of H are $(b \cos A, b \cos A \cot B)$. The coordinates of Y, the midpoint of BH, are therefore $(\frac{c}{2} + \frac{b \cos A}{2}, \frac{b \cos A \cot B}{2})$. Now, PY is a diameter of the nine-point circle, and hence its midpoint is the nine-point center N. The coordinates of N are therefore $(\frac{c}{4} + \frac{b \cos A}{2}, \frac{b \sin A}{4} + \frac{b \cos A \cot B}{4})$. As for the circumcenter O, its x-coordinate is that of Q since it is on the vertical line through Q. We find its y-coordinate as follows: O is on the perpendicular to BC through its midpoint M, by O's definition. This line has slope $\cot B$ since it is parallel to the altitude AJ. Hence, the rise from O to M is $\cot B$ times the run from O to M, which is just the difference in x-coordinates: $(\frac{c}{2} + \frac{b \cos A}{2}) - \frac{c}{2} = \frac{b \cos A}{2}$. Therefore, the rise from O to M is $\frac{b \cos A \cot B}{2}$, and the y-coordinate of O is that of M minus this

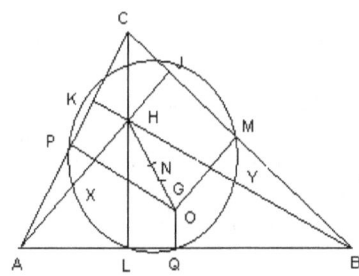

Fig. 5-48. The proof of the Euler line theorem. H is the orthocenter, N the nine-point center, G the centroid, and O the circumcenter.

rise. So O's coordinates are $(\frac{c}{2}, \frac{b \sin A}{2} - \frac{b \cos A \cot B}{2})$.

Introducing the symbols u, v, and w to represent the quantities $b \cos A$, $b \sin A$, and $b \cos A \cot B$ respectively, we simplify the coordinates of the centers to $O(\frac{c}{2}, \frac{v}{2} - \frac{w}{2})$, $G(\frac{c}{3} + \frac{u}{3}, \frac{v}{3})$, $N(\frac{c}{4} + \frac{u}{2}, \frac{v}{4} + \frac{w}{4})$, and $H(u, w)$. We now calculate the differences between these coordinates, which correspond to $x_2 - x_1$ and $y_2 - y_1$ in Sec. 5.7.2. If the four points are indeed on the same line, all the values of $x_2 - x_1$ and $y_2 - y_1$ will be proportional to each other. We find $G - O = (\frac{u}{3} - \frac{c}{6}, \frac{w}{2} - \frac{v}{6})$, $N - O = (\frac{u}{2} - \frac{c}{4}, \frac{3w}{4} - \frac{v}{4})$, and $H - O = (u - \frac{c}{2}, \frac{3w}{2} - \frac{v}{2})$. Since, for each coordinate, $H - O = 2(N - O) = 3(G - O)$, both parts of Th. 5-23 are confirmed at once.

5.8 Higher-Dimensional Geometries

So far in this chapter we have assumed that all points lie in one flat surface or plane. This type of geometry is called two-dimensional because two and only two lines can be drawn so that each is perpendicular to all the rest. However, we live in a world of three dimensions, as you can see by looking at any corner of the room you are in; except for unusually designed buildings, there are three lines, each clearly perpendicular to the other two. And there is no need to stop at three; mathematically, geometries exist in any number of dimensions. (Some of them have gained currency in recent physics theories!) I will end this chapter with a brief discussion of some of the properties of higher-dimensional geometries.

We begin by noting that the part of Th. 5-9 dealing with perpendiculars is no longer true; the mere fact that three or more lines can be mutually perpendicular bears this out. (The second half of Th. 5-9 does remain true.) In any n-dimensional geometry there are flat "hyperplanes" of dimensions from $n - 1$ down to 1, the latter being the ordinary straight line. Given such a geometry and n points in it, there is at least one $(n - 1)$-dimensional hyperplane containing all n points, and if there are more than one then there is at least one $(n - 2)$-dimensional hyperplane containing them as well.

Two lines can now fail to intersect without being parallel; the definition of parallel must be expanded to include the condition that parallel lines must lie in the same plane (2-dimensional hyperplane). In a geometry of $2n + 1$ dimensions, two n-dimensional hyperplanes can fail to intersect without being parallel. (The direction of a hyperplane is specified by a line called its *normal*; hyperplanes are parallel if and only if their normals are.) When hyperplanes do intersect, there is a minimum dimension of the figure that they intersect in; that dimension can be found by using the rule of the *codimension*. An m-dimensional hyperplane in an n-dimensional space has a codimension of $n - m$; the codimension of the intersection is the sum of the codimensions of the intersecting hyperplanes. Hence, two 4-dimensional hyperplanes in a 7-dimensional geometry must share at least a line $(7 - ([7 - 4] + [7 - 4]) = 1)$ unless they are parallel; they can share a plane or 3-dimensional hyperplane, but they cannot share exactly a point.

By analogy to polygons, higher-dimensional geometry has *polyhedra* (singular *polyhedron*); but where plane geometry contains an infinite variety of regular polygons, the regular (or *Platonic*) polyhedra are strictly limited: 5 in three-space and varying numbers in higher dimensions, but always at least two, which are analogues of the square and equilateral triangle.

The analogue of the circle is called the *hypersphere*; in three dimensions the "hyper" is left out. The intersection of a hypersphere with a (two-dimensional) plane is a circle; if the plane passes through the center of the sphere, the circle is called a *great* circle, otherwise, it is called a *small* circle. Great circles act like lines in non-Euclidean "elliptic" geometry; any two of them (on the same sphere) intersect, and the sum of the angles in a spherical "triangle" is greater than 180° - the bigger the triangle, the bigger the sum of its angles. In fact, a spherical triangle can have three right angles; such a triangle will cover one-eighth of the sphere's surface.

Higher-dimensional geometries can be non-Euclidean in their own right as well. While there are still just the two basic types, elliptic and hyperbolic, with larger numbers of dimensions a geometry can be elliptic along some of them, hyperbolic along others, and Euclidean along still others! In some of the most popular geometries that accompany physics theories, there are ten dimensions: three Euclidean ones that act like ordinary space, and seven elliptic ones which are so wound up that the largest structure in them is only about 10^{-35} meters across.

Coordinate geometry, too, can be used in more than two dimensions. In an n-dimensional geometry, n real numbers are required to specify the position of a point. An equation that is linear in n variables represents an $(n - 1)$-dimensional hyperplane. Quadratic equations represent surfaces (or hypersurfaces) called *quadric*; in general, the intersection of a plane and a quadric (hyper)surface is a conic section. An equation like $w^2 + x^2 + y^2 + z^2 = 4$ represents a (hyper)sphere, as the distance formula can be extended to any arbitrary number of dimensions.

Chapter 6   The Foundations of Arithmetic

In this chapter, we return to arithmetic, but at a deeper level. Just what is a number? How do we know that the laws in Sec. 3.1 are true? Why is 2 + 2 equal to 4 and not 3 or 5? How many kinds of numbers are there, and how many of each kind exist? This chapter will deal with these questions and similar ones, and introduce the concepts of a formal language and an algebraic system, as well as several types of such systems. It will turn out that one of these systems, the field, will describe the arithmetic of the real numbers.

6.1 The Domino Principles and the Peano Axioms

We begin by considering not numbers, but instead dominoes. Imagine a line of dominoes organized according to the following principles:

1. One of the dominoes is named Fred.
2. Each domino has exactly one domino to its immediate right.
3. Fred is not to the right (immediate or otherwise) of any domino.
4. Two different dominoes cannot have the same domino at their immediate right.
5. If the dominoes are arranged so that each one, in falling, will knock down the one to its immediate right, then if Fred is knocked over, all the dominoes will fall.

The last statement is called "the well-ordering principle for dominoes." The phrase "right of" shall imply "immediate" from here on, unless otherwise specified.

Now we will determine some of the properties of this row of dominoes. Principle 1 guarantees the existence of at least one domino – Fred. Then, since Fred is a domino, Principle 2 guarantees the existence of a second domino to Fred's (immediate) right. This domino cannot be Fred himself because nothing, dominoes included, can be to the right of itself, and anyway Principle 3 says that Fred is not to the right of any domino. Applying Principle 2 to this second domino (call it Greg), shows that a third domino must sit to Greg's right. This cannot be Fred because of Principle 3 and it certainly cannot be Greg himself. Let us call the third domino Henry and observe that, once again, Henry must have another domino to his right because of Principle 2. This domino cannot be Fred by principle 3, and cannot be Greg by Principle 4 as Greg is already to the right of Fred and therefore cannot serve as the domino to the right of Henry as well. The newest domino cannot be Henry either, as he is already placed to the right of Greg, and therefore must be different from all the previous dominoes whose existence has been proven.

We can proceed indefinitely in this fashion. Each time we deduce the existence of a new domino, Principle 2 demands the presence of a domino to occupy the space to its right. This domino cannot be Fred by Principle 3, nor can it be Greg, Henry, etc. by Principle 4 since they are already to the right of other dominoes. Hence, it must be an entirely new domino, to which Principle 2 applies, closing the cycle. The process of finding new dominoes can never end; there is no "last" domino in the same way that there is no "last" prime number (Sec. 4.4.4).

This proof of the endlessness of the line of dominoes uses only Principles 1-4. The role of Principle 5 is to clarify which dominoes are in the line and which are not. If Fred is knocked over under the assumption in Principle 5, he will topple Greg, who will in turn hit Henry, and so on, but there will always be just one domino falling at any one time. The only dominoes that will fall are Fred, Greg, Henry, the domino at Henry's right, the domino to the right of the domino at Henry's right, …, i. e., all dominoes that are to the right (immediate or otherwise) of Fred, as well as Fred himself. Since Principle 5 states that all the dominoes in the line fall, the only dominoes in the line are those characterized above.

The principles give Fred a role analogous to the origin of a ray: there is an endless row of dominoes to his right, but none to his left. The dominoes, however, are unlike the points in a ray in that they are discrete: we can talk about "immediate" right instead of just "right," which is all we can say for points.

In these three properties, endlessness, directionality, and discreteness, the line of dominoes resembles the group of objects at the foundation of arithmetic: the natural numbers. Indeed, the Domino Principles are merely adaptations of rules invented by the Italian mathematician Giuseppe Peano to describe the natural numbers and called the Peano axioms in his honor:

1. Zero is a number.

2. Every number has a unique successor, which is also a number.

3. Zero is not the successor of any number.

4. Different numbers have different successors.

5. If zero has a property, and the successor of every number with the property also has the property, then all numbers have the property.

The last axiom is the well-ordering principle. It is the basis of the method of mathematical induction discussed briefly in Sec. 4.3.

Since the Domino Principles and the Peano axioms are essentially the same, it should come as no surprise that the system of natural numbers established by the axioms has the same properties as the row of dominoes organized under the principles; i. e., the natural numbers are endless, directional, and discrete.

6.2 Formal Languages

For the further investigation of the properties of the natural numbers, we use a sophisticated form of an axiom system called a *formal language*. The use of the word "language" here should not lead to a comparison with English, French, or Latin; a computer language is a better comparison, but still not exact. A formal language consists of four lists: a list of the symbols used by the language; a list of rules for combining the symbols into statements of the language; an optional list of axioms, or starting material; and a list of *rules of inference* which are used to derive new statements from old ones. The object of studying a formal language is, as with any axiom system, to derive as many statements as possible. Note that I said that the list of axioms was optional; some formal languages have no axioms. These languages, and most others, have rules of inference that allow certain statements to be derived "out of thin air," so to speak, but the types of statements that can be so conjured up are strictly limited. It should also be noted that, while working with a formal language, we pretend not to assign any meaning to the symbols, or statements, that are merely manipulated according to the rules. I used "pretend" because in reality formal languages are invented with meanings in mind: *after* an interesting statement has been derived, this meaning can be attached to it.

The language to be used here is a sublanguage of "Typographical Number Theory," or "TNT," a language invented by Douglas Hofstadter in his book "Godel, Escher, Bach: An Eternal Golden Braid." Hofstadter would describe his book as "metamathematical"; he is fond of the prefix *meta-*. TNT is powerful enough to make any statement in number theory (once the proper meanings are attached to the symbols), and it can probably prove any provable statement in number theory. But it is exceedingly complex, especially in the large number of rules of inference. I have selected a sublanguage that is less complex, but still robust enough to prove the basic results of arithmetic. This language will be called "TNT Jr." to show that it is based on TNT.

The structure of TNT Jr. is given below.

> Symbols: 0 (the numeral zero)
> > The lower-case letters a through e
> > The capital letter S
> > The operators $\oplus$ and $\otimes$
> > Parentheses (and)
> > A right-pointing arrow $\rightarrow$
> > The equal sign $=$
>
> Rules for building up statements:
> > Numerals: 0 is a numeral, and any numeral preceded by an S is also a numeral.
> > Variables: Any lower-case letter is a variable.
> > Expressions: A variable or a numeral is an expression, and if x and y are expressions, then so are Sx, Sy, $(x \oplus y)$, and $(x \otimes y)$.
> > Statements: if x and y are expressions, then $x = y$ is a statement, and if p and q are statements, then $(p \rightarrow q)$ is a statement. In the statement $x = y$, any pair of parentheses that completely enclose either x or y may be dropped, and in the statement $(p \rightarrow q)$, the outermost pair of parentheses may be dropped.
> > Axioms: A1. $a \oplus 0 = a$.
> > > A2. $a \oplus Sb = S(a \oplus b)$.
> > > A3. $a \otimes 0 = 0$.
> > > A4. $a \otimes Sb = (a \otimes b) \oplus a$.
>
> > Rules of inference: Let $V(a)$ denote a statement containing the variable a (which in actuality can be

any of the lower-case letters), and $V(x)$ denote the same statement after all occurrences of a have been replaced with the expression x (with outermost parentheses, if any).

Specification: If $V(a)$ is a theorem, then so is $V(x)$ for any expression x.

Replacement: If $x = y$ is a theorem, then any theorem containing the expression x can have the x replaced with y, or if it contains y can have it replaced with x, and the resulting new statement will be a theorem.

Hypothesis: If, from a hypothetical statement p, we can derive a statement q, then $(p \to q)$ is a theorem. During the derivation of q from p, all statements used, including p and q, must be indented to show that they are not theorems. In deriving q from p, we may apply the rules of inference to any statements that are less indented than p but none that are more indented. Additional hypotheses used in deriving q must be further indented. No hypothetical statement can violate the Peano axioms. ($Sa = 0$ is an example of a statement that violates the Peano axioms, specifically axiom 3 in this example.)

Induction: If $V(0)$ is a theorem, and if $V(a) \to V(Sa)$ is a theorem, then $V(a)$ is a theorem.

In the rules of inference, the term "theorem" means either an axiom or a statement derived by applying one of the rules of inference to one or more previously proved theorems. Now, we will start deriving a few trivial results, then one with more importance.

6.2.1 Proving the "Obvious"

We begin by investigating the expression $(S0 \oplus S0)$. The statements of a proof are numbered for future reference, to justify later steps.

$L1.\ S0 \oplus S0 = S(S0 \oplus 0)$.

This is axiom A2, with a specified as $S0$ and b as 0.

$L2.\ (S0 \oplus 0) = S0$.

This is axiom A1, with a specified as $S0$. Note again that parentheses are optional around an expression that occupies all of the space on one side of the equal sign. Here we chose to include them.

$L3.\ S0 \oplus S0 = SS0$.

This is replacement, $L2$ in $L1$. The characters $(S0 \oplus 0)$ in $L1$ are replaced with $S0$, the other side of $L2$. If we attach the proper meanings to the symbols, we find that we have proven that $1 + 1 = 2$.

Now, we investigate the expression $S0 \otimes S0$.

$L1.\ S0 \otimes S0 = (S0 \otimes 0) \oplus S0$.	(A4, a specified as $S0$ and b as 0)
$L2.\ (S0 \otimes 0) = 0$.	(A3, a specified as $S0$)
$L3.\ S0 \otimes S0 = 0 \oplus S0$.	(replacement, $L2$ in $L1$)
$L4.\ 0 \oplus S0 = S(0 \oplus 0)$.	(A2, a and b both specified as 0)
$L5.\ (0 \oplus 0) = 0$.	(A1, a specified as 0)
$L6.\ 0 \oplus S0 = S0$.	(replacement, $L5$ in $L4$)
$L7.\ S0 \otimes S0 = S0$.	(replacement, $L6$ in $L3$)

We have proven that $1 \times 1 = 1$. So far, the results have been obvious, even trivial in nature, because they contain no variables. To prove theorems involving variables, we must bring in the more powerful rules of inference, hypothesis and induction.

Now, we will embark upon a long proof of something that is less obvious.

$L1.\ 0 \oplus 0 = 0$.	(A1, a specified as 0)
$L2.\quad (0 \oplus a) = a$.	(hypothesis)
$L3.\quad 0 \oplus Sa = S(0 \oplus a)$.	(A2, a specified as 0 and b as a)
$L4.\quad 0 \oplus Sa = Sa$.	(replacement, $L2$ in $L3$)
$L5.\ 0 \oplus a = a \to 0 \oplus Sa = Sa$.	(hypothesis rule, $L2$ and $L4$)
$L6.\ 0 \oplus a = a$.	(induction, $L1$ and $L5$)
$L7.\ a \oplus 0 = a$.	(A1)
$L8.\ a \oplus 0 = 0 \oplus a$.	(replacement, $L6$ in $L7$)

Note the indentation of $L2$ through $L4$, since they are used in a hypothesis, and the induction in $L6$. Here $V(a)$ is $0 \oplus a = a$, so $V(0)$ is $0 \oplus 0 = 0$ and $V(Sa)$ is $0 \oplus Sa = Sa$; therefore, $L1$ is $V(0)$ and $L5$ is $V(a) \rightarrow V(Sa)$. The induction rule can be used only in conjunction with the hypothesis rule, with the apparently paradoxical result that our hypothesis ($L2$) is what we wish to prove ($L6$). This explains the importance of indentation; we can drag lines in with impunity, but dragging them out is strictly prohibited. It should also be noted that, when we attach the intended meanings to the symbols, we have proven that zero commutes with all numbers under addition. This is only the first part of what we wish to prove; namely, that addition is commutative for any two numbers. So the proof continues.

$L9.\ Sa \oplus 0 = Sa.$	(A1, a specified as Sa)
$L10.\ a \oplus S0 = S(a \oplus 0).$	(A2, b specified as 0)
$L11.\ a \oplus S0 = Sa.$	(replacement, $L7$ in $L10$)
$L12.\ a \oplus S0 = Sa \oplus 0.$	(replacement, $L9$ in $L11$)
$L13.\ a \oplus SSb = S(a \oplus Sb).$	(A2, b specified as Sb)
$L14.\ Sa \oplus Sb = S(Sa \oplus b).$	(A2, a specified as Sa)
$L15.\quad a \oplus Sb = Sa \oplus b.$	(hypothesis)
$L16.\quad Sa \oplus Sb = S(a \oplus Sb).$	(replacement, $L15$ in $L14$)
$L17.\quad a \oplus SSb = Sa \oplus Sb.$	(replacement, $L16$ in $L13$)
$L18.\ a \oplus Sb = Sa \oplus b \rightarrow a \oplus SSb = Sa \oplus Sb.$	(hypothesis rule, $L15$ and $L17$)
$L19.\ a \oplus Sb = Sa \oplus b.$	(induction, $L12$ and $L18$)
$L20.\quad a \oplus b = b \oplus a.$	(hypothesis)
$L21.\quad Sb \oplus a = b \oplus Sa.$	($L19$, a specified as b, simultaneously b as a, and left and

right sides reversed)

We can reverse the left and right sides of a statement of equality because if $x = y$, then we can replace y with x in that statement itself to get $x = x$, and then replace the first x with y to get $y = x$.

$L22.\quad b \oplus Sa = S(b \oplus a).$	(A2, a specified as b and simultaneously b as a)
$L23.\quad Sb \oplus a = S(b \oplus a).$	(replacement, $L22$ in $L21$)
$L24.\quad a \oplus Sb = S(a \oplus b).$	(A2)
$L25.\quad a \oplus Sb = S(b \oplus a).$	(replacement, $L20$ in $L24$)
$L26.\quad a \oplus Sb = Sb \oplus a.$	(replacement, $L23$ in $L25$)
$L27.\ a \oplus b = b \oplus a \rightarrow a \oplus Sb = Sb \oplus a.$	(hypothesis rule, $L20$ and $L26$)
$L28,\ a \oplus b = b \oplus a.$	(induction, $L8$ and $L27$)

$L28$ is our goal; its "meaning" is that addition is commutative for all natural numbers. This fact is by no means obvious when the structure of TNT Jr. or TNT itself is considered without any meanings attached. In the natural numbers, the way addition is defined makes the commutative property obvious; $2 + 3 = 3 + 2$ because two apples and three oranges are as much fruit as three oranges and two apples. However, the axioms A1 and A2 in TNT Jr., which are also a part of TNT itself, and can be interpreted as an alternative definition of addition, seem asymmetric in that they define $a \oplus 0$ but not $0 \oplus a$, and similarly $a \oplus Sb$ but not $Sa \oplus b$. The proof above shows that $0 \oplus a$ and $Sa \oplus b$ are implicitly defined by the rules of inference and the axioms working in combination. It should be mentioned that if A2 were changed to $Sa \oplus b = S(a \oplus b)$, not only could we not prove that addition was commutative, but we would not even be able to calculate, say, $0 \oplus S0$.

We can prove, in similar and lengthier fashion, that addition is associative and that multiplication is commutative and associative. The vital distributive law can also be verified, although it would probably take several pages. This makes it possible to verify all the laws in Sec. 3.1 that do not involve subtraction or division (which are not defined in either TNT Jr. or TNT), with the exception of the multiplicative identity: $b \times 1 = b$. This is taken care of by defining the number $S0$ to be 1, so that $b \otimes S0 = (b \otimes 0) \oplus b = 0 \oplus b = b$. (A4, A3, and $L6$ of the above proof.)

Exponentiation can be defined in TNT Jr. by introducing its symbol, $\uparrow$, and two new axioms that serve to define it: $a \uparrow 0 = S0$ and $a \uparrow Sb = (a \uparrow b) \otimes a$. The first two laws of exponents in Sec. 3.1.3 can be derived, but not the third, since it involves subtraction and division.

6.3 What Is a Number?

The earliest concept of numbers was for counting; "four" is just the common attribute of the legs of a dog, the seasons of the year, and the Horsemen of the Apocalypse, while "five" is what is shared by the fingers of a hand, the arms of a starfish, and the acts of a Shakespearean play. The natural numbers are sometimes called *counting numbers* in this context.

With this idea of "number," the concept of negative numbers has no meaning. What could be less than nothing? Nevertheless, a few mathematicians advanced the idea. It took the Renaissance, with its more sophisticated credit system, to give negative numbers a concrete footing: a debt could be so represented. If you owed somebody 1,500 florins, your monetary position could be expressed as –1,500; furthermore, if you then received 1,500 florins from someone else (represented by a +1,500), you could pay off the debt, restoring your monetary position to zero. This motivated the final definition of a negative number: -a was the number that, when added to a, produced zero. In modern terminology, -a is the additive inverse of a.

From this definition of negative numbers, the sign laws and the rules for adding signed numbers can be worked out. Subtraction can now be defined, for all numbers, as in Sec. 3.1.4. The operation, of course, existed earlier, but was defined differently, as in Sec. 3.1: $a - b = c$ meant $c + b = a$. The fact that no "number" could be added to 4 to get 2 merely meant that there was no answer to $2 - 4$, and this gap is what led to the original creation of negative numbers.

The system of negative counting numbers, positive ones, and zero is called the *integers*. "Integer" means roughly "uncut."

Curiously, the concept of the fraction is much older than that of the negative integer. Egyptians of 2500 B.C. had a limited understanding of fractions; they could work with reciprocals of counting numbers like ½ and 1/10, and could split other simple fractions into sums of reciprocals. The Babylonians of a thousand years later knew much more. They could extend their marvelous (for 1500 B.C.) base-60 notation system into the equivalent of decimals, and knew, for example, that $1/8 = 7/60 + 30/3,600$ and $1/9 = 6/60 + 40/3,600$. The rules governing arithmetic with fractions must have been known from 1500 B.C. onward.

The irrational numbers were discovered by the Greeks. The Pythagorean theorem (Sec. 5.4) enabled geometers to construct line segments with lengths $\sqrt{2}$, $\sqrt{3}$, $\sqrt{5}$, and so on, which to the Greeks meant that $\sqrt{2}$, etc., were valid numbers. But then one of the members of Pythagoras' school made the mistake of discovering that $\sqrt{2}$ and the other square roots were irrational; that is, there did not exist a fraction equal to any of them. For the Pythagoreans, who believed that all things could be described with integers and ratios of integers (fractions), this had roughly the same effect that a proof that God did not exist would have had on the church in A.D. 500. One report says that the man was thrown off a cliff for his proof of the existence of irrational numbers. Nevertheless, irrational square and other roots gradually gained acceptance, and by 1400 were known to all mathematicians.

However, roots of algebraic equations which contain only integer coefficients are not the only kind of irrational number. Early in the 19th century, it was proven that π was not such a root. Certainly, a ratio as important as pi should be considered to be a valid number, so the concept of *transcendental* numbers was introduced. Late in the 19th century, mathematician George Cantor showed that there are infinitely more transcendental numbers than all other types combined. The so-called "real" number system includes all the transcendental numbers as well as the algebraic irrationals, the fractions, and the integers. It contains essentially every concept that can be used to measure a one-dimensional quantity.

But the fact that the square of every real number is positive, and therefore there is no real square root for $- 1$ and other negative numbers, left many mathematicians stumped. To give negative numbers square roots, the concept of "imaginary" numbers was developed, and it was followed shortly by the *complex* numbers. These new numbers are explained in Sec. 3.10.

There are other mathematical objects to which operations resembling arithmetic can be applied. For example, given two line segments, we can draw a third segment that is in a way the geometric mean between the first two; the construction is shown in Fig. 6-1, c being the third segment. There are mathematical objects called *matrices* on which "addition" and "multiplication" are defined, but do not have all the properties of normal addition and multiplication.

The history of mathematics has been characterized by a steady broadening

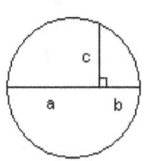

Fig. 6-1. The construction of the "geometric mean" between two segments.

of the concept of number and at the same time a continual increase in the level of abstraction. If the idea of holding -2 apples in your hand seems far-fetched, then holding $-3 + 3i\sqrt{3}$ apples must be truly outlandish. Each expansion of the number system comes with an increase of the symmetry of the system; the extra abstraction is a price that must be paid for the added symmetry. Without negative numbers, for example, there is an answer for $4 - 2$ but not $2 - 4$; without fractions, we can divide nine by three but not three by nine; and without complex numbers there is a square root of 16 but no square root for -16. The creation of the complex number system is the end of the line for generalizations of the counting numbers; no larger system with more symmetry than the complex numbers and containing the counting numbers can exist. Paradoxically, systems with *less* symmetry than the complex numbers and containing them have been invented; the quaternions, briefly mentioned in Sec. 3.10, are one such system.

6.4 Abstract Algebraic Systems

An algebraic system is a generalization of arithmetic to arbitrary objects whose properties are defined by their behavior under one or more binary operations which correspond to those of arithmetic. Formally, there are three lists: a list of one or more elements on which the operations are performed; a list of one or two operations; and a list of one or more laws which are obeyed by all the elements.

The most important type of abstract algebraic system is a *group*. In a group, the list of elements may be finite or infinite, and the group is labeled accordingly; there is one operation, symbolized by $*$, and the laws are:

L_1. Associativity: $a * (b * c) = (a * b) * c$.

L_2. Identity: one of the elements, denoted by e, has the property that $a * e = e * a = a$ for every element a in the group.

L_3. Inverses: For each element a, there is an element a', called the *inverse* of a, with the property that $a * a' = a' * a = e$.

Note that we do not require that the operation be commutative for all elements, just for the identity and inverse. If the operation is commutative for all elements, the group is called commutative, or *Abelian*, after Niels Abel, a pioneer in the study of groups.

Even though a group has only a rudimentary structure, the fact that the three laws must hold for all elements allows us to prove some theorems that must be true for all groups. For example, there can be only one identity element to a group. If there were two (call them e and f), then $e * f$ would be f since e is an identity, and at the same time $e * f$ would equal e since f is an identity. Since $e * f$ can have only one answer, e and f must be the same element. Similarly, any element can have only one inverse. If a had two inverses b and c, then $(b * a) * c = e * c = c$ and simultaneously $b * (a * c) = b * e = b$. Since the binary operation in a group is associative, $b * (a * c) = (b * a) * c$ and therefore $b = c$. We note that the identity e is its own inverse, and that if a is the inverse of b, then b is the inverse of a.

A group also features the important *Law of Cancellation*: if $a * b = a * c$ then $b = c$. This law can be proved by "multiplying" both sides of the equation $a * b = a * c$ on the left by a's inverse a': $a' * (a * b) = a' * (a * c)$. We then use the associative law, the inverse law, and the identity law to reduce the equation to $b = c$. We have to specify that a's inverse is multiplied by on the left because not all groups are Abelian.

We will denote the list of elements in a group G by $|G|$ and the number of elements in the list by $N(G)$ if that list is finite. The element e is its own inverse, and if e is the only element with that property (and $N(G)$ is finite), then $N(G)$ can be shown to be odd: we simply pair off each element other than e with its own inverse, thus producing a whole number of pairs, containing an even number of elements, with e itself still to be counted. If $|G|$ contains a sub-list $|H|$ that contains e, contains the inverses of all its elements, and contains $a * b$ and $b * a$ for every pair of elements a and b that it contains, then $|H|$ is part of a *subgroup* H of G.

We will now prove the famous Lagrange theorem about subgroups: If $N(G)$ is finite, then $N(H)$ is a factor of $N(G)$. This proof requires us to consider the *co-sets* of H in G. The co-set corresponding to an element a of G is the set of all elements $a * h_1$, $a * h_2$, $a * h_3$, etc., where h_1, h_2, h_3, etc., are the elements of H. The co-set corresponding to a is written $a * H$. We note the following: Since e is in H, $a * H$ contains a for every element a of G; thus all the elements of G are in some co-set. Also, if b is in $a * H$, then $b * H = a * H$. Since b is in $a * H$, there is an element h_1 in H such that $a * h_1 = b$. Then, any element in $b * H$ can be represented by $b * h_2$ for some h_2 in H, and $b * h_2 = (a * h_1) * h_2 = a * (h_1 * h_2) = a * h_3$ which is in $a * H$. Similarly, $b * h_1' = (a * h_1) * h_1' = a * (h_1 * h_1') = a * e = a$, and every element in $a * H$ must also be in $b * H$. The last two sentences mean that any two co-sets are either identical or totally different, with no two members in common. Finally, we note that every co-set contains precisely $N(H)$ elements: if $a * h_3 = a * h_5$ then $h_3 = h_5$ by the Law of Cancellation. Therefore, the co-sets constitute a division of $|G|$ into non-overlapping subsets with $N(H)$ elements each, and this can only happen if $N(H)$ divides $N(G)$. The number of co-sets is $N(G) \div N(H)$.

It is possible to define an operation on the co-sets that makes them into a group; denoting this operation by $\bullet$, we define $(a * H) \bullet (b * H)$ to be $(a * b) * H$. The identity element of this new group is the co-set containing e, which just happens to be $|H|$ itself. The new group defined in this way is called a *quotient* group, and is denoted by G/H. The curious will no doubt want to know about $G/(G/H)$, but I cannot help them; the elements of G/H are co-sets, not single elements, of G, so G/H is not a subgroup of G.

Examples of groups are everywhere in mathematics. Modular arithmetic, looking only at addition with any modulus greater than 1, provides us with groups. If the modulus is n, the group is denoted by Z_n and the inverse of a is $n - a$. The integers form a group with addition as the operation; this group is denoted by Z. (Z stands for *Zahl*, German for number, as in "Zahlvergnugen.") The non-zero rational numbers form a group if the operation is multiplication; we must exclude zero because it has no inverse. The set of all isometries (non-distorting motions) in the geometric plane is a group in which the operation consists simply of combining two motions; here the identity element is the "motion" where nothing moves. The possible rearrangements of n objects in a row form a group called the *symmetric* group on n elements and denoted S_n; S_n has $n!$ elements and the operation is applying one rearrangement to the results of another. These last two examples are non-Abelian, showing that it makes sense to talk about Abelian groups.

Two groups are *isomorphic* if we can match each element of one to a single element of the other in such a way that the operations in the two groups are preserved. If the matching runs $a \rightarrow x$, $b \rightarrow y$, and $c \rightarrow z$, then if $a * b = c$ in one group, we must have $x * y = z$ in the other, and the matching must also connect $a' \rightarrow x'$, $b' \rightarrow y'$, and $c' \rightarrow z'$. All finite groups are isomorphic to some subgroup of a symmetric group S_n.

Algebraic systems with two binary operations can be more varied than groups. The basic structure is called a *ring*. In a ring, the operations are analogous to addition and multiplication, and hence are symbolized by $+$ and $\times$. The laws of a ring are as follows:

A_1: Considering only addition, the ring is an Abelian group. The identity element is denoted by z, for zero, and the inverse of a is $-a$.

M_1: Multiplication is associative: $(a \times b) \times c = a \times (b \times c)$.

D_1: The distributive laws hold: $a \times (b + c) = (a \times b) + (a \times c)$ and $(a + b) \times c = (a \times c) + (b \times c)$.

In a ring, subtraction is defined by $a - b = a + (-b)$. Note that a ring need not have an identity element for multiplication, let alone a multiplicative inverse for every element.

Among the basic theorems for rings are the sign laws in Sec. 3.1 and the multiplicative property of zero: $a \times z = z \times a = z$ for all a in the ring.

The concept of a ring is mainly a starting point for more structured systems in which multiplication labors under additional constraints:

M_2: Multiplication is commutative.

M_3: There is a multiplicative identity, denoted by u.

M_4: The zero-product property (Sec. 3.6.2) holds.

M_5: Every element except z has a multiplicative inverse, denoted variously by a', a^{-1}, $\frac{1}{a}$, a^*, or other symbols.

M_5 implies M_4, but the other rules are independent of each other; in particular, M_4 does not imply M_5. (Of course, M_5 makes no sense without M_3.) M_4 is equivalent to the Law of Cancellation for multiplication.

If all of $M_2 - M_5$ hold, the structure is called a *field*; the rational numbers form a field. Together with M_1, $M_2 - M_5$ say that the elements, except for z, form an Abelian group relative to multiplication. If just $M_3 - M_5$ are used, and the non-z elements form a non-Abelian group with multiplication, we have a *division ring*. The system of quaternions is the basic example of a division ring. Finally, if only $M_2 - M_4$ are valid, we have an *integral domain*. The integers (naturally) are the prime example of an integral domain.

To any of these structures we may add the following, called the *order axiom*: The non-z elements can be split into two sets, P and N, such that (1) if a is in P, then $-a$ is in N; and (2) If a and b are in P, then $a + b$ and $a \times b$ are also. If the order axiom holds, the structure satisfying it is called an ordered ring, field, etc. The integers, rational numbers, and real numbers are ordered, but the complex numbers and quaternions are not.

In a field or division ring, the operation of division is defined as $a \div b = a \times b^{-1}$.

In an integral domain, an element a is called a *unit* if it has a multiplicative inverse. There are always at least two units, u and $-u$ (unless these two are the same, which is possible). Among the integers, 1 and -1 are the only units. I call an integral domain *closed* if u and $-u$ are the only units, and *open* if there are other units. One type of open integral domain is the set of all real numbers of the form $a + b\sqrt{n}$, where a, b, and n are integers and n is not a perfect square. (Each possible value of n produces a different integral domain.) Here, $a + b\sqrt{n}$ is a unit if and only if $a^2 - nb^2 = \pm 1$. As seen in Sec. 4.3, this Pell equation has an infinite number of solutions if n is not a square, so the related integral domain has an infinite number of units.

There are several specialized types of field as well. A field is called finite, or a *Galois* field after Evariste Galois, one of the founders of group theory, if it has only a finite number of elements. Modular arithmetic, to any prime modulus, produces a Galois field. A field is called *self-negative* if $a + a = z$ for any non-z element a in the field. If this holds for any non-z element, it must hold for all elements. A field is called *quadratic* if every element has a square root; that is, if for any element a, there exists a b such that $b \times b = a$. A field is called *algebraic* if the Fundamental Theorem of Algebra holds for it as it does for the complex numbers.

Some of these types are mutually exclusive: no ordered field can be finite, quadratic, self-negative, or algebraic; no finite field can be ordered or algebraic; and a finite field can be quadratic only if it is also self-negative. It is a conjecture of mine that the only algebraic field is the complex field; this may already have been proved or disproved by other mathematicians (of the professional variety). In Fig. 6-2 are the arithmetic tables for a four-element, self-negative, quadratic field; the reader may verify that the tables do indeed define a field, and that the polynomial $us^2 + us + u$ factors as $(us + a)(us + b)$ as well as two other possibilities but that $us^2 + us + a$ cannot be factored.

+	z	u	a	b
z	z	u	a	b
u	u	z	b	a
a	a	b	z	u
b	b	a	u	z

x	z	u	a	b
z	z	z	z	z
u	z	u	a	b
a	z	a	b	u
b	z	b	u	a

Fig. 6-2. The arithmetic tables for a finite field.

Just as a group may contain subgroups, so integral domains, rings, and fields may contain substructures. Since a ring includes a group (the Abelian group relative to addition), the Lagrange theorem applies to finite structures. For fields, we can say something more: If F is a finite field with a subfield F_1, then $N(F) = [N(F_1)]^k$ for some positive integer k. The reason behind this statement involves the idea of a *vector space*, another algebraic system which will be discussed in Chapter 10.

There is one more type of algebraic system which deserves mention in this section. It is called a *Boolean algebra* after George Boole. Like a ring, a Boolean algebra has two binary operations, but they are not closely analogous to addition and multiplication, and hence are denoted by different symbols, called *wedges* colloquially: $\wedge$ and $\vee$. The laws for a Boolean algebra are:

L_1: (Commutative laws) $a \wedge b = b \wedge a$ and $a \vee b = b \vee a$.

L_2: (Distributive laws) $a \vee (b \wedge c) = (a \vee b) \wedge (a \vee c)$ and $a \wedge (b \vee c) = (a \wedge b) \vee (a \wedge c)$.

L_3: (Identity elements) The Boolean algebra contains elements u and z such that $a \vee z = a$ and $a \wedge u = a$ for all elements a. u and z must be different elements.

L_4: (Complements) To each element a in the algebra there corresponds an element a' such that $a \vee a' = u$ and $a \wedge a' = z$.

Note carefully the direction of the wedges in L_4. The complement is not the same thing as a group-theoretical inverse. No algebraic structure can satisfy $L_1 - L_3$ above and the modified L_4: $a \wedge a' = u$ and $a \vee a' = z$. To prove this, we need only consider the trivial structure containing just the elements z and u. By L_3, $z \vee z = z$ and $z \vee u = u$. Hence, z' must be z by the modified L_4, and therefore $z \wedge z = u$. Similarly, we must have $u \wedge u = u$ and $u \wedge z = z$, so $u' = u$ and $u \vee u = z$. But now the distributive law L_2 says $u \wedge (u \vee u) = (u \vee u) \wedge (u \vee u)$ which reduces to $u = z$. Since z and u must be different elements, we have a contradiction, meaning that the structure cannot satisfy all of $L_1 - L_3$ and the modified L_4.

Since an actual Boolean algebra, with the unmodified L_4, does not contain a group, there is no cancellation law such as $a \wedge b = a \wedge c$ implies $b = c$. However, equality for both wedges does work: $a \wedge b = a \wedge c$ and $a \vee b = a \vee c$ together imply $b = c$.

The presence in the laws for a Boolean algebra of an extra distributive law that has no counterpart in ordinary arithmetic (certainly $4 + (6 \times 3)$ does not equal $(4 + 6) \times (4 + 3)$) means that there is no basic distinction between the operations $\wedge$ and $\vee$. In fact, exchanging the two wedges, as well as u and z, changes the first half of each of the laws into the second and vice versa. The result is called the *Principle of Duality*: any theorem in Boolean algebra, when modified by exchanging $\wedge$ and $\vee$, and u and z, becomes another theorem. Some of the basic theorems for Boolean algebras are: $a \wedge a = a \vee a = a$, the idempotency laws; $(a \wedge b)' = a' \vee b'$ and $(a \vee b)' = a' \wedge b'$, DeMorgan's laws; and the associative rules, $a \vee (b \vee c) = (a \vee b) \vee c$ and $a \wedge (b \wedge c) = (a \wedge b) \wedge c$.

The two most important examples of Boolean algebras are the "algebra" of sets and the propositional calculus (not to be confused with the infinitesimal calculus of the next chapter). In the algebra of sets, the elements are the sets, which contain mathematical or physical objects that share a common property. The operation $\vee$ is the *union* of two sets, which is defined as the set of all the objects that are in either or both of the first two sets. Union is denoted by $\cup$;

the symbol ∨ is a modified ∪. The operation corresponding to ∧ is *intersection*; the intersection of two sets is the set containing all objects that are in both of the sets. Hence, if A is the set of all Heisman Trophy winners, and B is the set of all people who have been Most Valuable Players of the major-league baseball All-Star Game, then $A \cap B$, the intersection, is the set whose only member is Bo Jackson, at least as of 2010. The ∩ for intersection is just an upside-down ∪, and the ∧ in Boolean algebra is a modified ∩. The complement of a set is another set containing all the objects that are in some larger set, called the *universal* set and denoted by I, but are not in the first set. The complement of I is the set with no elements, called the *empty* set and symbolized by ϕ. I and ϕ play the parts of u and z in a Boolean algebra.

The elements of the propositional calculus are propositions, statements (not necessarily mathematical) that are either true or false. The operations ∧ and ∨ are the words "and" and "or" respectively, the latter being used in its inclusive sense. The complement is the negation of a proposition, or the word "not." The element u is a statement that must be true, and z is a statement that must be false. (Example of z: "Raising taxes stimulates the economy to grow.") In abstract work, the propositions are denoted by lowercase letters p, q, r, and s; "and" is denoted by K, "or" by A, "not" by N, and the derived operations "implies" and "is equivalent to" by C and E respectively. The capital letters are written before the propositions they apply to; thus, "p and q" is written Kpq and "(r and s) is logically equivalent to (not-p and q)" is written $EKrsKNpq$. C is defined by $Cpq = ANpq$ and E by $Epq = KCpqCqp$. The notation is called operator prefix or Polish in honor of its inventor, Jan Lukasiewicz. Usually, a surname is used to honor a mathematician, but nobody knew how to pronounce "Lukasiewicz," so "Polish," after his country, Poland, was used instead.

6.5 The Real Number System

As mentioned above, the real numbers constitute an ordered field with respect to the operations of addition and multiplication. But actually *defining* what is meant by a real number is a surprisingly tricky process. Mathematicians have devised several ways of characterizing the real number system, and they are not necessarily equivalent. The simplest definition is an operational one: The real number system is completely defined as a field containing the rational numbers which satisfies the order axiom and the *Axiom of Completeness,* to be defined below.

An *upper bound* (abbreviated *ub*) of a set of numbers in an ordered algebraic system such as the real numbers is a number that is not exceeded by any number in the set. Thus, in the set {1/2, 2/3, ¾, 4/5, …} 10 serves as an upper bound because none of the numbers in the set can possibly get as large as 10. A *least upper bound* (*lub*) of a similar set is a number L such that no number in the set exceeds L, but for any positive number ϵ (Greek lowercase epsilon = e), there is a number in the set that exceeds $L - \epsilon$. The Axiom of Completeness says that any nonempty set of numbers that has an ub has a unique lub.

We can see that the Axiom of Completeness does not hold in the *rational* field by considering S, the set of all *rational* numbers whose squares are less than 2. Among the members of S are ½, 1, and the sequence $\frac{7}{5}, \frac{41}{29}, \frac{239}{169}, \dots,$ which can be obtained from the second table in Sec. 4.3 as $\frac{3+4}{5}, \frac{20+21}{29}, \frac{119+120}{169}, \frac{696+697}{985}, \dots,$ and is continued by applying to both numerator and denominator the rule $x_{n+1} = 6x_n - x_{n-1}$ as in that section. Each of the numbers in the sequence is larger than the one before it, but their squares are all less than 2, placing them in S. Now, 2 serves as an upper bound for S. in fact, any number whose square is *greater* than 2 (and is rational and positive) is an upper bound for S because, for positive numbers, the square is an increasing function; that is, $a^2 > b^2$ implies $a > b$ when a and b are both positive. [Let n be such a number (square greater than 2) and b any number in S; then we have $n^2 > 2 > b^2$ because b is in S, and hence $n > b$.] If the Axiom of Completeness held for rational numbers, S would have a unique *rational* lub. But there is a sequence $\frac{3}{2}, \frac{17}{12}, \frac{99}{70}, \frac{577}{408}, \dots,$ derivable from the first table in Sec. 4.3 by doubling each y value there, continuable by the same $x_{n+1} = 6x_n - x_{n-1}$ rule mentioned above, and such that each term is less than the one before it and the squares of all the terms are greater than 2. The amount by which the square of any term is greater than 2 is the reciprocal of the square of its denominator, and hence, for any rational number whatsoever whose square exceeds 2 (and thus might be the lub of S) we can find a number in this sequence whose square is less than that prospective lub. Hence, S has no *rational* lub, and, therefore, the Axiom of Completeness does not hold for rational numbers only. The actual lub of S, as you have no doubt guessed, is $\sqrt{2}$ itself, but the square root of two is not rational (Sec. 3.9).

A second characterization of real numbers makes use of geometry. An arbitrary infinite straight line is drawn, an arbitrary point is picked on that line to be the origin, and another arbitrary point on the line is chosen to serve as the "one." (See Fig. 6-3.) Now, every point on the line is identified with a real number in such a way that if the

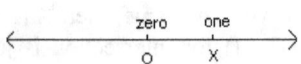

Fig. 6-3. The straight line as a model for the real number system. The segment OX is a unit segment.

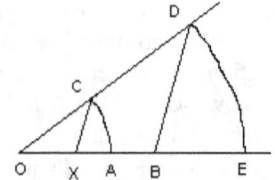

Fig. 6-4. "Multiplication" of two segments. OA times OB equals OE, when OX is a unit segment.

line segment AB is congruent with CD, then $|b - a| = |d - c|$, where the lowercase letters denote the real numbers assigned to the points named by the corresponding capital letters. Order is defined by which side of the origin the "one" point is. Geometric constructions are used to define addition and multiplication. In Fig. 6-4, multiplication is illustrated. O is the zero, X is the "one," $OC = OA$, BD is parallel to CX, and $OD = OE$. The number matched with E is defined as the product of those matched with A and B. It is not difficult to show that the number system defined in this way is a field.

The number-line interpretation also allows us to construct irrational numbers. Fig. 6-5 is drawn so that $XC = OX$, $AO = CO$, OX and DA are perpendicular to OA, and $OB = OD$. Therefore, by the preceding definition of multiplication, $b = a^2$. It can also be shown that $BX = OX$, and therefore $b = 2$; hence, $a = \sqrt{2}$, which is irrational. Square roots of numbers can also be constructed with the diagram in Fig. 6-6. Here, OA is a unit segment, OB is the segment whose square root is to be extracted, and the circle has AB for a diameter. OC, which is perpendicular to AB, has a length equal to the square root of OB.

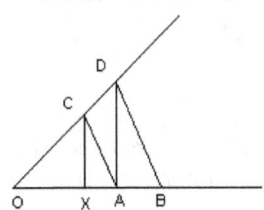

Fig. 6-5. One way to construct the square root of two; OA is the segment if OX is a unit segment. The angle at O is 45 degrees.

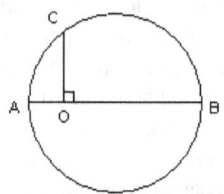

Fig. 6-6. The construction of a general square root; OC squared equals OB if OA is a unit segment.

Under this concept of real numbers, there are as many real numbers as there are points on a line. But we do not need to use an entire infinite straight line. Paradoxically, any line *segment*, however short, contains just as many points as a whole straight line, and, in a way, the segment has two other points besides! Fig. 6-7 sets up a correspondence between a segment and an entire line. The segment is AB. PQ is drawn parallel to and directly under AB so that $ABQP$ is a rectangle, and a semicircle with PQ as diameter and end-points is drawn. The infinite line XY is also parallel to AB and PQ. An arbitrary point C on the segment AB is picked, a perpendicular to AB through C is drawn and intersects the semicircle in one point, also labeled C, and then the ray from O (the center of the semicircle) through this point is drawn and extended to intersect XY in a third point labeled C. All these points C correspond to one another. To go in the other direction, we take D on XY, OD intersecting the semicircle in another D, and the line through this D perpendicular to PQ intersecting AB in a third D. Clearly, each point on the line matches exactly one point on the segment, and each point on the segment matches precisely one point on the line – with the exception of the segment's endpoints, A and B. A, for example, matches P on the

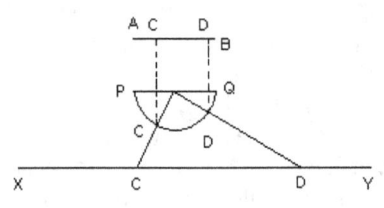

Fig. 6-7. The correspondence between an arbitrarily short line segment and an infinite line.

semicircle. But OP is parallel to XY and therefore does not intersect it in any point, so there is no point on the line that corresponds to A on the segment. Likewise, B has no counterpart on XY, so, in a sense, an arbitrarily short line segment actually has two *more* points than an infinitely long line! Of course, infinity plus two is still infinity; otherwise, the previous sentence would constitute a genuine paradox.

A third concept of real numbers expands on the set S of a couple of pages ago. Imagine all the *rational* numbers being split into two sets X and Y, such that every number in X is less than every number in Y; for example, X is the negative rational numbers and Y contains zero and the positive rationals. Each such split (called a *Dedekind cut* after the inventor of the concept) is defined as a real number. If there is either a largest rational in X or a smallest rational in Y, the real number defined by $X|Y$ is equal to that rational; otherwise, the real number defined by $X|Y$ is irrational. In the example just given, zero is the smallest rational in Y, so that cut is identified with the rational number zero. Note that if 0 were moved to X, the number defined by $X|Y$ would still be zero.

The sum of two reals, $A|B$ and $C|D$, is defined to be $E|F$, where E is the set of all rationals that can be expressed as the sum of a number from A and another from C, and F contains all other rationals. A real number $X|Y$ is positive if a positive rational is in X and negative if a negative rational is in Y. The product of two positive cuts $W|X$ and $Y|Z$ is defined as $P|Q$, where P contains all negative rationals, zero, and all rationals expressible as the product of a positive rational from W and another from Y, while Q contains all rationals not in P. The additive inverse of a real $X|Y$ is $Y'|X'$, where Y' contains the negatives of all the rationals in Y, and X' contains the negatives of all the rationals in X. With these definitions, we can extend the sign laws in Sec. 3.1 to apply to these cuts, and show that the real numbers thus defined are an ordered field by virtue of the corresponding theorem for the rationals. The definition of "greater than" is a little tricky: $A|B$ is greater than $X|Y$ if and only if at least <u>two</u> rationals in A are also in Y. By the definition of Dedekind cut, this means that an infinity of rational numbers must be in both A and Y.

The fourth characterization of rational numbers begins with the limit of a sequence of rationals (Sec. 3.9), and then makes the rational numbers involved be decimal fractions (Sec. 2.5). Since $\sqrt{2}$ is approximately 1.4142135, the sequence whose limit is $\sqrt{2}$ begins $\frac{1}{1}, \frac{14}{10}, \frac{141}{100}, \ldots, \frac{14,142,135}{10,000,000}, \ldots$ Any infinite decimal defines a sequence of rationals like this one which must have a limit. The proof uses the axiom of completeness: Any sequence of rationals corresponding to an infinite decimal is bounded and nondecreasing; the sequence for $\sqrt{2}$, for example, cannot possibly get as large as 2 or even 1.4142136 and can never backtrack because we are always adding non-negative terms of the form $\frac{d}{10^n}$, where d and n are non-negative integers with $0 \leq d \leq 9$. Now, the sequence has an upper bound (2 does well for $\sqrt{2}$'s sequence) and therefore must have a lub by the Axiom of Completeness. This lub is the limit of the sequence; call it L. If ϵ is any positive number, there must be a term of the sequence greater than $L - \epsilon$ or L would not be a lub. Once we find such a term, all those after it in the sequence must lie between L and $L - \epsilon$: greater than $L - \epsilon$ because the sequence is nondecreasing, and less than or equal to L because L is an ub. The last sentence fulfills the definition of the limit of a sequence and proves that L is that limit.

So any infinite decimal defines a real number. For finite (terminating) decimals like 0.5, we just attach an infinite string of zeros. We will now outlaw those decimals ending in an infinite string of 9's such as 0.499999... because they are equal to terminating decimals as will be shown. An infinite decimal with a repeating pattern such as 7.426262626... represents a rational number. We multiply the decimal by 10 raised to the power of the number of digits in the pattern: 7.42626... becomes 742.62626.... Then, we subtract the original number from the multiple: 742.62626... - 7.42626... = 735.20000..., where an infinite string of zeros appears because digits in the pattern are subtracted from matching digits in another repeat of the pattern. Since the difference between two multiples of a number is also a multiple of the same number, and a terminating decimal represents a rational number, the original infinite decimal must be a rational number. As for the infinite string of 9's, we perform the calculations as follows: Let $x = 0.499999...$. So $10x = 4.999999...$, and $9x = 10x - x = 4.500000...$ Since $9x = \frac{45}{10}$, $x = \frac{5}{10} = 0.500000...$. The numerator in the decimal fraction for $9x$ will always be a multiple of 9 because $ab...xyz9$ and $ab...xyz$ have the same remainder when divided by 9 (Sec. 1.1.2).

With the elimination of infinite strings of 9's, we can prove two statements about real numbers. First, between any two distinct real numbers, we can find an infinity of others. For two real numbers to be different, they must differ in at least one decimal place. We find the earliest difference, fill that place with any digit between the differing ones, and fill the others to the right any way we please (except for the law against the 9's), obtaining an infinite variety of numbers in between. If the differing digits are next to each other, say 3 and 4, we move over one place to the right. Unless we now have 39 and 40, there is a gap here that can be treated as above. If we do have 39 and 40, we continue moving to the right; unless we find 399 and 400, 3999 and 4000, ..., we will have a gap. But the infinite string of 9's has been outlawed, so eventually we will come across a digit in the smaller number that is 8 or less, and that will give us our gap.

The second statement is Cantor's theorem about how many real numbers there are. He showed that there can be no correspondence between all real numbers and all integers. Imagine that such a correspondence exists, and then define an infinite decimal as follows: Its k'th place is a 7 unless the k'th place in the real number corresponding to k is also a 7, in which case we define the k'th place of the new number to be 4. Assuming that a correspondence exists, we now have an infinite decimal that will look something like 0.77774774777777777477... and will represent a particular real number, which must have an integer corresponding to it. We call this integer r and ask: What is the r'th decimal place of our real number? From its definition, it is a 7 if and only if it is not a 7! This contradiction shows that no correspondence between integers and real numbers can exist. Since there are real numbers that are not integers, but no integers that are not real numbers, there must be *more* real numbers than there are integers.

But there are already an infinite number of integers! How can you have more than infinity? Cantor answered this paradox by saying that there are different kinds of infinity: an infinite number of them, in fact. The infinity of integers is just the "smallest" infinity, and is symbolized by $\aleph_0$ (read "aleph-null"); aleph being the first letter in the Hebrew alphabet and Cantor being Jewish. The next larger infinity is denoted by $\aleph_1$ ("aleph-one"), and was thought to be equal to c, the infinity of the real numbers, but it is now known that it is impossible to tell whether or not $\aleph_1$ and c are in fact the same. The statement that $c = \aleph_1$ is like the parallel postulate: you can accept or deny it without creating a logical contradiction.

Now is probably a good time to show that there are just as many rational numbers as there are integers, a fact which is by no means obvious. After all, you have $\aleph_0$ choices for the numerator of the fraction, and another $\aleph_0$ choices for the denominator. By elementary combinatorial rules, there are $\aleph_0 \times \aleph_0$ fractions, and certainly a number as big as "infinity" ought to get even bigger when squared. But there is a correspondence between fractions and integers. Considering only fractions with positive integers for both numerator and denominator, the fraction N/D is assigned the integer $\Delta(N + D - 2) + D$. This matches 1/1 with 1, 2/1 with 2, ½ with 3, 3/1 with 4, 2/2 with 5, 1/3 with 6, and so on. The correspondence the other way matches the integer K with the fraction $\frac{S+\Delta(S-2)-K}{K-\Delta(S-2)}$, where S is the integer part of $\frac{\sqrt{8K-7}+3}{2}$. This correspondence shows that $\aleph_0 \times \aleph_0 = \aleph_0$, but this equation is not like an ordinary algebraic equation in that you cannot "divide" both sides by $\aleph_0$ to get $\aleph_0 = 1$. Arithmetic with infinite numbers isn't the same as arithmetic with ordinary real numbers, and the alephs and c are not considered real numbers.

6.6 Modular Arithmetic – Advanced

In Chapter 4, the concept of modular arithmetic was introduced. Given any positive integer (except 1) as modulus, all integers can be reduced to an integer between zero and one less than the modulus. We now pretend that only this limited set of numbers exists, and we further require that the modulus be prime. In honor of this, we call the modulus p, not m. The integers between zero and $p - 1$ form a Galois field (Sec. 6.4) with respect to the operations of modular arithmetic. The proof of the field axioms is not very difficult, since ordinary addition and multiplication of integers obey most of them. To prove that a multiplicative inverse exists, we must show that for an arbitrary a ($0 < a < p$), there exists a b such that, in ordinary arithmetic, $ab = pk + 1$ for some integer k. If $a = 1$, there is no problem; we simply take $b = 1$ and $k = 0$. For other a's, we consider the sequence 0, a, $2a$, $3a$, …, $(p-1)a$ and their remainders after division by p. Each of the remainders must be between 0 and $p - 1$, and the zero must appear at the beginning of the sequence. If a remainder is repeated, say ma and na have the same remainder, then $(m - n)a$ is a multiple of p. But, since p is prime and m, n, and a are all less than p, this is impossible by the Fundamental Theorem of Arithmetic. Therefore, all the remainders must be distinct. There are p remainders and p possible values of remainders from 0 to $p - 1$, so every possible value for a remainder, including 1, must be represented. Once we find the remainder of 1, we have found our multiplicative inverse. This type of argument shows that, for finite structures, M_4 of Sec. 6.4 is equivalent to M_5, and therefore any finite integral domain is a field.

The field of integers modulo p is denoted Z_p. Once we have a and its multiplicative inverse a', the field axioms give us three more multiplicative inverses: $(a')' = a$; $(-a)' = -(a')$; and $(-a')' = -a$. However, these four numbers, a, a', $-a$, and $-a'$, are not always distinct. The possibility $a = -a$ and $a' = -a'$ is impossible unless $p = 2$, for it requires $a + a = 2a = p$. The possibility $a = a'$ and $-a = -a'$ is met by $a = 1$ or $p - 1$ but by no other numbers in any Z_p, as it requires $a^2 = kp + 1$ where a and k are both $< p$, and therefore $a^2 - 1 = (a + 1)(a - 1) = kp$. By the Fundamental Theorem of Arithmetic again, this means either $a + 1 = p$ or $a - 1 = 0$. The last possibility is $a = -a'$ and $a' = -a$. We can show that this must be satisfied whenever p is of the form $4n + 1$ by arranging the $p - 1 = 4n$ non-zero elements of Z_p by fours in the a, a', $-a$, $-a'$ pattern. Since 1 and $p - 1$ form an abnormal pair, and no other pairs of the types $a = a'$ or $a = -a$ are possible, we must have a second pair of numbers that must be of the type $a = -a'$. Now, $a = -a'$ requires $a(p - a) = kp + 1$, or $a^2 = pa - a(p - a) = (a - k)p - 1$. Hence, any prime of the form $4n + 1$ must have two multiples, less than its square, that are one more than squares. For $13 = 3 \times 4 + 1$, the multiples are $26 = 5^2 + 1$ and $65 = 8^2 + 1$; in Z_{13}, 5 and 8 are reciprocals and negatives of each other as $5 + 8 = 13$ and $5 \times 8 = 40 = 3 \times 13 + 1$.

Since Z_p is a field, we can solve any linear equation in it in the same way we solve such equations in the real field. The solution to $ax + b = 0$ is $x = -a'b$. The solution of quadratic equations is tougher. If you are given $ax^2 + bx + c = 0$ in Z_p, we can change it successively into (1) $x^2 + a'bx + a'c = 0$; (2) $x^2 + a'bx = -a'c$; (3) $x^2 + a'bx +$

$(2'a'b)^2 = (2'a'b)^2 - a'c$; and (4) $(x + 2'a'b)^2 = (2'a'b)^2 - a'c$, as if we were completing the square in an ordinary equation, but there is no easy way to tell whether $(2'a'b)^2 - a'c$ has a square root in Z_p. Exactly half of the non-zero elements in Z_p $(p > 2)$ have square roots, but which ones they are is far from obvious. Unlike the situation in real numbers, it is possible for neither or both of x and $-x$ to have square roots. In Z_5, for example, 1 and 4 have square roots while 2 and 3 do not. Modular square roots do resemble real ones in one respect; every number that has one has two that are negatives of each other. This property should not be surprising because it follows from the field axioms.

There is a theorem about the existence of square roots in modular fields. Given two primes p and q, p has a square root in Z_q if and only if q has a square root in Z_p, provided that p and q are not both of the form $4n - 1$, in which case one and only one of the square roots exists. This theorem, called *quadratic reciprocity,* can simplify the search for modular square roots. As an example, take $\sqrt{7}$ in Z_{19}. 7 and 19 are both of the form $4n - 1$, so $\sqrt{7}$ exists in Z_{19} if and only if $\sqrt{19}$ does not exist in Z_7. Now, $19 = 5 \pmod 7$ so we look for $\sqrt{5}$ in Z_7. Since 5 is not of the form $4n - 1$, $\sqrt{5}$ exists in Z_7 if and only if $\sqrt{7}$ exists in Z_5. $7 = 2 \pmod 5$, and we saw above that $\sqrt{2}$ does not exist in Z_5; therefore $\sqrt{5}$ does not exist in Z_7 and $\sqrt{7}$ does exist in Z_{19}. Knowing this, we examine 26, 45, 64, 83, etc., and find that $\sqrt{7} = 8$ and 11 in Z_{19}.

Chapter 7   An Introduction to Calculus

The first six chapters of this book have gone over arithmetic, algebra, geometry, trigonometry, and analytic geometry (plus other topics). These five branches of mathematics are sometimes lumped together as "pre-calculus" mathematics because they are prerequisites for the subject of this chapter. Called "calculus" for short, "the" calculus for long, and "the infinitesimal calculus" for longer, it is the mathematics of change, of movement, of dynamics. It is also the bulk of the section of math usually referred to as "analysis." Analysis deals with functions and their properties. Up until now, the word "function" has been used loosely, if at all, so this section will begin with a rigorous definition of a function.

7.1.1 Functions: Definitions

Given two sets A and B, we define their *Cartesian product* as the set of all ordered pairs (c, d) where c is an element of A and d is one of B. The Cartesian product of A and B is denoted $A \times B$. A subset of $A \times B$ is a function from A to B provided that each element of A appears exactly once as the first element of an ordered pair in the subset. In basic calculus, we consider only functions for which A and B are subsets of the set R of real numbers; advanced calculus, which will not be studied, can deal with some other types of functions.

A function from A to B is said to be *onto* B or *surjective* if every element of B is the second element of an ordered pair in the function. We can always restrict B so that a function becomes onto; the restricted B is called the *range* of the function, while A is called the *domain*. A function is called *one-to-one* or *injective* if no element of B appears more than once as the second half of an ordered pair. A function is called a one-to-one correspondence or *bijective* if it is both one-to-one and onto, or if it is both injective and surjective. Functions are denoted by letters like f or G; to show that (a, b) belongs to f, we write $f(a) = b$.

Functions can be thought of as black boxes with labels on them. The box for a function has an opening into which numbers are fed, and another opening from which the result comes out. One such black box spits out 9 when you insert 3, 15 when you insert 4, and 21 when you insert 5. If you put in the variable x, out comes the algebraic expression $6x - 9$. This can be used as a label for the function box. If we have already agreed to call this function G, we write $G(x) = 6x - 9$ to show the rule used for assigning elements of the range to elements of the domain.

7.1.2 Composite Functions

Function boxes can be hooked up to one another so that the output of one box becomes the input of the next. In terms of ordered pairs, we connect the function F from A to B with G from B to C as follows: Given x, a member of A, we find the one y in B such that (x, y) belongs to F, or $F(x) = y$. Then we find the z that makes (y, z) part of G, or $G(y) = z$. This procedure will give us exactly one z to any x, so the hooked-up functions combine to form another function, called the *composition* of F and G and written $G \circ F$. We have $G \circ F(x) = G(F(x))$. It should be noted that for composition to work, the range of F must overlap or lie within the domain of G.

A numerical example might be $F(x) = 2x + 1$ and $G(x) = x^2 - 2$. Therefore, $G \circ F(x) = G(F(x)) = G(2x + 1) = 4x^2 + 4x - 1$. Note that changing the order of the functions to get $F \circ G$ changes the result: $F \circ G(x) = F(G(x)) = F(x^2 - 2) = 2x^2 - 3$ which is not the same as $4x^2 + 4x - 1$. This is true in general: $F \circ G$ and $G \circ F$ are usually different. It is therefore important to note that the notation $F \circ G$ means that G is followed by F even though F is written first.

An example of restricted domains occurs when $F = 1 - x^2$ and $G = \sqrt{x}$. The domain of F is all real numbers, but the domain of G is only the non-negative real numbers because negative real numbers do not have real square roots. The range of F is the real numbers that are not greater than 1. Since the range of F overlaps the domain of G, the composition $G \circ F = \sqrt{1 - x^2}$ does exist, but its domain is not all real numbers, but rather the set of reals x for which $1 - x^2$ is non-negative. $1 - x^2 \geq 0$ requires $x^2 \leq 1$ and therefore $-1 \leq x \leq 1$, which is therefore the domain of $G \circ F$.

7.1.3 Inverse Functions

A bijective function f has an *inverse* function f^{-1} which satisfies, for all x in the domains, the equations $f^{-1}(f(x)) = x$ and $f(f^{-1}(x)) = x$; in other words, the compositions $f \circ f^{-1}$ and $f^{-1} \circ f$ are both equal to the identity function $I(x) = x$. We obtain f^{-1} from f by reversing all the ordered pairs in f: (a, b) is in f^{-1} if and only if (b, a) is in f. The domain of f is the range of f^{-1}, and the range of f is the domain of f^{-1}. Most functions can be restricted as to their domains and ranges so that they have inverses.

7.1.4 Implicit Relations

Some functions are defined by an equation involving two variables, rather than a formula containing only one. Instead of seeing $f(x) = 3 - x$, you may see $x + y = 3$. This equation can be solved algebraically for y (that is, y is isolated on one side of the equation in the same way that x is isolated in solving $2x + 5 = 9$) to give $y = 3 - x$. When y is isolated in this way, the equation is said to define y *explicitly* as a function of x. For any value of x, there is exactly one value of $3 - x$, and therefore y $(= 3 - x)$ is a function of x. The equation $x + y = 3$ is said to define y *implicitly* as a function of x. Sometimes, an implicit equation cannot be solved for y in terms of x, as when $\sin(x + y) + x = 3 \tan y$. In this case, the equation is called an *implicit relation*. For some or all values of x, one or more values of y will satisfy the equation, and hence y is defined implicitly as one or more functions of x.

7.1.5 Graphs of Functions

We draw the graph of a function $f(x)$ by plotting all of its ordered pairs $(x, f(x))$ on a Cartesian coordinate system. Of course, if either the domain or range of f is unbounded, we can draw only part of the graph on any finite sheet of paper. The graph of the function $f(x) = 1 - x^2$ is a parabola with the y-axis for its axis and opening downward. (See Fig. 7-2.) The graph of the function $g(x) = 3 - x$ or the implicit relation $x + y = 3$ is a straight line with slope -1 and y-intercept 3, as can be seen from the slope-intercept form of $y = (-1)x + 3$. (See Fig. 7-1.)

The graph of an inverse function is found by reflecting the original function's graph in the line $y = x$, as is made clear by Fig. 7-3. A curve is the graph of a function provided that no vertical line intersects it more than once, and the function is one-to-one provided that no horizontal line intersects the curve more than once, either.

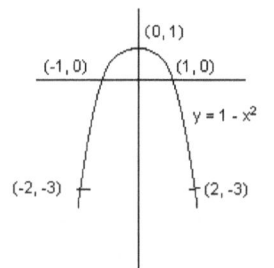

Fig. 7-2. A parabolic graph for the function $y = 1 - x^2$.

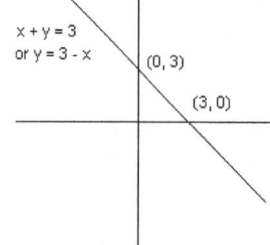

Fig. 7-1. The graph of the function $y = 3 - x$ or the implicit relation $x + y = 3$.

The graph of an implicit relation consists of all the points whose coordinates satisfy the defining equation. Since $x^2 + y^2 = 1$ is the equation for a unit circle, the implicit relation defined by $x^2 + y^2 = 1$ has the circle for its graph. Vertical lines will intersect the circle in two points, one above and one below the x-axis, so the implicit relation defines y as two separate functions of x, namely, $y = \sqrt{1 - x^2}$ and $y = -\sqrt{1 - x^2}$. Vertical lines through the lemniscate of Bernoulli (end of Sec. 5.7.3) likewise intersect it in two points (most of the time) and hence the lemniscate's equation defines y implicitly as two separate functions of x like the circle's equation does.

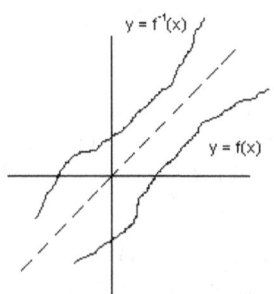

Fig. 7-3. The graph of a function and its inverse. The dashed line is $y = x$.

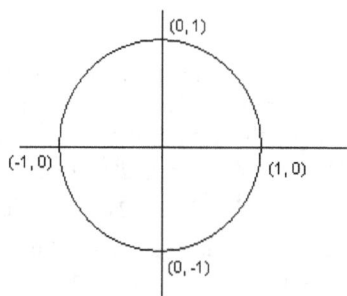

Fig. 7-4. The graph of the implicit relation $x^2 + y^2 = 1$, a unit circle.

One absolutely fundamental concept of calculus is that of the *limit* of a function as its *argument* (the x of $F(x)$) approaches, but does not necessarily reach, a certain value. We arrive at the definition of limit by means of an example.

Consider the function $F(x) = 2x - 1$ as the argument x gets close to 2. We compile a table of values of x and $F(x)$ as follows:

x	$2x$	$F(x)$	$3 - F(x)$
1	2	1	2
1.5	3	2	1
1.9	3.8	2.8	0.2
1.95	3.9	2.9	0.1
1.99	3.98	2.98	0.02
1.999	3.998	2.998	0.002
1.9999	3.9998	2.9998	0.0002
3	6	5	- 2
2.5	5	4	- 1
2.1	4.2	3.2	- 0.2
2.05	4.1	3.1	- 0.1
2.01	4.02	3.02	- 0.02
2.001	4.002	3.002	- 0.002
2.0001	4.0002	3.0002	- 0.0002

It is apparent from this table that the closer x gets to 2, the closer $F(x)$ gets to 3. For this reason, we wish to define limit in such a way that the limit of $F(x) = 2x - 1$ as x approaches 2 is precisely 3, which is symbolized $\lim_{x \to 2} 2x - 1 = 3$. To do this, we introduce the Greek letters δ (lowercase delta) and ϵ (lowercase epsilon) to represent the differences between x and 2, and $F(x)$ and 3, respectively. (A student of mathematics will become well acquainted with the Greek alphabet; we have already seen φ (phi) show up twice, and α (alpha), β (beta), γ (gamma), Γ (capital gamma), Δ (capital delta), θ (theta), and, indeed, almost every letter in that alphabet seems to denote some mathematical quantity or other.) From the table, we see that the smaller δ gets, the smaller ϵ gets as well: whenever δ is less than 0.01, ϵ is less than 0.02; whenever δ is less than 0.001, ϵ is less than 0.002; and so on. This motivates the following definition of limit: The limit of $F(x)$, as x approaches c, is L provided that, for any positive ϵ, there exists a positive δ such that, whenever the absolute value of $x - c$ is greater than zero but less than δ, the absolute value of $F(x) - L$ is less than ϵ. In symbols, we write $\lim_{x \to c} F(x) = L$ means that for each $\epsilon > 0$ there exists a $\delta > 0$ such that $0 < |x - c| < \delta$ implies $|F(x) - L| < \epsilon$.

To prove that $\lim_{x \to 2} 2x - 1 = 3$, we take $\delta = \epsilon/2$, so that if $\epsilon > 0$, then δ is positive as well, and note that $0 < |x - 2| < \frac{\epsilon}{2}$ can be transformed into $0 < |2x - 4| < \epsilon$ and then $0 < |(2x - 1) - 3| < \epsilon$. We then drop the left half of the compound inequality to get $|(2x - 1) - 3| < \epsilon$, which matches $|F(x) - L| < \epsilon$ when $F(x) = 2x - 1$ and $L = 3$.

The reader has no doubt noticed that the value of $F(x)$ when $x = 2$ is precisely equal to 3. But this fact cannot be used to prove that the limit is 3, since the compound inequality $0 < |x - c| < \delta$ means that we cannot take $x = c$ in determining the limit. What we lose in simplicity, however, we gain in flexibility. Because of that "$0 <$" in the delta inequality, the value – or lack of a value – of $F(c)$ does not affect the existence or value of the limit. For example, suppose instead of $F(x) = 2x - 1$, we had $G(x) = \frac{2x^2 - 5x + 2}{x - 2}$ to deal with. This function has no value when $x = 2$, since that makes the denominator $x - 2$ equal to zero. However, for all values of x *other* than 2, $G(x) = F(x)$ since $2x^2 - 5x + 2 = (2x - 1)(x - 2)$. Therefore, the algebraic steps involved in showing that $\lim_{x \to 2} F(x) = 3$ do the same for $G(x)$, thanks to the "$0 <$" in the delta inequality.

Similarly, we could have worked with the function $H(x)$, defined by $H(x) = 2x - 1$ when $x \neq 2$ and $H(x) = 7{,}654{,}321.098$ when $x = 2$. Again, since $H(x) = F(x)$ whenever x is *not* 2, $\lim_{x \to 2} H(x) = \lim_{x \to 2} F(x)$. The fact that, for $F(x)$, the value at 2 is equal to the limit there means that F is *continuous* at 2. We will have more to say about continuity later.

7.2.1 The Basic Theorem

The simplicity of the algebraic proof that $\lim_{x \to 2} 2x - 1 = 3$ is not usual. Most functions are more complicated than $2x - 1$, and it can become quite tedious to determine a limit directly from the definition. What is done instead is that the definition is used to prove a few simple limits, and then to prove the following four-part theorem, which is applied to more complex limits.

Theorem 7-1 (Basic Limit Theorem)
If $\lim_{x \to c} F(x) = L$ and $\lim_{x \to c} G(x) = M$, then:
I. $\lim_{x \to c} (F(x) + G(x)) = L + M$.
II. $\lim_{x \to c} (F(x) - G(x)) = L - M$.
III. $\lim_{x \to c} (F(x) \cdot G(x)) = LM$.
IV. $\lim_{x \to c} \left(\frac{F(x)}{G(x)} \right) = L/M$, provided that M is not zero.

The proof of I is simple enough: If $\lim_{x \to c} F(x) = L$, then for the positive number $\epsilon/2$ there is a δ_1 such that $0 < |x - c| < \delta_1$ implies $|F(x) - L| < \frac{\epsilon}{2}$, and for the same $\frac{\epsilon}{2}$ there is a δ_2 for $G(x)$ since $\lim_{x \to c} G(x) = M$. We let δ be the smallest of δ_1 and δ_2, so that $0 < |x - c| < \delta$ implies both $|F(x) - L| < \frac{\epsilon}{2}$ and $|G(x) - M| < \frac{\epsilon}{2}$, and therefore $|(F(x) + G(x)) - (L + M)| < \frac{\epsilon}{2} + \frac{\epsilon}{2} = \epsilon$. The proof of II is similar, but the proofs of III and IV are slightly more complicated and will not be shown.

The simple limits calculated directly from the definition are summarized below.

Theorem 7-2 (Simple Limits)
I. $\lim_{x \to c} x = c$.
II. If $F(x)$ is the constant k, then $\lim_{x \to c} F(x) = k$.
III. $\lim_{x \to c} ax + b = ac + b$.

Actually, we do not need part III of this theorem once we have parts I and II and parts I and III of Th. 7-1. But its proof is simple. To prove part I, we just take $\delta = \epsilon$, and for part II, we can use any positive δ we want. To prove part III, we let $\delta = \frac{\epsilon}{|a|}$ and proceed as we did in the proof that $\lim_{x \to 2} 2x - 1 = 3$.

Repeated applications of parts I, II, and III of Th. 7-1 to the simple limits in Th. 7-2, together with mathematical induction and one application of part IV of Th. 7-1 produces the following:

Theorem 7-3 (Continuity of Rational Functions)
· If $P(x)$ and $Q(x)$ are polynomials in x, and $Q(c)$ is not zero, then $\lim_{x \to c} \frac{P(x)}{Q(x)} = \frac{P(c)}{Q(c)}$.

A rational function is one that can be written as the quotient of two polynomials, and, as at the end of Sec. 7.1, continuity means that $\lim_{x \to c} F(x) = F(c)$. So this theorem is specified by saying that rational functions are continuous, except where a denominator is zero, which is why I have named it as I have.

7.2.2 How to Divide By Zero

The natural question now arises: What happens when $Q(c)$ *is* zero? More generally, how can one calculate $\lim_{x \to c} \frac{F(x)}{G(x)}$ when $G(c)$ is zero and F and G are any two functions of x, not just polynomials? There are two general strategies involved in this form of "division by zero": 1. Change the expression $F(x)/G(x)$ to a form which (a) is equivalent to $F(x)/G(x)$ for all values of x, except c, in an interval $|x - c| < h$, and (b) does not involve a denominator $D(x)$ that satisfies $D(c) = 0$, in which case the limit can usually be evaluated by replacing x by c in the new expression; or 2. Find functions $A(x)$ and $B(x)$ that satisfy $A(x) \le \frac{F(x)}{G(x)} \le B(x)$ for all values of x, except c, between $c - h$ and $c + h$ and, in addition, satisfy $\lim_{x \to c} A(x) = \lim_{x \to c} B(x) = L$, in which case $\lim_{x \to c} \frac{F(x)}{G(x)} = L$ as well.

Examples of both strategies occur in elementary calculus; the first strategy appeared a little earlier when $\frac{2x^2-5x+2}{x-2}$ was replaced with $2x - 1$, its equivalent for all x other than 2, to calculate the limit of the first expression as x approached 2 and therefore the denominator $x - 2$ approached zero. The second strategy, called the sandwich approach, will be used in determining the key limit $\lim_{x \to 0} \frac{\sin x}{x}$.

7.2.3 Infinity

There are many functions $\frac{F(x)}{G(x)}$ for which neither approach above will work. Alternatively, examination of the behavior of the function near c may make it clear that there is no number L that can satisfy the definition of limit in Sec. 7.2. An example of the latter occurs in trying to determine $\lim_{x \to 0} \frac{1}{x}$. As can be seen from the accompanying table, the values of the reciprocal function steadily increase as x gets closer to zero (or decrease if x is negative) and do not

x	$1/x$
0.01	100
0.001	1,000
0.0001	10,000

get close to any one number. In fact, for any large number M you might name, such as a million, a billion, a googol (10^{100}), or even a googolplex (10^{googol}), I can make $\left|\frac{1}{x}\right|$ be larger than M simply by keeping $|x|$ less than $\frac{1}{M}$, which remains positive no matter how big M is. Therefore, $\lim_{x \to 0} \frac{1}{x}$ is infinite by the following definition: $\lim_{x \to c} F(x) = \infty$ if and only if, for any positive, large M, there exists a δ such that $0 < |x - c| < \delta$ guarantees that $|F(x)| > M$. The symbol ∞, though read "infinity," is not to be confused with the alephs in Sec. 6.5. The alephs and c there are infinite cardinal numbers with which a sort of arithmetic can be performed, but this "infinity" has only the meaning given it by its use in the above definition.

In addition to infinite limits, there are also limits at infinity. These are defined as follows: $\lim_{x \to \infty} F(x) = L$ if and only if for any positive ϵ there exists an N such that $|x| > N$ implies $|F(x) - L| < \epsilon$. Thus, $\lim_{x \to \infty} \frac{1}{x} = 0$; we specify N as $\frac{1}{\epsilon}$ just as above we could use $\delta = \frac{1}{M}$. Similarly, $\lim_{x \to \infty} \frac{2x}{x+1} = 2$; we change $\left|\frac{2x}{x+1} - 2\right|$ to $\left|\frac{2x-(2x+2)}{x+1}\right|$ and finally to $\left|\frac{2}{x+1}\right|$, arriving at a situation similar to the previous one.

Finally, we have the case of an infinite limit at infinity, such as $\lim_{x \to \infty} x^2$. The definition here combines the infinite parts of the previous two: $\lim_{x \to \infty} F(x) = \infty$ means that for any large positive M there exists a finite N such that $|x| > N$ guarantees $|F(x)| > M$.

7.2.4 Right- and Left-Hand Limits

The absolute value signs in the definition of limit can cause lots of trouble, and some functions can behave quite differently on opposite sides of the value c that we are interested in. Consider the function $U(x)$, called the *unit step function*, defined as follows: $U(x) = 0$ if $x \le 0$, and $U(x) = 1$ if $x > 0$. Now ask: what is $\lim_{x \to 0} U(x)$? Since there are numbers arbitrarily close to 0 that satisfy $U(x) = 0$ and $U(x) = 1$, no single number can play the role of L in Sec. 7.2. If, however, we restrict our attention to positive values of x only, it is easy to see that the limit is 1; considering only negative values makes it clear that the limit is 0. Since the definitions of limit do not allow us to consider just positive or just negative numbers for x, we define a new kind of limit that involves just a one-sided approach. The so-called *right-hand* limit, $\lim_{x \to c+} F(x)$, is equal to L if, for any positive ϵ, there exists a positive δ such that $0 < x - c < \delta$ implies $|F(x) - L| < \epsilon$. Note the absence of the absolute value sign around $x - c$. The definition of the *left-hand* limit, $\lim_{x \to c-} F(x)$, is similar; the expression $x - c$ is replaced with $c - x$. With these definitions, it is easy to see that $\lim_{x \to 0+} U(x) = 1$ and $\lim_{x \to 0-} U(x) = 0$. The fact that the one-directional limits differ is sufficient to make the two-directional limit nonexistent.

When c is infinite, we write $\lim_{x \to +\infty} F(x)$ for the limit as x becomes large and positive, and $\lim_{x \to -\infty} F(x)$ for the limit as x becomes large and negative. For rational functions, these two limits are the same if they are finite, but other functions, such as the one symbolized by tanh, have different limits at opposite extremes if the number line. it happens that $\lim_{x \to +\infty} \tanh x = 1$, but $\lim_{x \to -\infty} \tanh x = -1$. The function tanh, read "hyperbolic tangent," is related to a hyperbola in a roughly similar way to that in which the ordinary tangent function is related to a circle.

7.2.5 Continuity

A function $F(x)$ is said to be *continuous* at a point c in its domain provided that $\lim_{x \to c} F(x) = F(c)$. It is

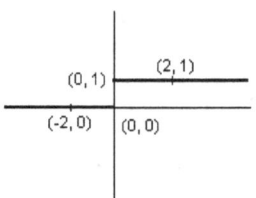

(0, 1)   (2, 1)

(-2, 0)   (0, 0)

continuous in an interval $a \leq x \leq b$ if it is continuous at all points in that interval. When you think of continuity, think "no holes or breaks in the function's graph." Note that in order for the limit $\lim_{x \to c} F(x)$ to even exist, both the left- and right-hand limits must exist, and they must be equal to each other. In Fig. 7-5, the graph of the unit step function $U(x)$ has a break when $x = 0$ because the left- and right-hand limits there are not equal. Therefore, this function is not continuous at $x = 0$.

It used to be thought that a function with an unrestricted domain could fail to be continuous at only a few isolated points, but consider the function $\varphi(x)$ defined as follows: $\varphi(x) = 1$ if x is rational, and $\varphi(x) = 0$ if x is irrational. Since there are both rational and irrational numbers arbitrarily close to any real number, the limit $\lim_{x \to c} \varphi(x)$ cannot exist for any value of c, and therefore $\varphi(x)$ is continuous *nowhere* despite the fact that it is defined everywhere. The graph of this function, called the *Dirichlet function* after its inventor, consists of two lines, $y = 0$ and $y = 1$, which are fuller of holes than the worst imaginable Swiss cheese. Since there are more irrational numbers than rational ones, the line $y = 0$ is more solid, but it still has no segment, however short, included anywhere.

The function $G(x) = \frac{2x^2 - 5x + 2}{x - 2}$ illustrates another, simpler, type of discontinuity. The function is not continuous at $x = 2$ since 2 is not in the function's domain. However, since $\lim_{x \to 2} G(x)$ does exist and is equal to 3, we can define a related function $G_1(x)$ as being equal to $G(x)$ when x is not 2 and to 3 when x is 2. This function $G_1(x)$ will be continuous at 2 since $G_1(2) = 3 = \lim_{x \to 2} G_1(x)$. In defining $G_1(x)$, we have removed the discontinuity of $G(x)$. We note that $G_1(x)$ is now identical to $F(x) = 2x - 1$ since it assigns the same values as $F(x)$ does to every real number. The graph of $G(x)$ has a hole: it is missing the point (2, 3) that is part of the graphs of $G_1(x)$ and $F(x)$.

7.2.6 The Intermediate Value Theorem

We will now prove a theorem that is crucial to more advanced results, including the Fundamental Theorem of Calculus.

Theorem 7-4 (The Intermediate Value Theorem)

Let $F(x)$ be continuous on the interval $a \leq x \leq b$, and let c be any number between $F(a)$ and $F(b)$. Then there is at least one number k, $a \leq k \leq b$, such that $F(k) = c$.

The geometric meaning of this theorem is that a horizontal line drawn between the lines $y = F(a)$ and $y = F(b)$ will intersect the graph of $F(x)$ at least once, provided that the graph is a continuous curve. This appears to be obvious (see Fig. 7-6), but conclusively proving it is surprisingly difficult. The proof here uses the Axiom of Completeness defined at the beginning of Sec. 6.5.

Assume that $F(a) < F(b)$ and consider the set S of all numbers between a and b whose function values are no larger than c. [If $F(a) > F(b)$, we exchange "greater" and "less" in certain parts of the proof, or work with $G(x)$, defined so that $G(x) = - F(x)$.] Since c is between $F(a)$ and $F(b)$, S has a for a member, and by its definition, S can have no members larger than b. Therefore, b is an upper bound for the nonempty set S, and, by the Axiom of Completeness, S has a lub which we denote by k.

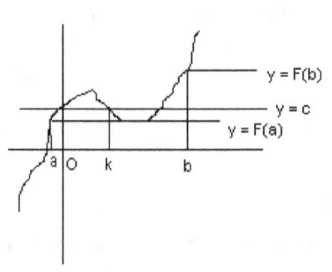

y = F(b)

y = c

y = F(a)

a   O   k   b

I now claim that $F(k)$ is precisely c. For, if $F(k) < c$, then let ϵ be the positive number $c - F(k)$. Since F is continuous on $a \leq x \leq b$ and therefore at k which must lie in this interval, $\lim_{x \to k} F(x) = F(k)$. By the definition of limit, there exists a positive δ such that, whenever $0 < |x - k| < \delta$, $|F(x) - F(k)| < \epsilon$, and therefore whenever $k < x < k + \delta$, $F(x) < F(k) + \epsilon$. But $F(k) + \epsilon = c$ by definition of ϵ, so there are values of x greater than k that satisfy $F(x) < c$, and hence are in S, contrary to the assumption that k is an upper bound for S.

On the other hand, if $F(k) > c$, then let ϵ be the positive number $F(k) - c$. As before, F is continuous at k, so there exists a δ such that whenever $0 < |x - k| < \delta$, $|F(x) - F(k)| < \epsilon$. Now we use the other half of the absolute value inequalities (remember that $|a| < b$ means $-b < a < b$) and conclude that $k - \delta < x < k$ implies $F(k) - \epsilon < F(x)$. But now $F(k) - \epsilon = c$, so $k - \delta < x < k$ implies $F(x) > c$. Hence no number between $k - \delta$ and k is in S, and therefore $k - \delta$ is just as good an upper bound for S as k is, again contrary to our choice of k as the <u>least</u> upper bound for S.

Therefore, by the definition of k, we can have neither $F(k) > c$ nor $F(k) < c$, and the only choice left is $F(k) = c$, Q.E.D.

The Intermediate Value Theorem can be used to justify the existence of solutions to equations that cannot be solved analytically, such as $x = \cos x$. In this equation, x is the radian measure of the angle whose cosine is x. To prove the existence of a solution, write $x = \cos x$ in the equivalent form $\cos x - x = 0$, and then consider $F(x) = \cos x - x$. Since $\cos 0 = 1$ and $\cos \frac{\pi}{2} = \cos 90° = 0$, $F(0) = 1$ and $F(\frac{\pi}{2}) = -\frac{\pi}{2}$. Now $-\frac{\pi}{2} < 0 < 1$, so the Intermediate Value Theorem guarantees the existence of a k satisfying $F(k) = 0$ and therefore being the solution to $x = \cos x$.

7.3 Differentiation and Derivatives

When a rock is dropped off a cliff, it falls under the influence of gravity. The distance, s, that it falls is a function of the amount of time, t, that it has been falling. If s is measured in meters and t in seconds, then $s = 4.903t^2$. Hence, the rock falls 4.903 meters in the first second, 14.710 in the second second, 24.517 in the third, etc., going faster and faster as it falls until it hits the ground at the bottom of the cliff. But the rock does not suddenly jump in speed from zero to 4.903 meters per second the instant it is released, then triple its speed to 14.710 m/s after precisely one second has elapsed. Indeed, by keeping track of s at intervals of, say, one tenth of a second, we find that the rock falls at an average speed of 0.490 m/s during the first tenth of a second, 1.471 m/s during the second tenth, and so on. By calculating s at shorter and shorter intervals, we find the rock's speed over shorter and shorter periods of time. What happens if we take this approach to the limit and try to calculate the rock's speed at a precise instant of time? The usual formula speed = distance/time cannot be used here; during an instant, no time at all passes and the rock falls through zero distance, giving the meaningless 0/0 as the speed.

The calculus comes to the rescue, however, via the idea of limit. We calculate s at a certain moment, after t seconds have passed, and then a short time later, after $t + \Delta t$ seconds have elapsed. (The symbol Δ is a Greek capital delta, and represents "change of.") During the additional Δt seconds, the rock falls an additional distance, called Δs. If we can find an expression for $\Delta s/\Delta t$, which gives the average speed over the interval of Δt seconds, and if this expression has a limit as Δt approaches (but does not reach!) zero, then this limit is defined to be the rock's instantaneous speed after t seconds. This instantaneous speed, denoted by v for velocity, is called the *derivative* of s at the point t, and the process of finding it is called *differentiation*.

An analogous problem crops up in analytic geometry, where *tangent* lines are concerned. A tangent (not to be confused with the trigonometric function) line to a curve at a point P is defined as the limit of lines called *secants* (again not to be confused with the trigonometric function) that cross the curve at P and a nearby point Q, as Q comes closer and closer to P. With P and the curve fixed, the secants are functions of the location of Q, or, more precisely, the slopes of the secants are functions of the x-coordinate of Q. (See Fig. 7-7.) The slope of the secant PQ is defined, as in any line, by $m = $ rise/run (Sec. 5.7.2). If the coordinates of P are (x_1, y_1) and Q are (x_2, y_2), then $m = \frac{y_2 - y_1}{x_2 - x_1}$. If this expression has a limit as Q approaches P, or, equivalently, x_2 approaches x_1, then this limit is defined as the slope of the tangent line at P. If we change notation slightly, so that the change in x between P and Q is called Δx (before it was $x_2 - x_1$) and similarly replace $y_2 - y_1$ with Δy, we have the same situation as we did with instantaneous speed: we then wanted $\lim_{\Delta t \to 0} \frac{\Delta s}{\Delta t}$, and now we want $\lim_{\Delta x \to 0} \frac{\Delta y}{\Delta x}$. As before, we call $\lim_{\Delta x \to 0} \frac{\Delta y}{\Delta x}$ the derivative of y with respect to x, and the process of finding it is called differentiation.

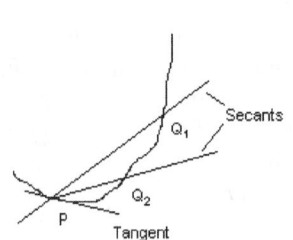

Fig. 7-7. The tangent line at P is the limit of secant lines, as Q gets closer and closer to P.

Other symbols for $\lim_{\Delta x \to 0} \frac{\Delta y}{\Delta x}$ include y', $\frac{dy}{dx}$, $\dot{y}$, and $D_x y$. The dy/dx was introduced by Leibniz, one of the inventors of calculus, and the less common $\dot{y}$ is due to Newton, the other inventor. Newton invented calculus first, but Leibniz published first; before Newton published, he used his new piece of mathematics to solve dozens of scientific puzzles and confirm his law of gravity.

7.3.1 Derivatives of Polynomials

We now consider a function $P(x) = x^n$, where n is a positive integer. We let x increase by a small amount Δx, and determine the amount by which $P(x)$ changes as a result. The binomial theorem (Sec. 4.7) says that $(x + \Delta x)^n = x^n + \binom{n}{1}\Delta x \cdot x^{n-1} + \binom{n}{2}(\Delta x)^2 x^{n-2} + \binom{n}{3}(\Delta x)^3 x^{n-3} + \cdots + \binom{n}{n-1}(\Delta x)^{n-1}x + (\Delta x)^n$. We subtract x^n to determine the *change* in $P(x)$ as a result of letting x increase to $x + \Delta x$, and then divide this change by Δx (we can do this as long as Δx is not zero). The result is

$$\frac{\Delta P(x)}{\Delta x} = \binom{n}{1}x^{n-1} + \binom{n}{2}\Delta x \cdot x^{n-2} + \binom{n}{3}(\Delta x)^2 x^{n-3} + \cdots + (\Delta x)^{n-1}.$$

As Δx approaches zero, so do all the terms that have Δx as a factor. The reason for this is part III of Th. 7-1, as $\lim_{\Delta x \to 0}(\Delta x) = 0$ (naturally!) Hence, $\lim_{\Delta x \to 0}\frac{\Delta P(x)}{\Delta x}$ consists of the one term above that does not have Δx as a factor, namely $\binom{n}{1}x^{n-1}$. Since $\binom{n}{1} = n$ for all positive integers n, we have $P'(x) = \frac{dP(x)}{dx} = \dot{P} = D_x P = nx^{n-1}$ for all positive integers n. Eventually, this formula will be shown to be valid for all real numbers n, provided that x is positive. We file this formula away as the first in a series of formulas for derivatives:

I. $\frac{d(x^n)}{dx} = nx^{n-1}$.

To use this formula to evaluate the derivative of any polynomial, we combine it with the following rules:

II. $\frac{d(F_1(x)+F_2(x))}{dx} = \frac{dF_1(x)}{dx} + \frac{dF_2(x)}{dx}$.

III. $\frac{d(cF(x))}{dx} = c\frac{dF(x)}{dx}$.

IV. $\frac{dc}{dx} = 0$.

In these formulas, F, F_1 and F_2 are functions of x and c is a constant. Rule II follows directly from part I of Th. 7-1 and the observation that $\frac{[F_1(x+\Delta x)+F_2(x+\Delta x)]-[F_1(x)+F_2(x)]}{\Delta x} = \frac{F_1(x+\Delta x)-F_1(x)}{\Delta x} + \frac{F_2(x+\Delta x)-F_2(x)}{\Delta x}$. Rule III is obvious once we notice that $\frac{cF(x+\Delta x)-cF(x)}{\Delta x} = c\frac{F(x+\Delta x)-F(x)}{\Delta x}$. As for Rule IV, we write $\frac{c(x+\Delta x)-c(x)}{\Delta x} = \frac{c-c}{\Delta x}$ (since c is a constant) $= 0$.

To show the use of the formulas to differentiate a polynomial, we work on $P(x) = 8x^4 - 5x^3 + 7x - 9$. Now, by Rule II,

$$\frac{dP(x)}{dx} = \frac{d(8x^4)}{dx} + \frac{d(-5x^3)}{dx} + \frac{d(7x)}{dx} + \frac{d(-9)}{dx}$$
$$= 8\frac{d(x^4)}{dx} + (-5)\frac{d(x^3)}{dx} + 7\frac{dx}{dx} + 0 \text{ (rules III and IV)}$$
$$= 8 \cdot 4x^3 - 5 \cdot 3x^2 + 7 \cdot 1x^0 \text{ (rule I)}$$
$$= 32x^3 - 15x^2 + 7.$$

This result tells us that the slope of the tangent to the curve $y = 8x^4 - 5x^3 + 7x - 9$ at an arbitrary point $P(x_0, 8x_0^4 - 5x_0^3 + 7x_0 - 9)$ on it is $32x_0^3 - 15x_0^2 + 7$. It gives us more information than that: It tells us how sensitive $P(x)$ is to small changes in its argument. Since $\lim_{\Delta x \to 0}\frac{\Delta P(x)}{\Delta x}$ is $32x^3 - 15x^2 + 7$, when Δx is small, the actual value of $\frac{\Delta P(x)}{\Delta x}$ will be close to $32x^3 - 15x^2 + 7$, and hence $\Delta P(x)$ itself will be close to $32x^3\Delta x - 15x^2\Delta x + 7\Delta x$. If we start at $x = 1$, for example, then $32x^3 - 15x^2 + 7 = 24$, and therefore any small change in x will change $P(x)$ by about 24 times as much. Since $P(1) = 1$, $P(1.01)$ will be close to 1.24. A calculation shows that $P(1.01)$ is actually 1.24332708, giving a good idea of how close an approximation calculus can produce. The value 24, the derivative at a particular point, is denoted $P'(1)$ or $\frac{dP(x)}{dx}\Big|_{x=1}$.

We note that the derivative of $P(x)$ is also a polynomial in x and therefore can be differentiated. The resulting derivative of a derivative is called the *second derivative*, symbolized by $P''(x)$ or $\frac{d^2P(x)}{dx^2}$. Similarly, we can find the third, fourth, ..., nth derivatives of a polynomial, and other functions as well once we figure out how to get their first derivatives. The results for our $P(x)$ are listed below:

$$P(x) = 8x^4 - 5x^3 + 7x - 9, \qquad\qquad P(1) = 1;$$

$$P'(x) = \frac{dP(x)}{dx} = 32x^3 - 15x^2 + 7, \qquad\qquad P'(1) = 24;$$

$$P''(x) = \frac{d^2P(x)}{dx^2} = 96x^2 - 30x, \qquad\qquad P''(1) = 66;$$

$$P^{(3)}(x) = \frac{d^3P(x)}{dx^3} = 192x - 30, \qquad\qquad P^{(3)}(1) = 162;$$

$$P^{(4)}(x) = \frac{d^4P(x)}{dx^4} = 192, \qquad\qquad P^{(4)}(1) = 192;$$

$$P^{(5)}(x) = \frac{d^5P(x)}{dx^5} = 0, \qquad\qquad P^{(5)}(1) = 0.$$

Each successive differentiation operation reduces the degree of the polynomial by 1, so an nth-degree polynomial has a constant for its nth derivative and therefore zero for its $(n+1)$st derivative. Additional differentiation produces no further change; since zero is a constant, it is its own derivative.

7.3.3. Derivatives of Rational Functions

We will now extend the list of functions that can be differentiated to rational functions (products and quotients of polynomials) by deriving the following two formulas:

$$\text{V. } \frac{d(F(x) \times G(x))}{dx} = F(x)\frac{dG(x)}{dx} + G(x)\frac{dF(x)}{dx}.$$

$$\text{VI. } \frac{d\left(\frac{F(x)}{G(x)}\right)}{dx} = \frac{G(x)\frac{dF(x)}{dx} - F(x)\frac{dG(x)}{dx}}{[G(x)]^2}.$$

Formula VI will allow us to extend Formula I to the case where n is a negative integer (at the moment n can only be a positive integer) by treating the function x^n as $\frac{1}{x^{-n}}$, where $-n$ is a positive integer.

To derive V, consider the function $H(x)$, defined as $H(x) = F(x) \times G(x)$ for all x. Now, give x an increment Δx, which produces changes in $F(x)$ and $G(x)$ given by $\Delta F(x)$ and $\Delta G(x)$. We calculate $H(x + \Delta x) = [F(x) + \Delta F(x)][G(x + \Delta G(x)] = F(x) \times G(x) + F(x) \times \Delta G(x) + G(x) \times \Delta F(x) + \Delta F(x) \times \Delta G(x)$. Since $H(x) = F(x) \times G(x)$, $\Delta H(x) = F(x) \times \Delta G(x) + G(x) \times \Delta F(x) + \Delta F(x) \times \Delta G(x)$. Dividing by Δx and taking the limit as $\Delta x \to 0$, we have $\frac{dH(x)}{dx} = \lim_{\Delta x \to 0} \frac{\Delta H(x)}{\Delta x} = \lim_{\Delta x \to 0}[F(x)\frac{\Delta G(x)}{\Delta x} + G(x)\frac{\Delta F(x)}{\Delta x} + \Delta F(x)\frac{\Delta G(x)}{\Delta x}]$. By means of parts I and III of Th. 7-1, we expand this to $F(x)\lim_{\Delta x \to 0}\frac{\Delta G(x)}{\Delta x} + G(x)\lim_{\Delta x \to 0}\frac{\Delta F(x)}{\Delta x} + \lim_{\Delta x \to 0}\Delta F(x)\frac{\Delta G(x)}{\Delta x}$. The definition of derivative enables us to simplify this expression to $F(x)\frac{dG(x)}{dx} + G(x)\frac{dF(x)}{dx} + \lim_{\Delta x \to 0}\Delta F(x) \cdot \frac{dG(x)}{dx}$. Finally, since F has a derivative, $\lim_{\Delta x \to 0}\Delta F(x)$ must be zero (otherwise, $\lim_{\Delta x \to 0}\frac{\Delta F(x)}{\Delta x}$ would not exist in the same way that $\lim_{x \to 0}\frac{1}{x}$ does not exist). Dropping the term with $\lim_{\Delta x \to 0}\Delta F(x)$ as a factor produces $\frac{dH(x)}{dx} = F(x)\frac{dG(x)}{dx} + G(x)\frac{dF(x)}{dx}$, which is Formula V.

To derive VI, we again consider $H(x)$, defined now as $F(x)/G(x)$. Then, $H + \Delta H = \frac{F + \Delta F}{G + \Delta G}$, where we use the standard abbreviations F for $F(x)$, etc. to save space, and the fact that x is the independent variable is understood. Then,

$$\Delta H = \frac{F + \Delta F}{G + \Delta G} - \frac{F}{G} = \frac{FG + G\Delta F - F\Delta G - FG}{G^2 + G\Delta G} = \frac{G\Delta F - F\Delta G}{G^2 + G\Delta G}.$$ We now divide by Δx, obtaining $\frac{\Delta H}{\Delta x} = \frac{G\frac{\Delta F}{\Delta x} - F\frac{\Delta G}{\Delta x}}{G^2 + G\Delta G}$. As we let $\Delta x \to 0$, $\frac{\Delta H}{\Delta x}$ becomes $\frac{dH}{dx}$, $\frac{\Delta F}{\Delta x}$ becomes $\frac{dF}{dx}$, $\frac{\Delta G}{\Delta x}$ becomes $\frac{dG}{dx}$, and ΔG itself approaches zero for the same reason as ΔF approached zero in the preceding paragraph. Since G is finite, $G^2 + G\Delta G$ becomes G^2, and we have $\frac{dH}{dx} = \frac{G\frac{dF}{dx} - F\frac{dG}{dx}}{G^2}$, which is Formula VI.

To see how to use these formulas to differentiate rational functions, we apply them to $F(x) = \frac{(x^2-1)(x^3+x^2)}{x^2+x+1}$. First, Formula VI is used to obtain $\frac{dF}{dx} = \frac{(x^2+x+1)\frac{d[(x^2-1)(x^3+x^2)]}{dx} - (x^2-1)(x^3+x^2)\frac{d(x^2+x+1)}{dx}}{(x^2+x+1)^2}$. It looks as if we have made backward progress, doesn't it? We now apply V to the first of the two derivatives in this expression to get $\frac{dF}{dx} = \frac{(x^2+x+1)[(x^2-1)\frac{d(x^3+x^2)}{dx} + (x^3+x^2)\frac{d(x^2-1)}{dx}] - (x^2-1)(x^3+x^2)\frac{d(x^2+x+1)}{dx}}{(x^2+x+1)^2}$. Now there are three derivatives to evaluate, but they are all polynomials. Applying Formulas I–IV to them produces $\frac{dF}{dx} = \frac{(x^2+x+1)[(x^2-1)(3x^2+2x) + 2x(x^3+x^2)] - (x^2-1)(x^3+x^2)(2x+1)}{(x^2+x+1)^2}$. It looks a lot more complicated than what we started with, but the lengthy numerator could be represented as a single 6th-degree polynomial, which is only slightly more complicated than the fifth-degree polynomial that our beginning numerator could be written as. In general, functions that are not polynomials become more complex when they are differentiated, although there are some notable exceptions, such as $\sin x$ and $\cos x$.

We will now use Formula VI to extend I to negative integers. Let y be the function $y = x^n$ where n is a negative integer; then we can also say ([3.10c], Sec. 3.8), that $y = \frac{1}{x^m}$, where $m = -n$ is a positive integer. By Formula VI, then,

$\frac{dy}{dx} = \frac{x^m \frac{d(1)}{dx} - 1 \cdot \frac{d(x^m)}{dx}}{(x^m)^2}$. Since 1 is a constant, $\frac{d(1)}{dx} = 0$ by Formula IV, and, since m is a *positive* integer, we can apply

Formula I to x^m, so $\frac{d(x^m)}{dx} = mx^{m-1}$. Hence, $\frac{dy}{dx} = \frac{-mx^{m-1}}{(x^m)^2} = \frac{nx^{m-1}}{x^{2m}}$ since $m = -n$ and $(x^m)^2 = x^{2m}$ by the Second Law of

Exponents in Sec. 3.1.3. Applying the Third Law of Exponents to this last fraction, we have $\frac{dy}{dx} = nx^{m-1-2m} = nx^{-n-1+2n} = nx^{n-1}$. So Formula I holds even when n is a negative integer. It also holds when n is zero: since $x^0 = 1$ is a constant, $\frac{d(x^0)}{dx} = 0 = 0x^{-1}$.

7.3.3 Derivatives of Composite Functions: The Chain Rule

In this section we derive the formula for the derivative of a composite function $G(F(x))$. First, we state it as Formula VII:

VII. $\frac{d(G \circ F)}{dx} = \frac{dF}{dx} \cdot \frac{dG(y)}{dy}\Big|_{y=F(x)}$.

The notation with G means we first find its derivative, using another letter like y to stand for the independent variable, and then after the differentiation replace that independent variable with $F(x)$. For example, $\frac{d(x^2+1)^3}{dx} = \frac{d(x^2+1)}{dx} \cdot \frac{dy^3}{dy}\Big|_{y=x^2+1} = 2x \cdot 3(x^2 + 1)^2$.

After deriving the chain rule, we will use it to extend Formula I to the case where n is any rational number, thus continuing to broaden the rule's scope. In particular, this will allow us to find the derivative of $\sqrt{x}$.

To derive Formula VII, we write $G(y) = G(F(x))$. Now, when we give x an increment Δx, this will cause F to change by an amount ΔF satisfying $\lim_{\Delta x \to 0} \frac{\Delta F}{\Delta x} = \frac{dF}{dx}$. At the same time, when we give F an increment ΔF, it will cause $G(F(x))$ or $G(y)$ to change by an amount ΔG satisfying $\lim_{\Delta F \to 0} \frac{\Delta G}{\Delta F} = \frac{dG}{dF}$. Now, $\frac{\Delta G}{\Delta x} = \frac{\Delta G}{\Delta F} \cdot \frac{\Delta F}{\Delta x}$ and hence $\lim_{\Delta x \to 0} \frac{\Delta G}{\Delta x} = \lim_{\Delta x \to 0} \frac{\Delta G}{\Delta F} \cdot \lim_{\Delta x \to 0} \frac{\Delta F}{\Delta x}$ (Th. 7-1, part III). Next, $\Delta x \to 0$ means $\Delta F \to 0$ since F has a finite derivative, so we have $\frac{dG}{dx} = \frac{dG}{dF} \cdot \frac{dF}{dx}$. Since we are using y and $F(x)$ to denote the same thing, we get $\frac{d(G \circ F)}{dx} = \frac{dG(y)}{dy} \cdot \frac{dF(x)}{dx}$, and y must be replaced with $F(x)$ after differentiation of G to get everything back in terms of one independent variable.

To use the chain rule, all we have to do is break the composite function down into its component functions. With $(x^2 + 1)^3$ above, we have $F(x) = x^2 + 1$ and $G(y) = y^3$ so that $G(F(x)) = [F(x)]^3 = (x^2 + 1)^3$. It is standard procedure to incorporate the chain rule into formulas for derivatives; under this practice Formula I does not involve just x^n but u^n, where u is an unspecified but differentiable function of x. Applying the chain rule to u^n produces the more general form of Formula I:

I. $\frac{d(u^n)}{dx} = nu^{n-1} \cdot \frac{du}{dx}$.

The other formulas we have so far do not involve single functions, but the use of u and v instead of F and G is standard. Rewriting the other formulas in terms of u and v produces:

II. $\frac{d(u+v)}{dx} = \frac{du}{dx} + \frac{dv}{dx}$.

III. $\frac{d(cu)}{dx} = c\frac{du}{dx}$.

IV. $dc/dx = 0$.

V. $\dfrac{d(uv)}{dx} = u\dfrac{dv}{dx} + v\dfrac{du}{dx}$.

VI. $\dfrac{d(\frac{u}{v})}{dx} = \dfrac{v\frac{du}{dx} - u\frac{dv}{dx}}{v^2}$.

VII. $\dfrac{dv}{dx} = \dfrac{dv}{du} \cdot \dfrac{du}{dx}$.

Although this form of the chain rule makes it look like $\dfrac{dv}{dx}$, etc., are ordinary fractions, they are not. dv/dx does not mean dv divided by dx any more than dx means d times x. Rather, $dv/dx = \lim_{\Delta x \to 0} \dfrac{\Delta v}{\Delta x}$, where $\dfrac{\Delta v}{\Delta x}$ is an ordinary fraction, but Δv does not mean Δ times v but the change in v associated with Δx, the change in x.

We will now use the chain rule to extend Formula I to the case where n is a rational number p/q and p and q are integers. The key is to view $v(x) = x^p$ as the composite of two other functions, $v(u) = u^q$ and $u(x) = x^{p/q}$. Then $v(u(x))$ = $v(x^{p/q}) = (x^{p/q})^q = x^p$. The chain rule enables us to write $\dfrac{dv}{dx} = \dfrac{dv}{du} \cdot \dfrac{du}{dx} = \dfrac{d(u^q)}{du}\Big|_{u=x^{p/q}} \cdot \dfrac{d(x^{\frac{p}{q}})}{dx}$. Since p and q are integers, the already known applicability of Formula I leads to $px^{p-1} = qu^{q-1} \cdot \dfrac{d(x^{\frac{p}{q}})}{dx} = q(x^{p/q})^{q-1} \cdot \dfrac{d(x^{\frac{p}{q}})}{dx}$. We solve this equation for $\dfrac{d(x^{\frac{p}{q}})}{dx}$ to obtain $\dfrac{d(x^{\frac{p}{q}})}{dx} = \dfrac{p}{q}x^{p-1-(q-1)(\frac{p}{q})} = \dfrac{p}{q}x^{(\frac{pq-q+pq+p}{q})} = \dfrac{p}{q}x^{(p-q)/q} = \dfrac{p}{q}x^{(\frac{p}{q})-1}$. Replacing p/q with its equivalent n, we get $\dfrac{d(x^n)}{dx} = nx^{n-1}$, where n is any rational number. In particular, when $n = \frac{1}{2}$, we have $\dfrac{d(\sqrt{x})}{dx} = \dfrac{1}{2\sqrt{x}}$, or, applying the chain rule, Formula VIII: $\dfrac{d(\sqrt{u})}{dx} = \dfrac{du/dx}{2\sqrt{u}}$. We state this as a separate formula despite the fact that it is just a particular case of Formula I due to the frequency with which square roots crop up in mathematics. As an example, take $F(x) = \sqrt{x^2 + 1}$. This is $\sqrt{u}$ with $u = x^2 + 1$, so Formula VIII tells us that $\dfrac{dF}{dx} = \dfrac{du/dx}{2\sqrt{u}} = \dfrac{2x}{2\sqrt{x^2+1}} = \dfrac{x}{\sqrt{x^2+1}}$.

7.3.4 Derivatives of Inverse Functions

Any bijective function f has an inverse function denoted f^{-1}. (The f^{-1}, of course, does not mean f to the minus-first power, that is, the reciprocal of f.) As mentioned briefly in Sec. 7.1.3, we find the ordered pairs of f^{-1} by switching members in the ordered pairs of f. Hence, to any secant line on the graph of f between the points (x_0, y_0) and (x_1, y_1) there corresponds a secant line on the graph of f^{-1} between the points (y_0, x_0) and (y_1, x_1). Since moving from f to f^{-1} exchanges rise and run for the secant lines, the slope of the secant line on the inverse function's graph is the reciprocal of the corresponding slope for the original function. As x_1 approaches x_0 and the secants become the tangent line for the original graph, the corresponding secants to the graph of f^{-1} must do likewise. Therefore, the slope of the tangent to the graph of f^{-1} at the point (y_0, x_0) is the reciprocal of the slope of the tangent to f at the point (x_0, y_0). In the language of calculus, this means that $\dfrac{d(f^{-1}(x))}{dx} = \dfrac{1}{\frac{df(y)}{dy}\big|_{y=f^{-1}(x)}}$, or, in u-v notation, $\dfrac{du}{dx} = \dfrac{1}{\frac{dx}{du}}$. As an example, we can consider $f(x) = 2x + 1$ and its inverse, $f^{-1}(x) = \dfrac{x-1}{2}$. We have $df/dx = 2$, so $\dfrac{d(f^{-1})}{dx}$ must be $\frac{1}{2}$. Since the derivative of f was a constant, no question about the argument can arise. For a more complicated example, consider $g(x) = x^2$ for which $g^{-1}(x) = \sqrt{x}$. Now, $\dfrac{d(g^{-1}(x))}{dx} = \dfrac{1}{2y}\Big|_{y=\sqrt{x}} = \dfrac{1}{2\sqrt{x}}$ in agreement with what we already know. Here, we replaced y with $g^{-1}(x) = \sqrt{x}$ after differentiating $g(x)$. The statement $x = g(y)$ is synonymous with $y = g^{-1}(x)$, so, if $x = y^2$, then $y = \sqrt{x}$. (At least, if y is positive. Technically, the function $y = x^2$ is not bijective, but we can restrict its domain to positive values of x so that it becomes bijective.)

The formula $\dfrac{du}{dx} = \dfrac{1}{\frac{dx}{du}}$ is listed as Formula IX on the list of derivative formulas.

7.3.5 Derivatives of Implicit Relations

Relations in which y is not given explicitly as a function of x have derivatives as well. To find them, we simply differentiate both sides of the defining equation, treating y as an unknown function of x, so that $\dfrac{d(y^3)}{dx}$, for example, becomes $3y^2 \cdot \dfrac{dy}{dx}$. Then we solve the resulting equation for $\dfrac{dy}{dx}$, which, in general, will come out in the form

$\frac{dy}{dx} = \frac{F(x,y)}{G(x,y)}$. When $G = 0$, we conclude that the derivative does not exist. As a good example, take the implicit relation $x^2 + y^2 = k$, where k is a constant. Differentiating, we find $2x + 2y\frac{dy}{dx} = 0$ (regardless of k), which leads to $\frac{dy}{dx} = -\frac{x}{y}$. The geometric meaning of this result is that the tangent to a circle at any point on it is perpendicular to the associated radius (Th. 5-14), as the slope from the line from (0, 0) to (x, y) is y/x; this is equal to tan φ in Sec. 5.7.2; perpendicular lines have inclination angles that differ by 90°; and $\tan(\varphi \pm 90°) = -1/\tan\varphi$. Note that this derivative does not exist when $y = 0$; this means that the tangent there is vertical.

The rule for finding dy/dx in implicit relations shows us another way to check the extension of Formula I into rational numbers. Instead of the function $y = x^{p/q}$, we may write $y^q = x^p$ and differentiate this equation: $qy^{q-1} \cdot \frac{dy}{dx} = px^{p-1}$ or $\frac{dy}{dx} = \frac{px^{p-1}}{qy^{q-1}}$. When we replace y with $x^{p/q}$ and simplify the exponent on x that results, we find $\frac{dy}{dx} = \frac{p}{q}x^{\left(\frac{p}{q}\right)-1}$.

7.3.6 Derivatives of Trigonometric Functions

We start with a function $y = \sin x$. To find its derivative, of course, we must evaluate $\lim_{\Delta x \to 0} \frac{\sin(x+\Delta x) - \sin x}{\Delta x}$ $= \lim_{\Delta x \to 0} \frac{\sin x \cos \Delta x + \cos x \sin \Delta x - \sin x}{\Delta x}$ (angle-sum formula, Sec. 5.6) $= \sin x \lim_{\Delta x \to 0} \frac{\cos \Delta x - 1}{\Delta x} + \cos x \lim_{\Delta x \to 0} \frac{\sin \Delta x}{\Delta x}$. To evaluate these last two limits, we must specify the way we measure angles in calculus. The use of radian measure leads to simpler values for both of the limits, so whenever trigonometric functions appear in calculus, the arguments are assumed to be radian measures of angles. When we write tan 1, we mean the tangent of an angle of one radian.

To actually determine $\lim_{\Delta x \to 0} \frac{\sin \Delta x}{\Delta x}$, we need the formula for the area of a piece of pie. The area for a whole circle, of course, is πr^2, r being the radius. It is apparent that the area of a pie slice is proportional to the square of the radius and to the size of the central angle. Since the whole circle corresponds to a central angle of 2π radians, the area of the pie slice with a central angle of θ radians is $\frac{1}{2}r^2\theta$. If we are dealing with a unit circle for which $r = 1$, then the area is just $\frac{1}{2}\theta$.

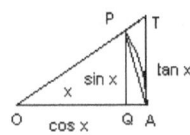

Fig. 7-8. Calculating the limit, as x approaches zero, of (sin x)/x, where x is measured in radians.

In Fig. 7-8, part of the "Trigonometry in Pictures" diagram in Sec. 5.6.4 is reproduced. We let x = the radian measure of the test angle AOP and read off $PQ = \sin x$, $OQ = \cos x$, and $AT = \tan x$. The area of the circular segment (pie slice) AOP is then $\frac{1}{2}x$, the area of the triangle AOP is $\frac{1}{2}\sin x$ (Th. 5-10B), and the area of triangle AOT is $\frac{1}{2}\tan x$. Because triangle AOP is included in pie slice AOP, which in turn is included in triangle AOT, we have $\frac{1}{2}\sin x < \frac{1}{2}x < \frac{1}{2}\tan x$. Dividing this inequality by $\frac{1}{2}\sin x$ results in $1 < \frac{x}{\sin x} < \sec x$, and taking reciprocals and reversing the direction of the inequality signs produces $\cos x < \frac{\sin x}{x} < 1$. It is evident that as x approaches zero, point Q approaches A, and therefore $\cos x = OQ$ approaches $OA = 1$. Therefore, by the sandwich process (Sec. 7.2.2), $\lim_{x \to 0} \frac{\sin x}{x} = 1$. From this limit, we calculate $\lim_{x \to 0} \frac{1-\cos x}{x}$ as follows: $\lim_{x \to 0} \frac{\sin x}{x} = 1$, so $\lim_{x \to 0} \frac{\sin^2 x}{x} = 1 \cdot 0 = 0$. Hence, $\lim_{x \to 0} \frac{1-\cos^2 x}{x} = 0$, and this is equal to $\lim_{x \to 0}(1 + \cos x) \cdot \lim_{x \to 0} \frac{1-\cos x}{x}$. Since the first of these limits is 2, the latter must be zero, and hence its negative, $\lim_{x \to 0} \frac{\cos x - 1}{x}$ must be zero as well.

These limits work just as well when Δx is used in place of x, and, placing them into the equation at the beginning of this section, we have $\frac{d(\sin x)}{dx} = \cos x$. Applying the chain rule to sin u, we have Formula X: $\frac{d(\sin u)}{du} = \cos u \cdot \frac{du}{dx}$. Since $\cos x = \sin(90° - x)$, or, since we are using radians, $\sin(\frac{\pi}{2} - x)$, $\frac{d(\cos x)}{dx} = \frac{d(\sin(\frac{\pi}{2}-x))}{dx} = \cos(\frac{\pi}{2} - x) \cdot \frac{d(\frac{\pi}{2}-x)}{dx} = \sin x \cdot (-1) = -\sin x$. Applying the chain rule again gives us Formula XI: $\frac{d(\cos u)}{dx} = -\sin u \cdot \frac{du}{dx}$. The derivatives for the other trigonometric functions can be found from these formulas and Formula VI, since all trig functions are either reciprocals or quotients of sine and cosine. For example, $\frac{d(\sec x)}{dx} = \frac{d(\frac{1}{\cos x})}{dx} = \frac{\cos x \frac{d(1)}{dx} - 1 \cdot \frac{d(\cos x)}{dx}}{\cos^2 x}$

$$= -\frac{-\sin x}{\cos^2 x} = \frac{\sin x}{\cos x} \cdot \frac{1}{\cos x} = \sec x \tan x.$$

7.4 Applications of Differentiation

We now have derivative formulas for all algebraic and trigonometric functions; in other words, for all the functions we have studied. The question is: what use are they? There are numerous applications of the derivative (otherwise calculus would not have been invented) and the next few sections will focus on some of them.

7.4.1 Max-Min Problems

One of the most frequent uses of the derivative is in finding the largest or smallest value of a function in a certain interval. For example, a widget maker can produce x widgets a week for the cost of $200 + 10x + 0.1x^2$ dollars. Because of market saturation he can only sell x widgets at a price of $30 - 0.2x$ dollars each. What is the optimum production level; that is, the one which maximizes his profit? To solve this problem through calculus, we use the theorem that follows.

Theorem 7-5. If a function has a extreme value (maximum or minimum) at an interior point of an interval, then the derivative of the function at that point is zero if it exists.

This theorem is proved by consideration of the left- and right-hand limits in the definition of the derivative; that is, $\lim_{\Delta x \to 0^-} \frac{f(x+\Delta x)-f(x)}{\Delta x}$ and $\lim_{\Delta x \to 0^+} \frac{f(x+\Delta x)-f(x)}{\Delta x}$. For the derivative to exist, these must be equal. Now, if f has a maximum or minimum, there is a number ϵ such that, whenever $|\Delta x| < \epsilon$, $f(x + \Delta x) - f(x)$ will have a certain sign (+ for a minimum) regardless of the sign of Δx. Therefore, one of the expressions above will approach some limit through negative numbers and the other will approach a limit through positive numbers. The only way that a negative number can get arbitrarily close to a positive number is for both of them to get arbitrarily close to zero, and this observation proves the theorem. The theorem mentions an interior point of an interval because if we are dealing with an endpoint, only one of the two directional limits will apply to the situation in question.

Th. 7-5 leads to the following strategy for solving max-min problems: Find the derivative, set it equal to zero, and solve the resulting equation. In the case of the widget manufacturer, we wish to maximize $N(x)$, his net profit. By basic accounting rules, $N(x) = R(x) - C(x)$, where $R(x)$ is his revenue from selling widgets and $C(x)$ is the cost of making them. Since revenue = unit sales times price, $R(x) = xP(x)$, where $P(x)$ is the price he gets for each widget. We are given $P(x) = 30 - 0.2x$, so $R(x) = 30x - 0.2x^2$. We are told that $C(x) = 200 + 10x + 0.1x^2$, so $N(x) = 20x - 0.3x^2 - 200$. To find an extreme value of $N(x)$, we take its derivative $N'(x) = 20 - 0.6x$. This is zero when $x = \frac{20}{0.6} = 33\frac{1}{3}$, so an extreme value of the profit is realized when 33 ⅓ widgets a week are produced. The net profit there, $N(33\frac{1}{3})$, is 666⅔ - 333⅓ - 200 = 133⅓ dollars a week.

But is this value a maximum or a minimum? Since $N(0) = -200$, it must be a maximum, but there is a way to determine which without evaluating the function elsewhere. This test involves the second derivative of $N(x)$. If N reaches a maximum, it must be rising before and falling after the maximum. Hence, N' goes from positive to negative, so it is declining, and therefore N'' is negative. For a minimum, N' rises from negative to positive as we cross the minimum, and therefore N'' is positive. In this example, $N'' = -0.6$, so the value of N when N' was zero is in fact a maximum.

Because of the conditions required in Th. 7-5, there are actually three types of places for a function to attain an extreme value in an interval:

(1) At an endpoint of the interval.
(2) At a point where the function's derivative does not exist.
(3) At a point where the function's derivative is zero; and, in this case, a positive second derivative means a minimum value for the function, and a negative second derivative means a maximum value.

Note that a function does not have to have a maximum or minimum value on an *open* interval (one that does not include the endpoints). For example, the function $f(x) = x^2$ has neither a maximum or a minimum on the open interval $1 < x < 2$ that does not include 1 and 2. For any positive ϵ, we can find a value that is less than $1 + \epsilon$ or greater than $4 - \epsilon$, but we cannot find 1 or 4 because the interval is open. For a closed interval such as $1 \le x \le 2$, there is a theorem stating that a maximum and minimum must exist, provided that the function is continuous everywhere in the interval.

7.4.2 Related Rates

Another common use of the derivative is in problems involving related rates. In these, there are two quantities, one of which is a function of time and the second a function of the first. You are asked to find the rate of change of the second quantity, given the rate of change of the first at a given instant. Symbolically, you are given $\frac{dP}{dt}$ and Q as a function of P, and the chain rule produces $\frac{dQ}{dt} = \frac{dQ}{dP} \cdot \frac{dP}{dt}$. These problems are non-symbolic for the most part (the word problems of calculus), and as a result the various derivatives involved have dimensional answers, e. g., dQ/dt may come out in square feet per second when dQ/dP is in square feet per inch and dP/dt is in inches per second. In this case, the numerical part of dQ/dt is simply the product of the numerical values of dQ/dP and dP/dt.

For a typical problem, take a circular patch of weed infestation. The diameter of the patch is increasing at the rate of one inch per day. How fast is its area growing at the moment when the diameter reaches five feet? We have $P = $ the diameter; we are given $dP/dt = 1$ inch per day. The second quantity, the area Q, is related to P by $Q = \frac{\pi P^2}{4}$ since the patch is a circle. Hence, $\frac{dQ}{dP} = \frac{\pi P}{2}$. If P is measured in inches, then Q is in square inches and dQ/dP is in square inches per inch. When the diameter is five feet, or 60 inches, $dQ/dt = (dQ/dP)(dP/dt) = \frac{\pi \cdot 60}{2} \times 1 = 30\pi$ square inches a day. To get $\frac{dQ}{dt}$ in square feet per day, we must divide by 144, so $\frac{dQ}{dt} = \frac{5\pi}{24}$, or about two-thirds of a square foot per day.

The major part of related-rate problems lies in finding Q as a function of P. Like word problems in algebra textbooks, where the difficulty is more in arriving at an equation than solving it once one is found, it is important to determine the precise relation between the two quantities involved. In the previous example, Q was the area of a circle with diameter P; in more advanced problems, the relation may not be so simple! Suppose we have a water tank in the shape of a right circular cone that is 200 feet tall and the circle on the bottom has a diameter of 100 feet. This tank is being filled with water at the rate of 15 cubic feet per second. How fast is the water rising when it is 20 feet deep? The formula for the volume of a right circular cone is $V = \frac{1}{3}\pi r^2 h$, where h is the height and r is the radius, not the diameter, of the circle at the base. For this particular tank, $r = h/4$ since $r = 50$ and $h = 200$ in the whole tank, and the empty portion will be similar to the whole tank. So we can say that the empty volume of the tank is $\frac{\pi h^3}{48}$ where $h = 200 - d$ is the depth that is *not* filled with water. The volume is P; we are given $dP/dt = $ -15 cubic feet per second (the minus sign appears because we are keeping track of the empty, not the filled, volume). The value of d, the depth, is our Q. We have $P = \frac{\pi(200-Q)^3}{48}$, and therefore $\frac{dP}{dQ} = -\frac{\pi(200-Q)^2}{16}$. We now use Formula IX to obtain $\frac{dQ}{dP} = -\frac{16}{\pi(200-Q)^2}$; when $Q = 20$, $\frac{dQ}{dP} = -\frac{16}{180^2 \cdot \pi}$. Multiplying this by $\frac{dP}{dt} = -15$ gives $\frac{dQ}{dt} = \frac{240}{180^2 \cdot \pi} = \frac{1}{135\pi}$ feet per second or $\frac{16}{3\pi}$ inches per minute, a barely noticeable rise. The numerical value of dQ/dt comes out in feet per second because dP/dt is given in cubic feet per second and Q is measured in feet.

7.4.3 Velocity and Acceleration

As mentioned in Sec. 7.3, the instantaneous velocity of an object that has covered $f(t)$ distance units in t time units is the derivative of $f(t)$. In general, the velocity will be a function of t as well, and its derivative is defined as the (instantaneous) acceleration of the object.

Take the example in Sec. 7.3, where $s = 4.903t^2$. Differentiating once produces the velocity $v = 9.806t$, and a second differentiation gives the acceleration $a = 9.806$. Since s is measured in meters and t in seconds, v is in meters per second and a is in meters per second per second, or meters per second squared. Since the formula for s applies to an object falling because of the earth's gravity, a is called the acceleration due to gravity (at the earth's surface), and is denoted by g. The actual value of g varies slightly as the observation is made at different points on the earth's surface because (1) the earth is not a perfect sphere, and (2) some of the force of gravity is canceled out by the effect of the earth's rotation.

From Newtonian mechanics comes the equation $F = ma$, where F is the net force on an object of mass m, and a is the resulting acceleration. Hence, once we know m and s as a function of t, we can find a and therefore F, the force. In the falling situation, we have $F = mg$ for an object of mass m kilograms, and F is measured in kilogram-meters per second squared, mercifully shortened to newtons (not Fig, but Sir Isaac!) We combine this with the general law of gravitation (also proposed by Newton) which states that the gravitational force between two bodies is given by $F = \frac{GM_1M_2}{r^2}$, where (capital) G is the universal constant of gravitation, M_1 and M_2 are the masses of the bodies, and r is the distance between their centers. If our falling body has mass M_1 and the earth has mass M_2, and the falling body is close to the earth's surface so that r is approximately the earth's radius of 6,378 kilometers, we have $F = mg = \frac{GM_1M_2}{r^2} = 9.806M_1 = \frac{GM_1M_2}{(6,378,000)^2}$ so that $GM_2 = 9.806 \times (6,378,000)^2$. Modern measurements show $G = 6.672 \times 10^{-11}$, and therefore the mass of the earth, M_2, is approximately 5.98×10^{24} kilograms. The extreme smallness of G made its measurement quite difficult, and there is still somewhat more uncertainty in its value than the number 6.672 indicates. This uncertainty produces an uncertainty in the mass of the earth – an uncertainty that can be determined by calculus.

7.4.4 Rolle's Theorem and the Mean Value Theorem for Derivatives

We now begin to build up toward the Fundamental Theorem of Calculus. The two theorems in the title of this section and Th. 7-4 in Sec. 7.2.6 constitute the first steps.

Theorem 7-6. (Rolle's Theorem)

If F is continuous on the closed interval $a \le x \le b$ and differentiable on the open interval $a < x < b$, and if $F(a) = F(b) = 0$, then there is a number c between a and b such that $F'(c) = 0$.

We consider two cases: (1) All the values of F between a and b are zero, and (2) F takes on some nonzero values between a and b. In case (1), F is a constant and hence its derivative everywhere between a and b is zero (Formula IV). As for (2), F has either a negative minimum value or a positive maximum value or both on the closed interval $a \le x \le b$, since it is continuous there. This minimum or maximum cannot be at either endpoint since F's values there are zero, so it must be at an interior point in the interval. By hypothesis, F is differentiable throughout the interior of the interval, so Th. 7-5 tells us that the value of F' at that extremum must be zero.

The geometric interpretation of this result is that a "smooth" curve connecting two points $(a, 0)$ and $(b, 0)$ on the x-axis, and not crossing any vertical line more than once, must have a horizontal tangent somewhere between a and b. The two curves in Fig. 7-9 both depict continuous functions satisfying $F(a) = F(b) = 0$, but the one on the left does not have the required horizontal tangent. It does not have to because it is not smooth; at the point c, where the curve has a "corner," the derivative does not exist. Hence, the function graphed does not meet the conditions of Rolle's theorem and hence need not have a horizontal tangent. The curve at the right, however, does meet the conditions of Rolle's theorem and hence must have a horizontal tangent.

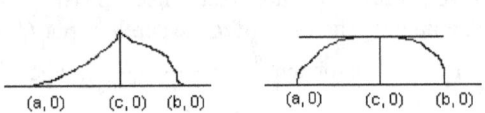

(a, 0)   (c, 0)   (b, 0)   (a, 0)   (c, 0)   (b, 0)

Fig. 7-9. The graphs of two continuous functions on the interval $a < x < b$. The function on the left need not have a horizontal tangent because it is not differentiable on the entire interval; at the point c, it has a corner and the derivative does not exist. The function on the right does have a horizontal tangent, as required by Rolle's theorem.

Looking at these diagrams should make it obvious that the fact that the x-axis connects the points $(a, F(a))$ and $(b, F(b))$ is immaterial. All that matters is that that line be horizontal – i. e., parallel to the tangent line. This observation is the basis for the Mean Value Theorem for Derivatives.

Th. 7-7. (The Mean Value Theorem for Derivatives)

If F is continuous on the closed interval $a \le x \le b$ and differentiable on the open interval $a < x < b$, then there is a number c between a and b such that $F'(c) = \frac{F(b)-F(a)}{b-a}$.

The first observation to be made is that Rolle's theorem is a special case of this theorem, where $F(a)$ and $F(b)$ both happen to be zero. The concept of generalization is critical to mathematical progress. The second is the geometric meaning of this theorem. A "smooth" curve, when any two points are joined by a chord, will have a point, on the portion cut off by the chord, where the tangent to the curve is parallel to the chord. Again, this meaning is a generalization of that of Rolle's theorem, where the chord was restricted to the x-axis.

We prove this theorem by introducing the auxiliary function G, defined as $G(x) = F(x) - F(a) - (x -$

a) $\frac{F(b)-F(a)}{b-a}$. Since a, b, $F(a)$, and $F(b)$ are constants, this function is equal to $F(x)$ minus a linear polynomial in x, and hence is continuous and differentiable wherever F is. Also, $G(a) = G(b) = 0$. For future reference, we calculate $G'(x) = F'(x) - \frac{F(b)-F(a)}{b-a}$. These observations mean that if F satisfies the conditions of Th. 7-7, then G satisfies those of Th. 7-6. Therefore, there is a number c such that $G'(c) = 0$ and therefore $F'(c) = \frac{F(b)-F(a)}{b-a}$, Q. E. D.

An immediate corollary of this theorem is that if a function's derivative is always zero, then the function is a constant. To prove this, take any two numbers a and b. Since differentiability implies continuity, a function with an identically vanishing derivative is continuous, and hence meets the requirements of Th. 7-7. Since the only possible value for $F'(c)$ in that theorem is zero, we conclude that $\frac{F(b)-F(a)}{b-a}$ is zero, and hence $F(b) - F(a)$ is zero for any two numbers a and b. This is synonymous with F being constant. In Sec. 7.3.1, the derivative of a constant was shown to be zero; this corollary shows that the converse is true as well.

An important second corollary states that if two functions have the same derivative, then they must differ by a constant. To prove this, we just apply the first corollary to the function defined as the difference of the other two functions.

7.5 Logarithmic and Exponential Functions

We are now ready to define a new kind of function; in fact, two new kinds. Some readers may remember the definition of a logarithm as being $\log x = y$ means $10^y = x$, and therefore $\log 1000 = 3$, $\log 100 = 2$, and $\log \sqrt{10} = \frac{1}{2}$. They may have wondered, when it was said that $\log 2 = 0.301$, how you could calculate $10^{0.301}$ to see whether or not it was 2? The definition of fractional exponents in Sec. 3.8 gives $10^{0.301} = \sqrt[1000]{10^{301}}$, which looks like a formidable task to calculate, and that does not finish the job, as $\log 2$ is not exactly 0.301 but an irrational number, $0.3010299957\ldots$, and irrational exponents are not defined as having any meaning in algebra. To be precise about logarithms, calculus must be used, which is why logarithms have been mentioned only in passing up until now.

There are two different methods for teaching logarithms in first-year calculus texts: one that pulls the function $\int_1^x \frac{dt}{t}$ out of a hat and then shows that it is a logarithm; and the one I am using, which assumes that a logarithm exists and then determines its properties, in particular its derivative. I prefer this second method both because it is the one my first-year calculus teacher used (although the textbook used the other one), and because it is less ad hoc and can be done earlier, before definite integrals are introduced.

The rigorous definition of a logarithmic function is that it is a function f satisfying, for all x and y in its domain, $f(xy) = f(x) + f(y)$. For reasons to be explained, we restrict the domain of logarithmic functions to the positive real numbers. An immediate consequence of this definition is that $f(1) = 0$, for $f(x) = f(x \cdot 1) = f(x) + f(1)$. From this, we have $f\left(\frac{1}{x}\right) = -f(x)$, as $\frac{1}{x} \cdot x = 1$. Zero cannot be in the domain of f, for if it were, we would have $f(0) = f(0 \cdot x) = f(0) + f(x)$, which can only hold if $f(x) = 0$ for all x, and a constant zero function would not be much use as a logarithm. If we let $y = x$ in the definition, we have $f(x^2) = 2f(x)$, and, by mathematical induction, $f(x^n) = nf(x)$ for all positive integers n.

7.5.1 Derivative of the Logarithmic Function

Now let f be a logarithmic function, and place it in the definition of a derivative: $\frac{df}{dx} = \lim_{\Delta x \to 0} \frac{f(x+\Delta x)-f(x)}{\Delta x}$. Since f is a logarithmic function, $f(x + \Delta x) - f(x) = f\left(\frac{x+\Delta x}{x}\right) = f(1 + \frac{\Delta x}{x})$. We now multiply the limit fraction by $\frac{1}{x}$ and divide its denominator by x to obtain $\frac{df}{dx} = \lim_{\Delta x \to 0} \frac{1}{x} \cdot \frac{f(1+\frac{\Delta x}{x})}{\frac{\Delta x}{x}}$. Since x is a constant during the limit operation, we can move $\frac{1}{x}$ outside the limit, and we replace $\frac{\Delta x}{x}$ with the new variable h. Since x is not zero, $\Delta x \to 0$ implies $h \to 0$. Our derivative equation is now $\frac{df}{dx} = \frac{1}{x}\lim_{h \to 0} \frac{f(1+h)}{h}$. We now subtract zero from the fraction's numerator in the form of $f(1)$, producing $\frac{df}{dx} = \frac{1}{x}\lim_{h \to 0} \frac{f(1+h)-f(1)}{h}$. The limit is now seen as equaling $f'(1)$, so $f'(x) = \frac{f'(1)}{x}$. We are free to choose any nonzero value for $f'(1)$, and one appears to be the most natural choice. The function resulting from the choice of $f'(1) = 1$ is called the *natural logarithm* of x, abbreviated in Romance-language style (adjective follows noun) as $\ln x$.

This definition of $\ln x$ shows that $\frac{d(\ln x)}{dx} = \frac{1}{x}$, and applying the chain rule to this produces Formula XII: $\frac{d(\ln u)}{dx} = \frac{du/dx}{u}$. If we let $u = \sin x$, then we have $\frac{d(\ln \sin x)}{dx} = \frac{\cos x}{\sin x} = \cot x$, and if we let $u = x^n$, then $\frac{d(\ln x^n)}{dx} = \frac{nx^{n-1}}{x^n} = \frac{n}{x}$. Note that this is n times the derivative of $\ln x$, so the functions $u = \ln x^n$ and $v = n \ln x$ must differ by a constant. When $x = 1$, both u and v are zero, so $\ln x^n = n \ln x$ for all positive x and all n for which x^n is defined.

7.5.2 The Values of $\ln x$

A logarithmic function has little practical use unless there is a way of calculating its values. We can at least estimate the values of $\ln x$ as follows: We know that $\ln 1 = 0$ because $\ln$ is a logarithm. When x is close to one, the derivative $1/x$ of $\ln x$ will be close to one, and therefore the difference quotient $\frac{\ln x - \ln 1}{x - 1}$ will be close to 1. Algebra shows that this implies $\ln x \approx x - 1$ when x is close to 1. To estimate $\ln 2$, we apply the Mean Value Theorem for Derivatives. The function $\ln x$ satisfies the conditions of Th. 7-7 on the interval $1 \leq x \leq 2$, so there must be a number c between 1 and 2 such that $f'(c) = \frac{f(2) - f(1)}{2 - 1} = f(2)$. Since f is $\ln$, $f'(c) = \frac{1}{c}$, and therefore $\ln 2$ must be between ½ and 1. We can refine this estimate by first considering $\ln 1\frac{1}{2}$. Th. 7-7 tells us that there is a c between 1 and 1½ such that $1/c = 2 \ln 1\frac{1}{2}$, and therefore $\frac{1}{3} < \ln 1\frac{1}{2} < \frac{1}{2}$. Then, applying Th. 7-7 to the interval $1\frac{1}{2} \leq x \leq 2$, we have a c in this interval such that $\frac{1}{c} = 2(\ln 2 - \ln 1\frac{1}{2})$. Combining this with the previous range, we have $\frac{7}{12} < \ln 2 < \frac{5}{6}$. Since $\ln 3 = \ln 2 + \ln 1\frac{1}{2}$, we also have $\frac{11}{12} < \ln 3 < 1\frac{1}{3}$.

These ranges are not very tight, but if we subdivide the interval between 1 and 2 into more parts, a more accurate value for $\ln 2$ can be computed. With ten parts, a calculation produces $0.66877 < \ln 2 < 0.71877$; taking the midpoint of this range gives $\ln 2 \approx 0.69377$, in good agreement with the actual five-place value of 0.69315. If we imagine the limit as more and more parts are used, we eventually get a value as close as we wish to $\ln 2$. The mathematical symbols are: $\ln 2 = \lim_{n \to \infty} (\frac{1}{n+1} + \frac{1}{n+2} + \cdots + \frac{1}{2n-1} + \frac{1}{2n})$.

Once we have $\ln 2$, we can determine $\ln 3/2$, $\ln 5/4$, $\ln 7/6$, etc. in similar fashion and then calculate $\ln 3 = \ln 2 + \ln 3/2$, $\ln 4 = 2 \ln 2$, $\ln 5 = \ln 4 + \ln 5/4$, and so on to compile a table of natural logarithms. For example, $\ln \frac{7}{6} = \lim_{n \to \infty} (\frac{1}{6n+1} + \frac{1}{6n+2} + \cdots + \frac{1}{7n-1} + \frac{1}{7n})$; a good value can be obtained by calculating the parenthesized expression for a moderate value of n and then adding $\frac{1}{84n}$ to the total. (The 84 comes from $(6 \times 7) \times 2$, and corresponds to the "splitting the difference" procedure used above for $\ln 2$.) There are somewhat easier methods of calculating logarithms that will be introduced in Sec. 7.10 on infinite series.

Since $\ln 2 > 0.5$, $\ln 2^n > 0.5n$, and there exist numbers with arbitrarily large natural logarithms. The reciprocals of these numbers will have arbitrarily large negative logarithms, so all real numbers are natural logarithms of some other number. In other words, the range of $\ln x$ is the entire set of real numbers. The domain, however, is restricted to positive numbers. To see why, let us apply the logarithm's definition to negative numbers. What is $\ln (-1)$? Because $(-1)^2 = 1$, $2 \ln (-1)$ must equal zero, so $\ln (-1) = 0$. Now consider $\ln (-a)$, where a is positive. Since $(-1)(-a) = a$, we have $\ln (-a) = 0 + \ln a$, or $\ln (-a) = \ln a$. But we wish to make $\ln x$ a one-to-one function so it will have an inverse that can be studied. Mathematics is still somewhat biased against negative numbers, so the negative half of the number line was excluded from the domain of $\ln x$.

7.5.3 Applications of the Logarithm

(Those who are already familiar with logarithms from recollections of high-school math can skip this section.)

The chief purpose of logarithms (at least in everyday math) is to expedite computations such as 1.573×0.927 or $(205)^{1.78}$. To accomplish this, the properties of logarithms are used: for 1.573×0.927, we write $\log P = \log 1.573 + \log 0.927$, P being the desired product. Tables are then used to determine the two logarithms, they are added, and the table referred to again to see what number has this sum for a logarithm. That number is P. For $(205)^{1.78}$, we write $\log Q = 1.78 \times \log 205$; this multiplication may itself be carried out by the use of logarithms. Of course, any logarithmic table will only have a finite accuracy, so there is a limit to the precision with which a calculation using logarithms can be done. But calculators and computers use logarithms to do their computations, so not much can be gained by resorting to them except speed.

The sound unit, the decibel, is a logarithm. Hence, multiplying the intensity of a sound just adds to its score in decibels. The decibel is so defined that multiplying a sound's intensity by a factor of 10 adds 10 decibels to its score, regardless of how loud the sound is. (The "bel" part of decibel, by the way, honors Alexander Graham Bell.)

Magnitudes of earthquakes are measured on a logarithmic scale as well. An increase of one point in the magnitude of an earthquake increases the intensity of its shaking by a factor of 10, no matter how strong or weak the quake is. Hence, a magnitude 8.8 quake such as the one that devastated Chile on February 27, 2010 is 65 times stronger than a 7.0 one such as the one that ruined Haiti just a few weeks earlier on January 12.

7.5.4 The Exponential Function and the Number e

We now consider the inverse function of the natural logarithm. (Recall that above we restricted the domain of ln to the positive numbers so it would have a unique inverse.) This function, sometimes referred to as the *antilogarithmic* function, is usually called *the exponential* function, emphasis on "the" because it is one of many exponential functions, but the only one directly associated with the natural logarithm. It is denoted sometimes by "exp x," but usually, for reasons that will be seen, e^x.

The functions exp and ln are inverses, so exp $(\ln x) = x$ and ln (exp x) $= x$. Also, the domain and range of exp are the range and domain respectively of ln. Because of this, the properties of exp are the reverse of those of ln: exp $(a + b) = $ exp a x exp b and exp $(na) = (\exp a)^n$. It should be noted that, since the range of ln is all real numbers, the domain of exp is all real numbers.

The number whose natural logarithm is 1 is the second most important irrational number in mathematics, after pi. It is denoted by e, and its value is approximately 2.71828. Since e is defined so that ln $e = 1$, we have exp $1 = e$. Hence, exp $a = $ exp $(1a) = (\exp 1)^a = e^a$. Now, exp a is defined for all real a, and this equation is taken as defining e^a for all real a, including irrational numbers. We rewrite the properties of exp in terms of e to get $e^{a+b} = e^a \cdot e^b$ and $e^{na} = (e^a)^n$, which match the laws of exponents in Sec. 3.1.3, showing that the definition of e^a as exp a is in fact consistent with those laws. In particular, $e^0 = 1$ since ln $1 = 0$ and $e^{-x} = 1/e^x$ since ln $1/x = -$ ln x.

The derivative of the exponential function can now be calculated using Formula VIII since it is the inverse of a function with a known derivative. We write $\dfrac{d(\exp x)}{dx} = \dfrac{1}{\frac{d(\ln y)}{dy}\big|_{y=\exp x}} = \dfrac{1}{\frac{1}{y}} = y = \exp x$. This function is its own derivative – it is immune to differentiation! The operation of differentiation has no more effect on it than rubbing with steel wool has effect on a diamond. Combining this remarkable fact with the chain rule produces Formula XIII: $\dfrac{d(e^u)}{dx} = e^u \cdot \dfrac{du}{dx}$. Therefore, $\dfrac{d(e^{ax})}{dx} = ae^{ax}$, $\dfrac{d(e^{\ln x})}{dx} = \dfrac{1}{x} \cdot e^{\ln x} = 1$ since $e^{\ln x} = x$, and $\dfrac{d(e^{\sin x})}{dx} = \cos x \, e^{\sin x}$.

7.5.5 General Exponential Functions

Because of the relationship between ln and exp (or e^x), we have $e^{\ln a} = a$ for all positive a. We now raise both sides of this equation to the power x to get $a^x = (e^{\ln a})^x$. The second law of exponents says $(e^{\ln a})^x = e^{x \ln a}$. We extend the concept of exponent to irrational numbers by extending that law to the case where x is irrational and therefore *defining* a^x to be equal to $e^{x \ln a}$. With this definition, we can at last extend Formula I to irrational values of the exponent n by writing $x^n = e^{n \ln x}$ and therefore $\dfrac{d(x^n)}{dx} = \dfrac{d(e^{n \ln x})}{dx} = \dfrac{n}{x} e^{n \ln x} = \dfrac{nx^n}{x} = nx^{n-1}$. When the independent variable x is the exponent rather than the base, we write $\dfrac{d(a^x)}{dx} = \dfrac{d(e^{x \ln a})}{dx} = \ln a \cdot e^{x \ln a} = a^x \ln a$. The laws of exponents for irrational powers can all be derived from the properties of exp. For example, $a^m a^n = a^{m+n}$ because $e^{m \ln a} \cdot e^{n \ln a} = e^{(m+n) \ln a}$.

7.5.6 General Logarithmic Functions

We now define the general logarithmic function, $\log_a x$, by saying that $\log_a x = y$ means simply that $a^y = x$. Applying the definition above, we have $e^{y \ln a} = x$, or, taking ln of both sides, ln $x = y$ ln a or $y = \dfrac{\ln x}{\ln a}$. Therefore, the definition of $\log_a x$ can be replaced by the definition $\log_a x = \dfrac{\ln x}{\ln a}$. The number a is called the *base* of the logarithm $\log_a x$. If we let e be the base, we have $\log_e x = \dfrac{\ln x}{\ln e} = \ln x$ since ln e equals one. Therefore, we say that e is the base of the system of natural logarithms. As to common logarithms, the base for them is 10, so we have $\log_{10} x = \frac{\ln x}{\ln 10} \approx \frac{\ln x}{2.30259} \approx 0.43429 \ln x$. Similarly, we can convert a common logarithm back into a natural one by multiplying it by $\ln 10 \approx 2.30259$.

The derivative of $\log_a x$ is readily computed from its definition as $\frac{\ln x}{\ln a}$ to be $\frac{1}{x \ln a}$. We can even consider the base to be the independent variable: $\frac{d(\log_x a)}{dx} = \frac{d(\frac{\ln a}{\ln x})}{dx} = -\frac{\ln a}{x(\ln x)^2}$. (In working this out for yourself, remember that $\ln a$ is a constant.)

7.5.7 Logarithmic Differentiation

With irrational exponents defined, it is now possible to consider functions such as x^x or $(\sin x)^{\cos x}$, or, more generally, u^v, where both u and v are functions of x, and find derivatives for them. There are two approaches to the differentiation of these functions, but both lead to the same calculations, and, of course, the same result. The first relies on the definition of u^v as $e^{v \ln u}$ and then calculates the derivative of that function as $\frac{d(v \ln u)}{dx} \cdot e^{v \ln u}$ according to Formula XIII. The second, called *logarithmic differentiation*, begins by taking natural logarithms in the equation $y = u^v$, getting $\ln y = v \ln u$, and then differentiating that equation implicitly (Sec. 7.3.5) to arrive at $\frac{1}{y} \cdot \frac{dy}{dx} = \frac{u}{v} \cdot \frac{du}{dx} + \ln u \cdot \frac{dv}{dx}$. Multiplying this by y and then replacing y with u^v produces Formula XIV: $\frac{d(u^v)}{dx} = vu^{v-1}\frac{du}{dx} + u^v \ln u \frac{dv}{dx}$. Applying this formula to x^x, where $u = x$ and $v = x$, produces $\frac{d(x^x)}{dx} = x^x(\ln x + 1)$.

The advantage of logarithmic differentiation is that it can be applied to other types of functions as well. Take, for instance, the example in Sec. 7.3.2: $F(x) = \frac{(x^2-1)(x^3+x^2)}{x^2+x+1}$. We take natural logarithms here, substituting y for $F(x)$, to get $\ln y = \ln(x^2 - 1) + \ln(x^3 + x^2) - \ln(x^2 + x + 1)$. Differentiating produces $\frac{dF}{dx} = F \cdot (\frac{2x}{x^2-1} + \frac{3x^2+2x}{x^3+x^2} - \frac{2x+1}{x^2+x+1})$. The result is just as complicated as in Sec. 7.3.2, but is obtained much more rapidly.

This technique can also be used to extend Formula V for products to any finite number of functions. Let $y = u_1 u_2 u_3 \ldots u_n$, where all the u's are functions of x. Then $\ln y = \ln u_1 + \ln u_2 + \cdots + \ln u_n$ and $\frac{dy}{dx} = y \left(\frac{\frac{du_1}{dx}}{u_1} + \frac{\frac{du_2}{dx}}{u_2} + \cdots + \frac{\frac{du_n}{dx}}{u_n} \right)$. When y is replaced with the product of the u's and all fractions are cleared, the result is $\frac{d(u_1 u_2 \ldots u_n)}{dx} = \frac{du_1}{dx} \cdot u_2 u_3 \ldots u_n + u_1 \cdot \frac{du_2}{dx} \cdot u_3 \ldots u_n + u_1 u_2 \cdot \frac{du_3}{dx} \cdot \ldots u_n + \cdots + u_1 u_2 \ldots u_{n-1} \cdot \frac{du_n}{dx}$. In other words, to differentiate a product, find the derivative of each factor in turn and multiply it by the other factors, undifferentiated. Then add all the resulting products and the total is the desired derivative.

7.6 Indefinite Integration

Throughout mathematics, reversing a process is as important (or more so) than carrying out the process itself. Differentiation is no exception to this "rule." The process of answering the question "What function has this as a derivative?" is called *antidifferentiation*, or *indefinite integration*. The word "indefinite" is used because there is also a process called "definite integration" that at first seems totally unrelated to indefinite integration, but will eventually be revealed to be quite intimately connected to it.

We begin to explore integration by asking the question "What function has $2x$ as its derivative?" Before we answer, the notation for integrals should be pinned down. The function with $2x$ as derivative is denoted $\int 2x \, dx$, where $\int$ is called the *integral sign*, and the dx serves to remind us that x is the variable which the derivative is taken with respect to. The $2x$ is called the *integrand*; that is, the function being integrated.

From Formula I for derivatives, we have $\frac{d(x^2)}{dx} = 2x$, so $\int 2x \, dx = x^2$. But x^2 is not the only function whose derivative is $2x$; there are also $x^2 + 1, x^2 - 5, x^2 + \frac{\sqrt{7}}{3}, x^2 - \ln 4$, etc. In fact, $x^2 + C$, for any value of the constant C, has $2x$ as its derivative. The second corollary to Th. 7-7 shows that all functions whose derivative is $2x$ are given by $x^2 + C$. We therefore write $\int 2x \, dx = x^2 + C$, and, in general, add a "+ C" to all answers to indefinite integrals.

7.6.1 Basic Formulas

Reversing various derivative formulas gives a short list of basic integral formulas. Reversing Formula III, for example, produces the important $\int cf(x)dx = c\int f(x)dx$, where c is any constant. We reverse Formula I to get $\int nx^{n-1}\,dx = x^n + C$; replacing n with $n+1$ and then dividing by $(n+1)$ via the above rule results in $\int x^n\,dx = \frac{x^{n+1}}{n+1} + C$. If $n+1 = 0$, of course, this rule does not apply. To integrate $1/x$, we reverse Formula XII to get $\int \frac{1}{x}\,dx = \ln x + C$. (Note that this integral is usually written $\int \frac{dx}{x}$; it is standard practice to write the dx in the numerator of a fraction being integrated if that numerator would otherwise be 1.) From Formulas X and XI, we get the integrals $\int \sin x\,dx = -\cos x + C$ and $\int \cos x\,dx = \sin x + C$. From Formula XIII, $\int e^x\,dx = e^x + C$. The chain rule, Formula VII, can be reversed to produce the integration formula $\int F(G(x)) \cdot \frac{dG(x)}{dx}\,dx = \int F(u)\,du$, where, after performing the second integration, we replace u with $G(x)$. This rule is especially valuable if $G(x) = cx$ for some constant c, since then $\frac{dG(x)}{dx} = c$, which can be "factored out" of an integral by using the first integral formula here. Examples of what might be called "the chain rule for integrals" include $\int \sin ax\,dx = -\frac{1}{a}\cos ax + C$, where $G(x) = ax$; $\int e^{ax}\,dx = \frac{1}{a}e^{ax} + C$, where $G(x)$ is also ax; $\int \sin^2 x \cos x\,dx = \frac{1}{3}\sin^3 x + C$, where $G(x) = \sin x$; and $\int \cot x\,dx = \int \frac{\cos x}{\sin x}\,dx = \ln \sin x + C$, where $G(x)$ is $\sin x$. To evaluate $\int (3x+2)^3\,dx$, we first multiply the inside of the integral by 3 ($= \frac{d(3x+2)}{dx}$) and multiply the outside by $1/3$ to cancel the 3, obtaining $\frac{1}{3}\int (3x+2)^3 \cdot 3\,dx$. This integral can now be evaluated by the chain rule, with $F(u) = u^3$ and $G(x) = 3x+2$, to get a final result of $\frac{1}{12}(3x+2)^4 + C$.

7.6.2 The Guessing Game

Some integrals that cannot be evaluated easily lend themselves to a sort of "guessing game" approach in which one first guesses an answer, then differentiates the guess to see how close he is to the real answer. The guess is then modified according to the result of the differentiation and the process repeated until the correct answer is found. A good example of this is the integral $\int xe^x\,dx$. Based on Formula XIII, $xe^x + C$ is guessed, but its derivative is $xe^x + e^x$, not just xe^x. To correct this, we reduce our guess by an amount whose derivative is e^x – which is e^x itself. The new guess is $xe^x - e^x + C$, which turns out to be correct, the extra $-e^x$ canceling out the unwanted e^x in the derivative of the first guess. A similar example is the integral $\int x \sin x\,dx$ in which the first guess of $-x\cos x$ produces a derivative of $x \sin x - \cos x$. To remove the $-\cos x$, we add to the guess a function whose derivative is $\cos x$; that is, $\int \cos x\,dx = \sin x + C$. This yields the correct answer, $\int x \sin x\,dx = \sin x - x \cos x + C$.

7.6.3 Substitution in Integrals

Most integrals do not yield easily to the guessing game. The integral $\int \frac{dx}{x^2 + a^2}$ is a good example; even the nature of a guess is up in the air, and any reasonable guess is apt to be wide of the mark, with the "residual" being even more bothersome than the original integrand. Many of these integrals can be evaluated by a change of variable: a substitution of the form $x = f(u)$ or $u = f(x)$, or perhaps several such substitutions.

The rules for substitutions in integrals are as follows:

(1) When substituting $x = f(u)$, remember to replace the dx with $\frac{df(u)}{du}\,du$, as well as replacing all other x's in the integrand with $f(u)$.

(2) When substituting $u = f(x)$, you must have $\frac{df(x)}{dx}$ present in the integrand to become part of du.

(3) The function used must have a unique inverse; if it does not, its domain must be restricted so as to give the function an inverse, but no further than necessary to do so.

(4) After evaluating the transformed integral, you must undo the substitution(s) to place the result in terms of the original variable.

In the case of $\int \frac{dx}{x^2 + a^2}$, the difficulty stems from having two terms in the denominator of the integrand. One way to turn the sum of two squared terms into a single squared term is to use the Pythagorean trigonometric identities in Sec. 5.6. Since $1 + \tan^2\theta = \sec^2\theta$, $a^2 + a^2\tan^2\theta = a^2\sec^2\theta$, which means that substituting $x = a\tan\theta$ replaces $x^2 + a^2$ with $a^2\sec^2\theta$. Also, $\frac{d(a\tan\theta)}{d\theta} = a\sec^2\theta$, so that the total effect of $x = a\tan\theta$ on $\int\frac{dx}{x^2+a^2}$ is to transform it into $\int\frac{a\sec^2\theta}{a^2\sec^2\theta}d\theta = \frac{1}{a}\int d\theta$. There is a slight problem, however. The function $a\tan\theta$ does not have an inverse because it is not one-to-one ($\tan\frac{\pi}{4} = \tan-\frac{3\pi}{4} = \tan\frac{5\pi}{4}$, etc.), so its domain must be restricted. Fortunately, there is an easy way to do this restriction: the domain is reduced to $-\frac{\pi}{2} < \theta < \frac{\pi}{2}$, on which $\tan\theta$ (and therefore $a\tan\theta$) takes on all real values exactly once. The resulting inverse function, denoted by $\tan^{-1}x$ in keeping with Sec. 7.1.3, fulfills the inverse requirement in (3) above.

With the inverse requirement taken care of, we return to the transformed integral $\frac{1}{a}\int d\theta$. The "missing" integrand can be treated as the constant one, so $\int d\theta = \theta + C$ and $\frac{1}{a}\int d\theta = \frac{1}{a}\theta + C$. The inverse substitution is determined by "solving" $x = a\tan\theta$ for θ: $\frac{x}{a} = \tan\theta$, so $\theta = \tan^{-1}\frac{x}{a}$; that is, $-\frac{\pi}{2} < \theta < \frac{\pi}{2}$ and θ is the number (radian measure of an angle) whose tangent is $\frac{x}{a}$. Replacing θ with $\tan^{-1}\frac{x}{a}$ gets us back to x and therefore gives the final answer to $\int\frac{dx}{x^2+a^2}$ as $\frac{1}{a}\tan^{-1}\frac{x}{a} + C$.

There are two other types of trigonometric substitutions to match the two other Pythagorean identities in Sec. 5.6: the substitution $x = a\sin\theta$ replaces $a^2 - x^2$ with $a^2\cos^2\theta$, and $x = a\sec\theta$ replaces $x^2 - a^2$ with $a^2\tan^2\theta$. These are particularly effective when the square root of $x^2 + a^2$, $a^2 - x^2$, or $x^2 - a^2$ appears in the integrand. For example, in $\int\sqrt{a^2 - x^2}\,dx$, we substitute $x = a\sin\theta$ and $dx = a\cos\theta\,d\theta$ to remove the square root sign and transform the integral into $a^2\int\cos^2\theta\,d\theta$. This integral is in turn evaluated by squaring the half-angle formula (also in Sec. 5.6) $\cos\frac{A}{2} = \sqrt{\frac{1+\cos A}{2}}$, replacing A with 2θ, and placing the resulting formula into the integral: $a^2\int\cos^2\theta\,d\theta$ becomes $\frac{a^2}{2}\int(1 + \cos 2\theta)d\theta$ which equals $\frac{a^2}{4}(2\theta + \sin 2\theta) + C$ or $\frac{a^2}{2}(\theta + \sin\theta\cos\theta) + C$. To perform the inverse substitution, we have $\theta = \sin^{-1}\frac{x}{a}$, $\sin\theta = \frac{x}{a}$, and $\cos\theta = \sqrt{1 - \sin^2\theta} = \frac{\sqrt{a^2-x^2}}{a}$. The final result is $\int\sqrt{a^2 - x^2}\,dx = \frac{a^2}{2}\sin^{-1}\frac{x}{a} + \frac{x}{2}\sqrt{a^2 - x^2} + C$. The complicated result is quite normal; even simple functions can have complicated integrals, especially if they involve square roots.

7.6.4 Algebraic Transformation of Integrals

Some integrals can be evaluated by manipulating the integrand algebraically, without substituting a new variable. One good example of this is $\int\frac{x^3}{(x^2+4)^3}\,dx$. Here we first split the x^3 into x times x^2, and then rewrite the x^2 as $(x^2 + 4) - 4$. The $x^2 + 4$ can now be used to cancel out one of the factors of $x^2 + 4$ in the denominator, transforming the integral into the sum of $\int\frac{x}{(x^2+4)^2}\,dx$ and $\int\frac{-4x}{(x^2+4)^3}\,dx$. These integrals can be multiplied and divided by constants to get $2x$, the derivative of $x^2 + 4$, into their numerators. This step results in $\int\frac{x^3}{(x^2+4)^3}\,dx = \frac{1}{2}\int\frac{2x}{(x^2+4)^2}\,dx - 2\int\frac{2x}{(x^2+4)^3}\,dx$. These last two integrals can be evaluated by the chain rule, with $u = x^2 + 4$, to give an answer of $-\frac{1}{2x^2+8} + \frac{1}{(x^2+4)^2} + C$.

A trigonometric integral such as $\int\sin^m x\cos^n x\,dx$, where at least one of m or n is an odd integer, can be evaluated in similar fashion. In $\int\sin^n x\cos^3 x\,dx$, we replace $\cos^2 x$ with its equivalent $1 - \sin^2 x$ to transform the integral into $\int\sin^n x\cos x\,dx - \int\sin^{(n+2)}x\cos x\,dx$. Since $\cos x$ is the derivative of $\sin x$, these two integrals can be immediately evaluated to produce $\frac{\sin^{(n+1)}x}{n+1} - \frac{\sin^{(n+3)}x}{n+3} + C$. If both m and n are even, we first transform $\int\sin^m x\cos^n x\,dx$ into an integral that involves only one of the functions. We then use the half-angle formulas in Sec. 5.6 in the form $\sin^2 x = \frac{1}{2}(1 - \cos 2x)$ or $\cos^2 x = \frac{1}{2}(1 + \cos 2x)$. For a large example, we take $\int\sin^4 x\cos^4 x\,dx$. We remove the cosine by writing $\cos^4 x = (1 - \sin^2 x)^2 = 1 - 2\sin^2 x + \sin^4 x$, transforming the integral into $\int(\sin^4 x - 2\sin^6 x + \sin^8 x)dx$. Replacing $\sin^2 x$ with $\frac{1}{2}(1 - \cos 2x)$ results in $\frac{1}{4}\int(1 - 2\cos 2x + \cos^2 2x)\,dx - \frac{1}{4}\int(1 - 3\cos 2x + 3\cos^2 2x - \cos^3 2x)dx + \frac{1}{16}\int(1 - 4\cos 2x + 6\cos^2 2x - 4\cos^3 2x + \cos^4 2x)dx$. We have reduced the largest exponent from 8 to 4. Collecting the powers of cos $2x$ reduces this expression to $\int \left(\frac{1}{16} - \frac{1}{8}\cos^2 2x + \frac{1}{16}\cos^4 2x\right) dx$ as the first and third power terms conveniently drop out. The other terms are dealt with as before: we replace $\cos^2 2x$ with $\frac{1}{2}(1 + \cos 4x)$, transforming the integral further into $\int \left(\frac{1}{64} - \frac{1}{32}\cos 4x + \frac{1}{64}\cos^2 4x\right) dx$, and replacing $\cos^2 4x$ with $\frac{1}{2}(1 + \cos 8x)$ removes all exponents larger than one, giving $\int(\frac{3}{128} - \frac{1}{32}\cos 4x + \frac{1}{128}\cos 8x) dx$. This can now be evaluated to produce the final answer of $\frac{3x}{128} - \frac{\sin 4x}{128} + \frac{\sin 8x}{1,024} + C$.

7.6.5 Applications of Indefinite Integration

The primary purpose of indefinite integration is as the first step toward obtaining a numerical answer for a definite integral (see Sec. 7.7.2). Leaving this aside, indefinite integration is used to solve differential equations and related problems called initial value problems.

A *differential equation* is an equation containing an algebraic combination of two variables and one or more derivatives of the unknown function connecting the variables. A typical example is $x^2 \frac{dy}{dx} + 3xy = 0$. This is called a *first-order* equation because only the first derivative of y with respect to x appears.

The solution of general differential equations is beyond a mere "introduction" to calculus, but certain easy ones fall within this chapter's scope. If the equation has the form $\frac{dy}{dx} = F(x)$ or $\frac{dy}{dx} = \frac{F(x)}{G(y)}$, it can be solved by integration. We multiply the equation by dx and $G(y)$ to change it into $G(y)\ dy = F(x)\ dx$, and then apply indefinite integration to both sides to get $P(y) + C_1 = Q(x) + C_2$, or $P(y) = Q(x) + C$, since the difference of two constants is another constant.

The example given just above, $x^2 \frac{dy}{dx} + 3xy = 0$, is not in the form $\frac{dy}{dx} = \frac{F(x)}{G(y)}$, but it can be transformed into such an equation as follows: subtract $3xy$ and then divide by x^2 to get $\frac{dy}{dx} = -\frac{3y}{x}$, and then use the fact that $\frac{1}{a} \div \frac{1}{b} = \frac{b}{a}$ to change the equation to $\frac{dy}{dx} = \frac{-1/x}{1/3y}$. We change to the form $G(y)\ dy = F(x)\ dx$ to get $\frac{dy}{3y} = -\frac{dx}{x}$, and then integrate to obtain $\frac{1}{3}\ln y = -\ln x + C = \ln\frac{C''}{x}$ where $\ln C' = C$. To get this into the form $y = f(x)$ (and most differential equation solutions can't be placed into $y = f(x)$ form), we multiply by 3 and take exponentials. The result is $y = C'x^{-3}$.

If, in addition to the differential equation, we are also given an initial condition such as $y = 10$ when $x = 1$, the problem becomes an *initial value problem*. The calculus is not affected; the solution $y = C'x^{-3}$ still holds. However, the condition $y = 10$ when $x = 1$ allows us to determine the value of C': since $10 = C'(1)^{-3}$, the constant must be 10 and the final solution to the initial value problem is $y = 10x^{-3}$.

Another application of indefinite integration concerns velocity and distance. As velocity is the derivative of distance, so distance is the antiderivative – that is, the indefinite integral – of velocity. If a particle falls under the influence of gravity so that its velocity after t seconds is $v(t) = 9.806t$ meters per second, then the distance it has covered in t seconds is $\int v(t)dt = 4.903t^2 + C$ meters. We determine C by assuming that the particle has covered no distance in zero seconds; therefore $0 = 4.903(0^2) + C$ and $C = 0$. If we are given that the particle starts at a height of 1 kilometer, then its height decreases as it falls, meaning $v(t)$ is negative and therefore $h(t) = \int v(t)dt = C - 4.903t2$. Since $h(0) = 1000$, $C = 1000$ and the particle's height after t seconds is $1000 - 4.903t^2$. This means that the particle will reach sea level ($h = 0$) in a little more than 14 seconds. Gravity is a strong force!

Velocity is an indefinite integral of acceleration in the same way as distance is an indefinite integral of velocity. If the acceleration is given by $a(t)$, then the velocity $v(t) = \int a(t)dt$ and the distance $s(t) = \int v(t)dt = \int(\int a(t)dt)dt$, which is usually written $\iint a(t)dt^2$. If $s(t)$ is any function satisfying $\frac{d^2s}{dt^2} = a(t)$, then the most general such function is given by $s(t) + C_1 t + C_2$, where C_1 and C_2 are arbitrary constants. In an initial value problem with a second-order equation such as $\frac{d^2s}{dt^2} = a(t)$, we are given two conditions that together allow the determination of C_1 and C_2.

7.7 Definite Integration

It is possible to view the relation between velocity and distance in another way. If a particle is traveling with a velocity $v(t)$, then between t_0 and t_1 it covers a distance approximately equal to $v(t_1)\Delta_1 t$, where $\Delta_1 t = t_1 - t_0$. Between times t_1 and t_2 it travels a further distance about $v(t_2)\Delta_2 t$, where $\Delta_2 t = t_2 - t_1$. The particle continues in this way until it reaches its goal at time t_n, having traveled a distance approximately equal to $v(t_1)\Delta_1 t + v(t_2)\Delta_2 t + v(t_3)\Delta_3 t + \cdots + v(t_n)\Delta_n t$, which is compressed with the use of *sigma* notation into $\sum_{i=1}^{n} v(t_i)\Delta_i t$. The Σ is a capital sigma (yet another Greek letter), which stands for "sum." It and the associated "$i = 1$" and "n" mean that, in what follows, the letter i is to be replaced by each integer from 1 to n in turn, and the resulting terms are to be added up. Thus, $\sum_{i=1}^{4} 2^i = 2^1 + 2^2 + 2^3 + 2^4 = 30$, and $\sum_{i=1}^{4} i^2 = 1^2 + 2^2 + 3^2 + 4^2 = 30$. (The coincidence is just that.)

The sum $\sum_{i=1}^{n} v(t_i)\Delta_i t$ is only an approximation to the actual distance covered by the particle between t_0 and t_n. This approximation improves as the intervals $\Delta_i t$ are made smaller and the number of intervals increases. In the limit, all the intervals are infinitely small (the width of the largest one approaches zero), the number of intervals increases indefinitely, and the approximation becomes the exact distance. Symbolizing the largest $\Delta_i t$ by $[\Delta]$, we have: Exact distance $= \lim_{[\Delta] \to 0} \sum_{i=1}^{n} v(t_i)\Delta_i t$. Provided that the function $v(t)$ is continuous for $t_0 \leq t \leq t_n$, this limit will exist, and it is defined to be the *definite integral* of $v(t)$ from t_0 to t_n, written $\int_{t_0}^{t_n} v(t)dt$. The symbols for definite integration closely resemble those for indefinite integration; the only difference is the addition of t_0 and t_n, called the *limits* of the definite integral. Do not confuse this use of the word "limit" with its ordinary mathematical meaning! The reason for the similarity is the close relation between the two types of integration, as stated in the Fundamental Theorem of Calculus to be proved below.

By the way, definite integrals are by no means limited to velocity and distance. Any quantity that can be subdivided in a way that corresponds to a subdivision of space or time can, at least in principle, be measured with the help of a definite integral. In practice, we need a continuous function relating the quantity to time or space, and there may not be a function available. However, there is, in this situation, a set of observational values that can be used to approximate a continuous function and its definite integral.

7.7.1 The Mean Value Theorem for Integrals

To prove the Fundamental Theorem of Calculus, we start with a fairly trivial observation.

Th. 7-8. If, for all x in the interval $a \leq x \leq b$, $m \leq F(x) \leq M$, then $m(b - a) \leq \int_a^b F(x)dx \leq M(b - a)$.

This theorem is true because if F is bounded by m and M, then all the values $F(x_i)$ that can appear in the approximating sum $\sum_{i=1}^{n} F(x_i)\Delta_i x$ are also between m and M, and, of course, the individual values $\Delta_i x$ must add up to $b - a$ since they are determined by a partition of $a \leq x \leq b$. Hence, all approximating sums $S = \sum_{i=1}^{n} F(x_i)\Delta_i x$ must lie between $m(b - a)$ and $M(b - a)$, and therefore their limit, $\int_a^b F(x)dx$, must as well.

This theorem is a necessary preliminary step for the following more advanced result.

Th. 7-9. (The Mean Value Theorem for Integrals)
If F is continuous for $a \leq x \leq b$, then there is some number c between a and b such that $F(c) \cdot (b - a) = \int_a^b F(x)dx$.

Proof: Since F is continuous on the closed interval, it must have a minimum value m and a maximum value M on that interval. Therefore, it satisfies the conditions for Th. 7-8, and $m(b - a) \leq \int_a^b F(x)dx \leq M(b - a)$. Since $\int_a^b F(x)dx$ is some number, we divide it by $b - a$ to come up with a k such that $k(b - a) = \int_a^b F(x)dx$. Naturally, k must be between m and M. Now let y be a value of x such that $F(y) = m$ and $a \leq y \leq b$, and Y be a similar value for M. Since F has m and M as minimum and maximum values, both y and Y must exist, and, furthermore, the interval between y and Y must be a subset of the interval between a and b. We apply the Intermediate Value Theorem to $F(x)$, k, y, and Y and arrive at a c between y and Y (and therefore between a and b) such that $F(c) = k$. Substituting this into the defining equation for k produces the conclusion of this theorem.

This theorem is called the Mean Value Theorem for Integrals because multiplying the equation in Th. 7-7 by $(b - a)$ and then "integrating" each side so that $F'(c)$ becomes $F(c)$ and $F(b) - F(a)$ is replaced with $\int_a^b F(x)dx$ transforms that equation into the one in Th. 7-9.

The number $\frac{\int_a^b F(x)dx}{b-a}$ is defined as the *average* value of F on the interval $a \leq x \leq b$. With this definition and Th. 7-9, it becomes clear that at some point in any interval, a continuous function must take on its average value in that interval.

7.7.2 The Fundamental Theorem of Calculus

Before we can use Th. 7-9 to prove the Fundamental Theorem of Calculus, we must prove the following about definite integrals.

Th. 7-10. $\int_a^b f(x)dx + \int_b^c f(x)dx = \int_a^c f(x)dx$.

This is true because "melding" a partition of $a \le x \le b$ with one of $b \le x \le c$ produces one of $a \le x \le c$, and, therefore, adding any approximating sum for $\int_a^b f(x)dx$ to a similar sum for $\int_b^c f(x)dx$ will produce an approximating sum for $\int_a^c f(x)dx$. Hence, the total of the limits of the first two sums will be the limit of the third sum, Q. E. D.

Th. 7-10 looks rather obvious, and its proof is short. It has two important corollaries: $\int_a^a f(x)dx = 0$ and $\int_b^a f(x)dx = -\int_a^b f(x)dx$, which are proven from this theorem in the same way that the results ln $1 = 0$ and $\ln\frac{1}{x} = -\ln x$ are proven from the definition of logarithms in Sec. 7.5.

We will now prove the Fundamental Theorem itself.

Th. 7-11. (The Fundamental Theorem of Calculus)

If $f(x)$ is a continuous function on $a \le x \le b$, and $F(x)$ is any indefinite integral of $f(x)$, that is, $\frac{dF(x)}{dx} = f(x)$, then $\int_a^b f(x)dx = F(b) - F(a)$.

To prove this theorem, we will change the notation slightly and write $\int_a^x f(t)dt$. Now, since f is continuous on $a \le t \le b$, the integral $\int_a^x f(t)dt$ exists for all x between a and b inclusive (Sec. 7.7) and therefore determines a function of x; call it $Q(x)$. We note for future reference that $Q(a) = 0$ as shown above.

We study $Q(x)$ by calculating its derivative with respect to x. This requires consideration of $\lim_{\Delta x \to 0} \frac{Q(x+\Delta x)-Q(x)}{\Delta x} = \lim_{\Delta x \to 0} \frac{\int_a^{x+\Delta x} f(t)dt - \int_a^x f(t)dt}{\Delta x} = \lim_{\Delta x \to 0} \frac{\int_x^{x+\Delta x} f(t)dt}{\Delta x}$ (Th. 7-10). Now, by Th. 7-9, there must be a c between x and $x + \Delta x$ such that $f(c) \cdot \Delta x = \int_x^{x+\Delta x} f(t)dt$, or $\frac{\int_x^{x+\Delta x} f(t)dt}{\Delta x} = f(c)$. Therefore, $\frac{dQ(x)}{dx} = \lim_{\Delta x \to 0} f(c)$, where c lies between x and $x + \Delta x$. As Δx approaches zero, $x + \Delta x$ of course approaches x, and therefore c which is sandwiched in between must approach x as well. Therefore, we can replace the "$\Delta x \to 0$" in the expression for $Q'(x)$ with "$c \to x$", and, since f is continuous, $\lim_{c \to x} f(c) = f(x)$, so $\frac{dQ(x)}{dx} = f(x)$.

Now any indefinite integral $F(x)$ of $f(x)$ can differ from $Q(x)$ only by a constant since F and Q both have f as a derivative (second corollary to Th. 7-7). We determine this constant by using the fact that $Q(a) = 0$: $F(x) - Q(x) = C$; $F(a) - Q(a) = C$; $F(a) - 0 = C$; $C = F(a)$; $F(x) - Q(x) = F(a)$; $Q(x) = F(x) - F(a)$. Taking x to be b, we have $Q(b) = \int_a^b f(t)dt = F(b) - F(a)$, Q. E. D. Note that the value of a definite integral is not affected by the choice of the letter we use to denote the variable of integration: $\int_a^b f(x)dx = \int_a^b f(t)dt = \int_a^b f(y)dy$, etc.

Definite integrals are difficult, and sometimes impossible, to evaluate directly from their definition as the limit of approximating sums. The Fundamental Theorem makes things much easier. For example, let us try to evaluate $\int_0^1 x\, dx$ from the definition: We set up a partition of $0 \le x \le 1$ by choosing $x_0 = 0$, $x_1, x_2, \ldots, x_{n-1}, x_n = 1$ in increasing order, and then make the inspired choice of an x in each interval exactly halfway between its endpoints. The approximating sum $\sum_{i=1}^n f(x_i)\Delta_i x$ then becomes $\frac{x_1}{2} \cdot x_1 + \frac{x_1+x_2}{2} \cdot (x_2 - x_1) + \frac{x_2+x_3}{2} \cdot (x_3 - x_2) + \cdots + \frac{x_{n-1}+x_n}{2} \cdot (x_n - x_{n-1})$ which equals $\frac{x_1^2}{2} + \frac{x_2^2-x_1^2}{2} + \frac{x_3^2-x_2^2}{2} + \cdots + \frac{x_n^2-x_{n-1}^2}{2}$, where the only uncanceled term is $\frac{x_n^2}{2}$ or ½. Therefore, $\int_0^1 x\, dx = \frac{1}{2}$. Using the Fundamental Theorem, $\int_0^1 x\, dx = \frac{1^2}{2} - \frac{0^2}{2} = \frac{1}{2}$.

The bar and subscript notation $F(x)\big|_a^b$ is used to mean $F(b) - F(a)$ just as $F(x)\big|_a$ means $F(a)$. Using this notation, we write $\int_0^1 x\,dx = \left.\frac{x^2}{2}\right|_0^1$.

The estimation of ln 2 in Sec. 7.5.2, as well as the formulas given there for ln 2 and ln 7/6, are in fact the estimation of and formulas for the values of definite integrals. Since $\int \frac{dx}{x} = \ln x$, $\int_a^b \frac{dx}{x} = \ln b - \ln a$, and, in particular, $\int_1^a \frac{dx}{x} = \ln a$ since ln 1 = 0. In fact, the use of the Mean Value Theorem for Derivatives in deriving the approximations and formulas there is an exact analogue of the use of the Mean Value Theorem for Integrals in proving Th. 7-11.

A few examples of definite integrals evaluated by Th. 7-11 follow.

1. $\int_0^2 x^2\,dx = \left.\frac{x^3}{3}\right|_0^2 = \frac{8}{3}$.

2. $\int_0^{\pi/2} \sin x\,dx = -\cos x\big|_0^{\pi/2} = -(0-1) = 1$.

3. $\int_0^{\pi/2} \sin 2x\,dx = -\left.\frac{\cos 2x}{2}\right|_0^{\pi/2} = -\frac{1}{2}(\cos\pi - \cos 0) = -\frac{1}{2}((-1)-1) = 1$.

4. $\int_1^2 \ln x\,dx = x\ln x - x\big|_1^2 = (2\ln 2 - 2) - (1\ln 1 - 1) = 2\ln 2 - 1 \approx 0.38629$.

5. $\int_0^1 e^x\,dx = e^x\big|_0^1 = e - 1 \approx 1.71828$.

6. $\int_0^1 \frac{dx}{x^2+1} = \tan^{-1} x\big|_0^1 = \frac{\pi}{4}$.

7.7.3 Theories of Integration

The definition of the definite integral presented here (called the Riemann integral) is not the only one that mathematicians have cooked up. In addition to Riemann, there are also the Lebesgue, Cauchy, and other definitions, each one named after the first-rank mathematician who invented it. For continuous functions – the only ones we have studied, for the most part – these definitions are all equivalent. But non-continuous functions are common in real life, and they are also created by mathematicians, such as Dirichlet's function (Sec. 7.2.5), and for these, the various definitions are not equivalent. For example, $\int_0^1 \phi(x)\,dx$ does not exist if we use the Riemann definition, since proper choices of the x's in the partition intervals can give us a sum of 0 or 1 (or anything in between) regardless of the value of $[\Delta]$, but $\int_0^1 \phi(x)\,dx$ exists and is zero using the Lebesgue definition.

7.8 Applications of the Definite Integral

As mentioned at the beginning of Sec. 7.7, a definite integral can in principle measure any quantity that can be subdivided in a way that corresponds to a subdivision of space or time. More precisely, if Q is the quantity that we wish to measure, and the entire quantity corresponds to an interval $a \leq x \leq b$ for some variable x, with the amount of Q corresponding to a subdivision Δx of x being approximately $f(x)\Delta x$ for any x in the subdivision, then the total Q is given by $\int_a^b f(x)\,dx$, provided that the approximation $f(x)\Delta x$ becomes exact as Δx approaches zero; that is, $\lim_{\Delta x \to 0} \frac{\Delta Q}{f(x)\Delta x} = 1$, where ΔQ is the actual amount of Q corresponding to Δx. Usually, a "natural" choice of f will satisfy the requirement, but there are a few applications where care is required. These will be mentioned as they come up.

7.8.1 Area

The earliest mathematical precursors to the definite integral were formulas computing area. Simple geometry suffices for polygons, but areas bounded by non-straight curves require a little thinking. The last group (c. 150 B.C.) of classical Greek mathematicians were the first to make progress beyond the circle – in fact, they came within an eyelash of inventing the definite integral! Archimedes managed the extraordinary trick of measuring the area under a parabola:

his result is equivalent to our $\int_0^a x^2 dx = \frac{a^3}{3}$.

Area also motivated the ultimate definition of the definite integral itself. What is the area bounded by the curve $y = f(x)$, the x-axis, and the lines $x = a$ and $x = b$? The area is naturally associated with the interval $a \le x \le b$. To a subdivision Δx of this interval is associated that part of the area lying directly over it. (See Fig. 7-10.) We approximate this part with a rectangle having width Δx and height $f(x)$ for

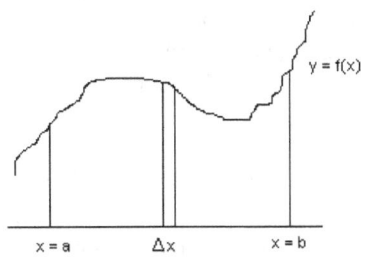

Fig. 7-10. The area under a curve can be computed by using a definite integral.

an arbitrary x in the interval. The area of the approximation is $f(x)\Delta x$, and the approximation becomes exact as Δx approaches zero because the difference is a shape with two small dimensions, while the approximation itself has only one small dimension. (See Fig. 7-11.) In mathematical language, the error in the approximation is an *infinitesimal of the second order*, while the approximation itself is an infinitesimal of the first order. The approximation $f(x)\Delta x$ meets the requirement for a definite integral if and only if the error is an infinitesimal of the second or higher order.

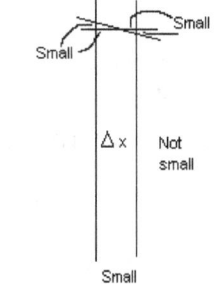

Figure 7-11. Infinitesimals of the first and second order.

Therefore, the answer to the question at the top of this page is $\int_a^b f(x)dx$. The area under one arch of the sine curve, for example, is given by $\int_0^\pi \sin x \, dx$, which equals $-(\cos \pi - \cos 0)$ or 2. If we want the area bounded by the sine curve and the x-axis between $x = 0$ and $x = 2\pi$, we integrate $\int_0^{2\pi} \sin x \, dx$ and get zero. Certainly the area should be greater than two because we now have two lens-shaped areas being counted instead of just one. It seems as if the second one has negative area! Well, from the point of view of the definite integral it does. All the values of $\sin x$ between π and 2π are negative (Sec. 5.6.4), so all the values of $(\sin x_i)\Delta_i x$ in an approximating sum for $\int_\pi^{2\pi} \sin x \, dx$ are as well. Therefore, the integral itself has a negative value. If we wish to add areas above and below the x-axis without cancellation, we must find separately each area above and each below, and then strip the minus signs from the "negative" areas before adding. (See Fig. 7-12.) This is done by integrating between successive zeros of $f(x)$. Thus, the total area of the two lenses is $\int_0^\pi \sin x \, dx + \int_\pi^{2\pi} |\sin x| dx$ which is 4. (We could have reached this answer by noting that the lenses are congruent figures since $\sin(\pi + x) = -\sin x$, and just doubling the earlier answer of two.)

We are not limited to areas bounded by straight lines and one curve. For example, what is the area of one shape that is between the sine and cosine curves? An approximating rectangle for part of the area, shaded in Fig. 7-13, has height $\sin x - \cos x$, so the area is given by $\int_{\pi/4}^{5\pi/4} (\sin x - \cos x)dx$. This evaluates as $-(\sin x + \cos x)\big|_{\pi/4}^{5\pi/4}$ or $2\sqrt{2}$. (Remember that $\sin\frac{\pi}{4} = \cos\frac{\pi}{4} = \frac{\sqrt{2}}{2}$ and $\sin\frac{5\pi}{4} = \cos\frac{5\pi}{4} = -\frac{\sqrt{2}}{2}$.)

In general, the area bounded by $y = f(x)$, $y = g(x)$, $x = a$, and $x = b$ is given by $\int_a^b [f(x) - g(x)] \, dx$, but, if the two curves cross, we must break up the integration at each point where they cross and keep track of which curve is on top.

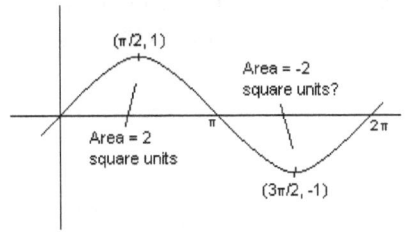

Fig. 7-12. The sine curve and areas bounded by it and the x-axis. To get the total area bounded by it and the x-axis between 0 and 2π, we must integrate the absolute value of $\sin x$.

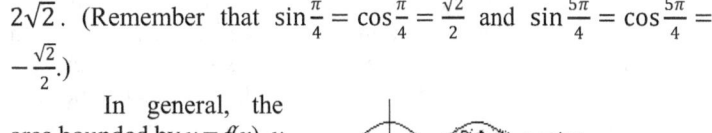
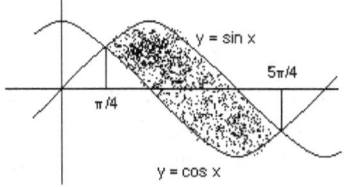

Fig. 7-13. The area between the sine and cosine curves can be calculated with a definite integral.

7.8.2 Volume

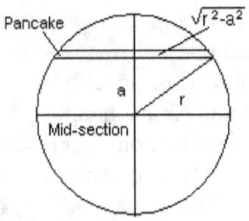

Fig. 7-14. The volume of a sphere determined by calculus.

Archimedes, already mentioned, was the first to determine that the volume of a sphere with radius r is $\frac{4}{3}\pi r^3$. To discover this fact with calculus, we slice the sphere into a stack of pancakes of various sizes. The radius of each pancake is $\sqrt{r^2 - a^2}$, where a is the distance of the pancake from the sphere's midsection. (See Fig. 7-14.) The thicknesses of the pancakes determine a subdivision of $-r \le a \le r$, and each pancake's volume is approximately $\pi(r^2 - a^2)\Delta a$. Hence, the sphere's volume is $\int_{-r}^{r} \pi(r^2 - a^2)da = \pi(ar^2 - \frac{a^3}{3})\Big|_{-r}^{r} = \pi\left(2r^3 - \frac{2}{3}r^3\right) = \frac{4}{3}\pi r^3$.

In general, if the cross-sectional area of a solid is given by $f(x)$ and the solid is included between $x = a$ and $x = b$, then the volume is given by $\int_a^b f(x)dx$, just like in determining area. The formula for the volume of a pyramid is determined this way: the base, which has area B, is taken to represent $x = h$ and the vertex to represent $x = 0$. Any slice parallel to the base between $x = 0$ and $x = h$ intersects the pyramid in a polygon similar to the base in the ratio $x:h$. By Th. 5-13(B), the area of this slice is $\left(\frac{x}{h}\right)^2 \cdot B$, and hence the volume of the pyramid is given by $\int_0^h \left(\frac{x}{h}\right)^2 \cdot B\, dx = \int_0^h \frac{B}{h^2} x^2 dx = \frac{1}{3} B \frac{h^3}{h^2} = \frac{1}{3}Bh$. Thus, the volume of any pyramid is one-third that of a prism with the same base and height, an extension into three dimensions of Th. 5-10(B) about the areas of triangles. Generally, in n dimensions, the volume of a "pyramid" is $1/n$ times the volume of a matching "prism."

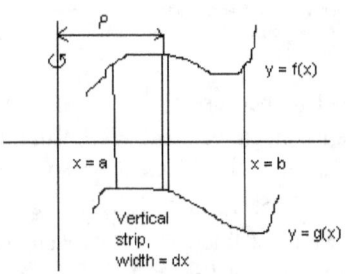

Fig. 7-15a. An area rotated around a vertical line. The vertical strip has height equal to f(x) - g(x) and width equal to dx.

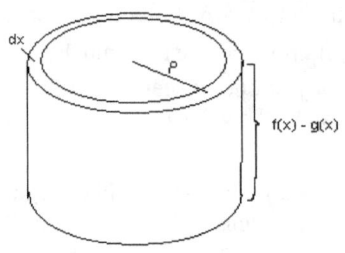

Fig. 7-15b. The vertical strip, when rotated about the axis, generates a cylindrical shell.

Not all solids can be cut into slices with easily measurable cross-sectional areas. For most of these, advanced methods such as triple integration must be used, but certain shapes can be measured with elementary methods. One such shape is that generated when an area bounded by $y = f(x)$, $y = g(x)$, $x = a$, and $x = b$ is rotated around a vertical line that does not intersect the area. (See Fig. 7-15a.) To determine the volume in this case, the area which sweeps through the volume is cut into thin vertical strips, each of which generates a thin cylindrical shell during the rotation. (See Fig. 7-15b.) A shell can be cut open along a line parallel to the axis of ratation and flattened out into a thin rectangular slab whose width is the height of the vertical strip of area, which is $f(x) - g(x)$. The length of the slab is the distance through which the strip traveled, which is 2π times the distance between the rotation axis and the strip. This distance, symbolized by ρ (lower-case rho, which, despite its resemblance to an English p, is actually equivalent to English r), will depend on the position of the axis of rotation and of the area strip. The third dimension of the slab, its thickness, is equal to the width of the strip, which we take to be $\Delta_i x$ in some partition of $a \le x \le b$. (See Fig. 7-15c.) The volume of the slab associated with the subinterval $\Delta_i x$ is therefore $2\pi\rho(f(x) - g(x))\Delta_i x$, and therefore the total volume generated by the area is $\int_a^b 2\pi\rho\big(f(x) - g(x)\big)dx$. In this integral, ρ is usually $x - c$ for some constant c.

As an example, we determine the volume of the *torus* (doughnut) generated by rotating the circle centered at the origin with radius r about the line $x = - R$ ($R > r$). The circle's equation is $x^2 + y^2 = r^2$ or $y = \pm\sqrt{r^2 - x^2}$. thus, we can treat the circle as lying between the lines $x = - r$ and $x = r$, being bounded above by the semicircle $y = \sqrt{r^2 - x^2}$, and bounded below by the semicircle $y = -\sqrt{r^2 - x^2}$. Hence,

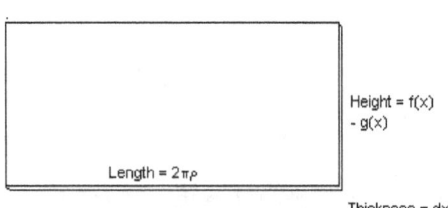

Fig. 7-15c. The cylindrical shell, when unrolled and flattened out, becomes a rectangular slab whose dimensions are as shown. This slab is an element of volume for the shape generated when the area is rotated about the axis.

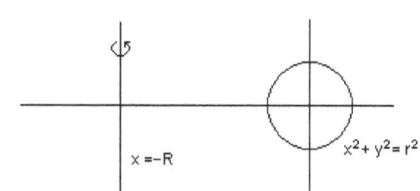

Fig. 7-16a. The circular region bounded by x²+ y²= r² is rotated about the line x = - R to generate a volume.

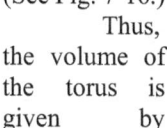

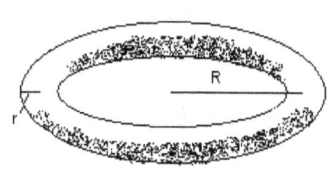

Fig. 7-16b. The torus (doughnut) generated when the circle of Fig. 7-16a is rotated about the line x = - R.

$f(x) - g(x)$ here is $2\sqrt{r^2 - x^2}$. The value of ρ is $x + R$ since the axis of rotation is the line $x = -R$. (See Fig. 7-16.) Thus, the volume of the torus is given by $4\pi \int_{-r}^{r} (R + x)\sqrt{r^2 - x^2}dx$. We split this integral up into $4\pi R \int_{-r}^{r} \sqrt{r^2 - x^2}dx$ and $2\pi \int_{-r}^{r} 2x\sqrt{r^2 - x^2}dx$. To the former, we apply the formula derived in Sec. 7.6.3 for $\int \sqrt{a^2 - x^2}dx$, and to the latter, we apply the chain rule for integrals (Sec. 7.6.1) with $F(u) = \sqrt{u}$ and $G(x) = r^2 - x^2$. The results are a value of $2\pi^2 R r^2$ for the first integral and a big fat zero for the second. Hence, the volume of the torus is $2\pi^2 R r^2$.

The zero result for the second integral is an example of the fact that if $f(x)$ is an *odd* function $(f(-x) = -f(x))$, then its integral over any interval symmetric about the origin vanishes. This is apparent from Fig. 7-17, as well as from the definition of the definite integral: to any term $f(x_i)\Delta_i x$ in the approximating sum, there corresponds a term $f(-x_i)\Delta_i x = -f(x_i)\Delta_i x$, if the partition is symmetrical. Therefore, this particular type of approximating sum vanishes, and so must the limit of all such sums. If f is continuous, all possible approximating sums must have the same limit (Sec. 7.7), and therefore this limit must vanish as well. The result comes in handy in variuos definite-integral applications, as it did in the calculation of the volume of the torus. It can save a lot of effort!

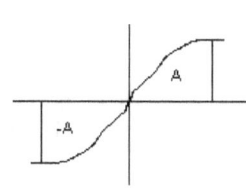

Fig. 7-17. The integral of an odd function over any interval symmetric about the origin vanishes.

7.8.3 Arc Length

To measure the lengths of straight lines, we use rulers, or the distance formula if the lines are in a Cartesian plane. But there are no curved rulers, and the distance formula applies only to straight line segments. Calculus gets around this problem by chopping a curve into infinitely many, infinitely short segments that can be considered straight.

In Fig. 7-18, a segment of a curve is shown. The whole curve can be considered to extend from P_0 to P_n and have length s. The portion of s accounted for between P_i and P_{i+1} is $\Delta_i s$, which is approximately the hypotenuse of a right triangle having sides $\Delta_i x$ and $\Delta_i y$. Hence, $\Delta_i s$ is approximately $\sqrt{(\Delta_i x)^2 + (\Delta_i y)^2}$. To change this into the form $f(x_i)\Delta_i x$, we divide inside the radical by $(\Delta_i x)^2$ and then multiply outside the radical by $\Delta_i x$. The result is $\sqrt{1 + \left(\frac{\Delta_i y}{\Delta_i x}\right)^2} \Delta_i x$. Now, as the segments $\Delta_i s$ approach zero in length, not only does this approximation become exact, but also the quotient $\frac{\Delta_i y}{\Delta_i x}$ becomes $\frac{dy}{dx}$. Therefore, the length of arc, s, is equal to $\int_{P_1}^{P_2} \sqrt{1 + (dy/dx)^2}\, dx$, where P_1 denotes the value of x at one end of the curve and P_2 that value at the other end. If the curve's equation is given by $x = f(y)$, the value of s is given by $\int_{P_1}^{P_2} \sqrt{1 + (dx/dy)^2}\, dy$, where P_1 and P_2 are now y values.

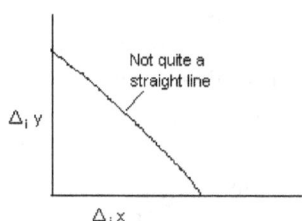

Not quite a straight line

$\Delta_i y$

$\Delta_i x$

Fig. 7-18. An approximatinon to the arc length of a curve is obtained by chopping the curve up into very many, very short segments.

As an example, let us determine the length of arc of that portion of the circle $x^2 + y^2 = 1$ that lies in the first quadrant. In the first quadrant, $y = \sqrt{1 - x^2}$, and thus $\frac{dy}{dx} = \frac{-x}{\sqrt{1-x^2}}$. Then,

$\sqrt{1 + (dy/dx)^2} = \sqrt{1 + \frac{x^2}{1-x^2}} = \sqrt{\frac{1}{1-x^2}} = \frac{1}{\sqrt{1-x^2}}$. The ends of the curve are at $x = 0$ and $x = 1$, so the integral to

evaluate is $\int_0^1 \frac{dx}{\sqrt{1-x^2}}$. We make the substitution $x = \sin\theta$ (Sec. 7.6.3), which rather demolishes this integral,

transforming it into just $\int_0^{\pi/2} d\theta$. The limits on this integral must be stated in terms of the new variable; when $x = 0$, θ
$= 0$ since $\sin 0 = 0$, and when $x = 1$, $\theta = \pi/2$ since $\sin \pi/2 = 1$. Alternatively, we could evaluate the indefinite integral
$\int d\theta = \theta + C$ and then perform the inverse substitution $\theta = \sin^{-1} x$ to get $\int_0^1 \frac{dx}{\sqrt{1-x^2}} = \sin^{-1} x|_0^1$. Either way, the

value of the integral is $\frac{\pi}{2}$, which should come as no surprise since it is the length of one-quarter of a circle which must
have a circumference of 2π.

The formula for arc length can be easily modified to fit the case where both x and y are functions of a third
variable such as t or θ. The third variable is called a *parameter*, and equations in terms of a parameter are called
parametric equations. Instead of dividing and multiplying $\sqrt{(\Delta_i x)^2 + (\Delta_i y)^2}$ by $\Delta_i x$, we use $\Delta_i t$ or $\Delta_i \theta$. The result is

$s = \int_a^b \sqrt{\left(\frac{dx}{dt}\right)^2 + \left(\frac{dy}{dt}\right)^2}\, dt$, where $t = a$ provides one end of the curve and the other is reached when $t = b$. An example

using this formula is the *astroid*, a four-pointed curve (shown in Fig. 7-19) whose equations are $x = \cos^3 \theta$, $y = \sin^3 \theta$.
(The curve's equation can be written with just two variables as $x^{2/3} + y^{2/3} = 1$, but try determining the arc length
from this equation and you will quickly find out why the parameter θ is usually used.)

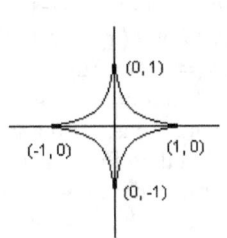

(0,1)

(-1,0)   (1,0)

(0,-1)

Fig. 7-19. The astroid, a
curve that is best studied
with parametric equations.

Using derivative formulas I, X, and XI, we have $\frac{dx}{d\theta} = -3 \cos^2\theta \sin\theta$ and $\frac{dy}{d\theta} =$

$3 \sin^2\theta \cos\theta$. Hence, $\left(\frac{dx}{d\theta}\right)^2 + \left(\frac{dy}{d\theta}\right)^2 = 9 \sin^2\theta \cos^2\theta (\sin^2\theta + \cos^2\theta)$ and

$s = \int_0^{2\pi} 3 \sin\theta \cos\theta\, d\theta = \frac{3}{2}\int_0^{2\pi} \sin 2\theta\, d\theta$ ... - NOT! The limits on the integral
enable θ to vary through the curve's corners, which are points where s is not
continuous. In general, arc length integrals must be evaluated *between* corners of the
curve. Here, we take advantage of the curve's symmetry and say that the total length is
four times the length in the first quadrant, which lies entirely between corners. Hence,
$s = 4 \cdot \frac{3}{2} \cdot \int_0^{\pi/2} \sin 2\theta\, d\theta = 6$. As a check, we note that this is slightly more than the
perimeter of a square with the same four corners, as one would expect. (To tell the
truth, the reason why the integral $\frac{3}{2}\int_0^{2\pi} \sin 2\theta\, d\theta$ does not give the correct arc length
of the astroid is that a square root is always supposed to be positive. If the absolute

value of $\sin 2\theta$ had been used in the integral, it would have produced the correct answer, despite the integration being
across the corners. However, in other problems, the integration across corners will produce wrong answers. It's better
to be safe than sorry, especially in mathematics.)

7.8.4 Mass

The density of an object is defined as the mass contained in a unit volume. The formula Density = Mass ÷
Volume can be turned around to produce Mass = Volume x Density, which in turn can be used to calculate the mass of
an object of known size and density. For example, the mass of a three-foot-deep hemispherical bowl filled with water,
which has a density of 62.4 pounds per cubic foot, is $\frac{4}{3}\pi(3)^3 \times \frac{1}{2}$ (since we are dealing with a hemisphere) x 62.4 =
3,530 pounds. (That's just the mass of the water, not its container.)

Just as with Distance = Time x Velocity, the formula Mass = Volume x Density can only be used if the
density is a constant. If our bowl were filled with a compressible fluid so that the density was, say, 20 pounds per cubic
foot times the depth in feet of the spot at which we are measuring the density, then we have a different problem. The
way this problem is solved is to determine the mass of an infinitesimally thin slab of fluid at a certain depth x, then add
up (integrate) the slabs for all the values of x between 0 and 3.

From Sec. 7.8.2, the volume of a slab of thickness dx at a depth of x feet is $\pi(9 - x^2)dx$ cubic feet, and
therefore its mass is $20\pi x(9 - x^2)dx$ pounds. Therefore, the total mass is $\int_0^3 20\pi x(9 - x^2)dx = 20\pi(\frac{9}{2}x^2 -$
$\left.\frac{x^4}{4}\right)\Big|_0^3 = 405\pi$ or about 1,272 pounds. The large difference from the homogeneous hemisphere with about the same
maximum density is because most of the volume of the hemisphere is at shallow depths.

In general, the mass of an object that occupies a volume V and has a density of δ is given by $\int_V \delta\, dV$. The symbol $\int_V \ldots$ means that the limits of the integral must describe the volume V. For most volumes, V must be described in two or three variables, and δ may be a function of more than one variable. This requires multiple or iterated integration, as in $\int_0^1 \int_0^{1-x} \int_0^{1-x-y} (x + y + z)\, dz\, dy\, dx$, which gives the mass of the pyramid cut out of the *first octant* by the plane $x + y + z = 1$, in which the density at any point (x, y, z) is $x + y + z$. (The first octant is the portion of Cartesian 3-space in which all three coordinates are positive. See Sec. 5.8.) As this is only an introduction to calculus, multiple integration is beyond the scope of this book. In the earlier problem, both δ and V could be described in one variable, so multiple integration was not needed.

In mathematics and physics, it is frequently possible to treat three-dimensional objects as though they were only two- or one-dimensional. Thus, we speak of the mass per square foot of a metal plate or the mass per foot of a rope even though metal plates and ropes have thickness. We write formulas like $M = \int_R \delta\, dA$ or $\int_C \delta\, ds$ to calculate the mass of such bodies. As in the case of volume, $\int_R \ldots$ and $\int_C \ldots$ mean that the limits of the integral must conform to the boundaries of the area R or curve C. As for dA and ds, they can frequently be reduced to the form $f(x)\, dx$ as with the hemisphere example above. If, for example, the equation of the curve C is given by $x = f(t)$ and $y = g(t)$, then $ds = \sqrt{\left(\frac{df}{dt}\right)^2 + \left(\frac{dg}{dt}\right)^2}\, dt$ as with the astroid in Sec. 7.8.3. Similarly, if R is bounded by the lines $x = 0$, $y = a$, $y = b$, and the curve $x = f(y)$, then $dA = f(y)\, dy$, and the limits on the integral are a and b.

The density function δ may also refer to more than one variable. Occasionally, a problem seeming to require multiple integration actually can be solved with a single integral by way of a judicious change of variable. For example, take the mass of the square whose corners are $(0, 0)$, $(2, 0)$, $(0, 2)$, and $(2, 2)$, where the density per unit area is $x + y + 2$. Since δ contains two variables, it appears as if multiple integration is required. However, if we are clever and introduce a "k-axis" along the line $y = x$, then strips of area perpendicular to this "axis" will have a constant density. (See Fig. 7-20.) Since we should make the units for k be the same length as those for x and y, the density along any strip is $2 + k\sqrt{2}$. (At the point $(1, 1)$, say, the value of k is $\sqrt{2}$, while that of $x + y$ is 2.) The length of a strip is $2k$ if k is between 0 and $\sqrt{2}$ and $4\sqrt{2} - 2k$ if k is between $\sqrt{2}$ and $2\sqrt{2}$. Therefore, the total mass of the square is $\int_0^{\sqrt{2}} (2 + k\sqrt{2})(2k\, dk) + \int_{\sqrt{2}}^{2\sqrt{2}} (2 + k\sqrt{2})(4\sqrt{2} - 2k)\, dk = \int_0^{\sqrt{2}} (2\sqrt{2}k^2 + 4k)\, dk + \int_{\sqrt{2}}^{2\sqrt{2}} (8\sqrt{2} + 4k - 2\sqrt{2}k^2)\, dk = \left(\frac{2}{3}\sqrt{2}k^3 + 2k^2\right)\Big|_0^{\sqrt{2}} + \left(8\sqrt{2}k + 2k^2 - \frac{2}{3}\sqrt{2}k^3\right)\Big|_{\sqrt{2}}^{2\sqrt{2}} = \frac{8}{3} + 4 + 32 + 16 - \frac{64}{3} - \left(16 + 4 - \frac{8}{3}\right) = 32 - \frac{48}{3} = 16.$

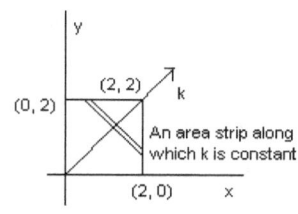

Fig. 7-20. A mass problem seeming to require multiple integration can be reduced to one that doesn't by introducing a third variable.

(0, 2)

(2, 2)

(2, 0)

y

k

x

An area strip along which k is constant.

7.8.5 Moments, Centroids, and Centers of Gravity

If you are old enough to remember the see-saw, you probably remember the solution to the problem of balancing a fat kid with a skinny one, or a kid with an adult, was to have the heavier person sit closer to the center of the see-saw. The mechanical principle behind this is known as the *law of the lever* (a see-saw is one of the myriad disguises for a lever), and is stated as $m_1 d_1 = m_2 d_2$, where m_1 and m_2 are the masses of the children balancing on the see-saw and d_1 and d_2 are the distances of the children from the see-saw's center, the balancing point (known formally as the *fulcrum*). It is obvious from the form of this law that if m_2 is bigger than m_1, then d_2 must be smaller than d_1; i. e., the heavier person sits closer to the fulcrum.

The quantities $m_1 d_1$ and $m_2 d_2$ are called the *moments* of each child relative to the fulcrum. Additional children getting on the see-saw have their own moments, and the balance condition is that the total of all the moments on the left side of the fulcrum must equal the total moments on the right side. If we adopt the convention that distance to the left of the fulcrum is negative, the condition becomes that the total of all moments, positive and negative, is zero. Formally, we have $\sum_{i=1}^n m_i d_i = 0$.

Now imagine several children on one side of the see-saw and one adult who weighs as much as all the children together. Where should he sit to exactly balance the moments of the children? A little algebra shows that

$d_A = \frac{\sum m_i d_i}{\sum m_i}$, where d_A is the proper distance for the adult (insert a minus sign for directed distances) and $\sum m_i$ is the mass of all the children, and hence the adult.

It is obvious that this adult can be balanced by a copy sitting opposite, that is, on the side that had the children, and the same distance from the fulcrum. In other words, the combination of children can be replaced by a single adult with the same total mass without affecting the see-saw's balance, provided the adult's distance satisfies $d_A = \frac{\sum m_i d_i}{\sum m_i}$. For this reason, the point corresponding to d_A is called the *center of gravity* of the mass of the children.

Where does calculus come in to all this? Well, so far we have considered a discrete arrangement of mass in the form of children. If instead we have a continuous distribution of mass along the see-saw, we cannot form a sum $\sum m_i d_i$. Instead, we have mass equal to $m(x)$ at a distance x from the fulcrum. Then, assuming the mass is distributed between a and b units from the fulcrum, the total mass is $\int_a^b m(x)dx$, and the moment associated with that mass is $\int_a^b x\, m(x)dx$. Thus, the location of the center of gravity is given by $\frac{\int_a^b xm(x)dx}{\int_a^b m(x)dx}$, because a single particle with mass equal to $\int_a^b m(x)dx$ situated at the center of gravity will have the same moment as the distribution of mass between a and b.

For mass distributions in two or more dimensions, we calculate separately the moments relative to the x-axis and y-axis, or the three or more coordinate planes or hyperplanes. The equations are similar to the one above, except that there is one for each coordinate, and, as in the calculation of mass itself, the limits of the integrals must describe the volume (or area) through which the mass is distributed.

if the density $m(x)$ or δ is a constant, we can move it across the integral sign and then cancel it out, since it appears in both the numerator and denominator of each center-of-gravity fraction. The integrals then refer only to the geometric object involved. The center of gravity of a homogeneous geometric object is called the *centroid* of the object. If an object has an axis of symmetry, the centroid must lie on that axis. We use this fact to find the centroid of a quarter-circle as follows: By the symmetry about the line $y = x$ (dotted in Fig.7-21), the two coordinates of the centroid must be the same. The x-coordinate of the centroid is given by $\bar{x} = \frac{\int_0^r x\sqrt{r^2-x^2}dx}{\int_0^r \sqrt{r^2-x^2}dx}$. We skip the detailed evaluation of the denominator, since it must be the area $\frac{\pi r^2}{4}$ of the quarter-circle. As for the numerator, we multiply by -2 inside and divide by -2 outside the integral sign to get $-\frac{1}{2}\int_0^r -2x\sqrt{r^2-x^2}dx = -\frac{1}{2}\cdot\frac{2}{3}(\sqrt{r^2-x^2})^3\big|_0^r = -\frac{1}{3}(0-r^3) = \frac{r^3}{3}$. Dividing this by $\pi r^2/4$ gives the x-coordinate of the centroid as $4r/3\pi$, approximately $0.4244r$. Note that in determining the moment relative to the y-axis, we integrate along strips parallel to that axis, so that the distance from that axis is constant along their entire length.

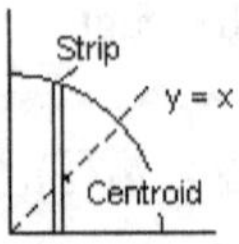

Fig. 7-21. The centroid of a quarter circle must lie on the axis of symmetry of the area, which here is the line y = x. Strips parallel to the y-axis, such as this one, are at a constant distance from that axis.

7.9 Approximations to Definite Integrals

In many practical situations that call for a definite integral, we do not have a continuous function to integrate, but only a set of experimental or measured values corresponding to discrete values of the variable of integration. Alternatively, we may have a continuous function, but one that has no 'elementary" indefinite integral, such as $\sin(x^2)$ or $\sqrt{1+x^4}$. (All finite combinations of algebraic, logarithmic, and trigonometric functions or their inverses are referred to as "elementary functions.")

For either of these cases, only a numerical approximation to the integral can be determined. We may form an approximating sum as in Sec. 7.7, but a better estimate can usually be made with one of two standard approximation formulas: the *trapezoidal rule* and *Simpson's rule*.

The trapezoidal rule, as might be guessed, approximates the area under a function's graph with a series of trapezoids, while Simpson's rule approximates with parabolic segments. In general, Simpson's rule is more accurate, but the uncertainty in many experimental measurements frequently overwhelms any accuracy gain. Both rules require evenly spaced function values; in addition, Simpson's rule requires an odd number of observations. Letting h denote the interval between observations, y_0, y_1, etc., the observations themselves, and T and S the trapezoidal and Simpson's rule approximations, $T = \frac{h}{2}(y_0 + 2y_1 + 2y_2 + \cdots + 2y_{n-2} + 2y_{n-1} + y_n)$ and $S = \frac{h}{3}(y_0 + 4y_1 + 2y_2 + 4y_3 + 2y_4 + \cdots + 2y_{2n-2} + 4y_{2n-1} + y_{2n})$.

To illustrate the rules, we take the example of an accelerating car. It starts with zero velocity at zero time, and accelerates as shown in the table. How much ground does it cover in ten seconds? The interval between measurements

Seconds	MPH
0	0
1	18
2	30
3	39
4	47
5	54
6	60
7	65
8	70
9	74
10	78

is one second, so the trapezoidal rule gives $T = \frac{1}{2}(0 + 36 + 60 + 78 + 94 + 108 + 120 + 130 + 140 + 148 + 78) = 496$. To get the answer in miles, we must divide by 3600; the car travels approximately 0.1378 miles or 727½ feet. For Simpson's rule, $S = \frac{1}{3}(0 + 72 + 60 + 156 + 94 + 216 + 120 + 260 + 140 + 296 + 78) = 497.3$, equivalent to 0.1381 miles or 729.4 feet. Rounding off to two significant figures (the accuracy of the speed measurements) makes both rules yield the same answer.

7.10 Infinite Sequences and Series

Back in Sec. 4.1, the formula for the sum of a geometric progression $a_1 + ra_1 + r^2a_1 + \cdots + r^{n-1}a_1$ was shown to be $a_1 \cdot \frac{1-r^n}{1-r}$. It was also stated that if $|r| < 1$, then the sum of the <u>infinite</u> progression $a_1 + ra_1 + r^2a_1 + \cdots$ is defined and equal to $\frac{a_1}{1-r}$. An infinite geometric progression is an example of an *infinite sequence*, and its indicated sum is an example of an *infinite series*. Both concepts, particularly the series, play a key role in the calculus of calculations, such as computing ln 2 or cos 1, and the series is important in the study of certain differential equations, such as $(x^2 - 1)\frac{d^2y}{dx^2} + 2x\frac{dy}{dx} + ny = 0$.

Since an infinite series is studied, at least at first, through an associated infinite sequence, we begin this section with sequences.

7.10.1 Functions as Infinite Sequences

The simplest example of an infinite sequence is a differentiable function restricted in its domain to the non-negative integers. To obtain the limit of such a sequence, we can apply the definitions in Sec. 7.2.3 and/or the following two theorems.

Theorem 7-12 (L'Hopital's rule, infinite case).
If $f(x)$ and $g(x)$ are everywhere differentiable, and both $\lim_{x\to a} f(x)$ and $\lim_{x\to a} g(x)$ are infinite, then $\lim_{x\to a} \frac{f(x)}{g(x)} = \lim_{x\to a} \frac{f'(x)}{g'(x)}$. ($a$ here may be finite or infinite.)

Theorem 7-13 (Functions applied to infinite sequences).
If $\lim_{n\to\infty} a_n = L$ and f is continuous at L, then $\lim_{n\to\infty} f(a_n) = f(L)$.

The infinite sequence itself is denoted by a_n, and its limit by $\lim_{n\to\infty} a_n$. For the rest of Section 7.10, the "$n \to \infty$" will be omitted except where its omission may cause confusion.

7.10.2 Common Infinite-Sequence Limits

The following infinite sequences arise often enough to warrant a tabulation of their limits:

A. $\lim \frac{1}{n} = 0$.
This limit was evaluated in Sec. 7.2.3.

B. $\lim \frac{n-1}{n} = 1$.

We write this as $1 - \frac{1}{n}$ and apply Th. 7-1, part I, and the previous result.

C. $\lim x^{1/n} = 1$ (x positive).

We rewrite $x^{1/n}$ as $e^{\ln(x^{\frac{1}{n}})} = e^{\frac{\ln x}{n}}$, then apply Th. 7-13 (which is legal regardless of the value of the limit, as e^x is continuous everywhere) to get $\lim x^{1/n} = e^{\lim \frac{\ln x}{n}}$, then apply Th. 7-1, part III, as $\ln x$ is a constant, and finally result A above.

D. $\lim n^{1/n} = 1$.

We use the same strategy as for result C, except that in evaluating $\lim_{n \to \infty} \frac{\ln n}{n}$, we now apply Th. 7-12, which produces $\lim_{n \to \infty} \frac{\ln n}{n} = \lim_{n \to \infty} \frac{1/n}{1}$, which equals zero by result A again.

E. $\lim \left(1 + \frac{1}{n}\right)^n = e$.

To calculate this limit, we must use the fact that $\lim_{x \to 0} \frac{\ln(1+x)}{x} = 1$. This is obvious once we notice that that fraction is simply $\frac{d(\ln x)}{dx}$ when $x = 1$, as was done in Sec. 7.5.1. It can also be proven from the zero case of L'Hopital's rule, whose statement is obtained from the infinite case merely by replacing $\lim_{x \to a} f(x) = \infty$ and $\lim_{x \to a} g(x) = \infty$ with $\lim_{x \to a} f(x) = 0$ and $\lim_{x \to a} g(x) = 0$ respectively. Once the value of $\lim_{x \to 0} \frac{\ln(1+x)}{x}$ is obtained, we replace x with $1/n$ to get $1 = \lim_{n \to \infty} n \ln(1 + \frac{1}{n}) = \lim_{n \to \infty} \ln(1 + \frac{1}{n})^n$, and the rest follows as in results B and C.

F. $\lim x^n = 0$ if $|x| < 1$.

This can be calculated directly from the definition in Sec. 7.2.3: we take N to be $\frac{\ln \epsilon}{\ln |x|}$. Both of these logarithms will be negative, so N is positive and if $n > N$, then $\ln|x^n| = n \ln|x|$ will be more negative than $\ln \epsilon$.

G. $\lim \frac{P(n)}{x^n} = 0$ if P is any polynomial and $x > 1$.

This limit was the origin of Malthus' saying that "exponential growth, however slow, always betters, in the long run, arithmetic growth, however fast." Even if the degree of P is one trillion, and x one and one trillionth, after one trillion applications of Th. 7-12, the numerator in the limit will be a constant, while the denominator will still approach infinity as n increases without bound. Malthus believed that humanity's food production could increase only "arithmetically," that is, polynomially, while the population could increase exponentially. Therefore, the above limit seemed to imply that eventually everyone would starve to death – unless we limit our population-increase capability by having fewer babies. Since everyone has not starved to death yet, it appears that Malthus was wrong, but you have to remember that he mentioned "the long run," not just the next few centuries.

7.10.3 Recurrence Relationships as Infinite Sequences

The second important type of infinite sequence is one generated by a recurrence relationship. Simple progressions usually do not do anything interesting: an arithmetic progression never has a finite limit, a harmonic progression always has a zero limit (not counting the "null" progressions where $d = 0$), and a geometric progression with $|r| > 1$ never has a finite limit. If $|r| < 1$, result F above shows that the limit is zero. If $r = -1$, the sequence has no limit (unless $a_1 = 0$) despite being bounded; its terms oscillate back and forth between a_1 and $-a_1$ without approaching any *one* number.

More complex recurrence relationships lead to more interesting results. For example, consider the sequence $a_1 = 1, a_2 = \frac{1}{2}, a_{n+1} = \frac{1}{2}(a_{n-1} + a_n)$; that is, each term is the average (rather than the sum as in a Fibonacci series) of the previous two. The sequence continues with $a_3 = \frac{3}{4}, a_4 = \frac{5}{8}, a_5 = \frac{11}{16}, a_6 = \frac{21}{32}, \ldots$ Evidently, the oscillations get steadily smaller, so the sequence must have some limit, but what is it? Looking at decimal values, we find $a_4 = .625, a_5 = .6875, a_6 = .6563, a_7 = .6719,$ and $a_8 = .6641$. The numbers are "zeroing in" on 2/3, and we can prove this by considering the related sequence $b_n = a_n - 2/3$: $b_1 = 1/3, b_2 = -1/6, b_3 = 1/12, b_4 = -1/24,$ etc. The b's clearly form a geometric progression with $r = -1/2$; since $|r| < 1$, the limit of the b's is zero and therefore $\lim a_n = 2/3$. In fact, $a_n = \frac{2}{3} - \frac{(-1)^n}{3 \cdot 2^{n-1}}$, which has a limit of 2/3 by result F and Th. 7-13.

The strategy of finding a closed-form expression for a_n (i. e., one that does not involve the other a's) is generally applicable to recurrence relationships. The strategy involved is to first determine some pattern (here the terms are approaching 2/3) and then subtract the pattern from the terms, leaving a residual that is (we hope) simpler than the terms themselves. This process is repeated until we have a tentative closed-form expression for a_n, which can be verified by mathematical induction (See Secs. 4.3 and 6.1.). The verification for the current sequence goes as follows: 1. The formula $a_n = \frac{2}{3} - \frac{(-1)^n}{3 \cdot 2^{n-1}}$ is true for $n = 1$, as it says $a_1 = \frac{2}{3} - \frac{-1}{3} = 1$. It is also true for $n = 2, 3, 4, 5$, and 6, as a little arithmetic will show. Now, if it holds for n and $n + 1$, where n is a positive integer, then $a_n = \frac{2}{3} - \frac{(-1)^n}{3 \cdot 2^{n-1}}$ and $a_{n+1} = \frac{2}{3} - \frac{(-1)^{n+1}}{3 \cdot 2^n} = \frac{2}{3} + \frac{(-1)^n}{3 \cdot 2^n}$. Then $a_{n+2} = \frac{a_n + a_{n+1}}{2} = \frac{\frac{4}{3} - \frac{(-1)^n}{3 \cdot 2^n}}{2} = \frac{2}{3} - \frac{(-1)^{n+2}}{3 \cdot 2^{n+1}}$. Since this is just the formula for a_n when n is replaced with $n + 2$, the formula holds for $n + 2$ if it holds for $n + 1$ and n. Therefore, it holds for all positive n by mathematical induction.

7.10.4 Infinite Series

Sequences based on recurrence relationships are particularly important because they can be useful in working with infinite series – which are indicated sums of infinitely many terms. The terms of an infinite series are denoted by a_n, and the series itself by $\sum_{n=1}^{\infty} a_n$. (See the beginning of Sec. 7.7 for what sigma notation means.) With each infinite series is associated an infinite sequence of *partial sums*, s_n, defined by $s_1 = a_1, s_2 = a_1 + a_2 = s_1 + a_2, s_3 = a_1 + a_2 + a_3 = s_2 + a_3, s_4 = s_3 + a_4$, and, in general, $s_n = s_{n-1} + a_n$. Now, if the sequence s_n has a limit S, we say that the series $\sum a_n$ *converges* and that its sum is S. If s_n fails to have a limit, we say that $\sum a_n$ *diverges*. (We leave off the "$n = 1$" and the "∞" from the sigmas just as we leave off the "$n \to \infty$" from the limit notation: where doing so can't cause confusion.)

Our first theorem on infinite series follows directly from this definition.

Th. 7-14 (*n*th-term test for infinite series).
$\sum_{n=1}^{\infty} a_n$ converges only if $\lim_{n \to \infty} a_n = 0$.

Proof: If $\sum a_n$ converges, it has a sum S, which must be the limit of the associated sequence of partial sums s_n. Hence, for any number $\frac{\epsilon}{2} > 0$, there exists an N such that, for all $n > N$, $|S - s_n| < \frac{\epsilon}{2}$. Now, if $n > N$, then $n + 1$ is as well, so for all $n > N$, $|S - s_n| < \frac{\epsilon}{2}$ and $|S - s_{n+1}| < \frac{\epsilon}{2}$. Writing these inequalities as compound ones without absolute value signs, $-\frac{\epsilon}{2} < S - s_n < \frac{\epsilon}{2}$ and $-\frac{\epsilon}{2} < S - s_{n+1} < \frac{\epsilon}{2}$. We now drop the left side of the first compound inequality and the right side of the second, multiply the second by -1 (reversing the remaining inequality sign there), and add the two to get $s_{n+1} - s_n < \epsilon$, valid for all $n > N$. But, by the definition of the s's, $s_{n+1} - s_n = a_{n+1}$. Therefore, for all $n > N$, $a_{n+1} < \epsilon$. Repeating this argument with the other halves of the compound inequalities produces $a_{n+1} > -\epsilon$; combining this with the above result gives $|a_{n+1}| < \epsilon$ for all $n > N$, Q. E. D.

Note that this theorem says "only if," not "if and only if." $\lim_{n \to \infty} a_n = 0$ is a necessary, but not sufficient, condition for the convergence of $\sum a_n$. To see why, consider the series $a_1 = \frac{1}{2}, a_2 = a_3 = \frac{1}{4}, a_4 = a_5 = a_6 = a_7 = \frac{1}{8}, a_8$ through $a_{15} = \frac{1}{16}, a_{16}$ through $a_{32} = \frac{1}{32}$, etc. Obviously, $\lim a_n = 0$, but it is just as obvious that each group of terms with the same value has a sum of ½. Since there are an infinite number of such groups, the partial sums cannot have a finite limit, and the series diverges. (We have $s_{2^n - 1} = \frac{n}{2}$.)

For a geometric series, however, $\lim a_n = 0$ is necessary and sufficient. Provided that $a_1 \neq 0, \lim_{n \to \infty} a_n = 0$ is the same as $|r| < 1$. As in Sec. 4.1, we have $s_n = a_1 \frac{1 - r^n}{1 - r}$, and by result F in Sec. 7.10.2 and Th. 7-13, $\lim_{n \to \infty} s_n = \sum_{n=1}^{\infty} a_n = \frac{a_1}{1 - r}$ if $|r| < 1$.

Note that if a_n is replaced with ca_n, then S is replaced with cS. This is true in general, and is stated as the next theorem.

Th. 7-15 (Multiples of series).

If $\sum a_n$ converges and has as sum S, then $\sum ca_n$ converges and has as sum cS. If $\sum a_n$ diverges, then $\sum ca_n$ diverges unless $c = 0$.

Proof: Replacing a_n with ca_n multiplies all partial sums by c, and therefore by Th. 7-13 multiplies their limit S by c as well. On the other hand, $\sum ca_n$ cannot converge if $\sum a_n$ diverges, unless $c = 0$. For otherwise, $\frac{1}{c}$ exists and $\sum \frac{1}{c} \cdot ca_n$ would converge, by the first half of this theorem. But $\frac{1}{c} \cdot ca_n = a_n$.

Infinite repeating decimals are infinite geometric series in disguise. For example, $0.33333333\ldots$ means $\frac{3}{10} + \frac{3}{100} + \frac{3}{1000} + \frac{3}{10000} + \cdots = \sum_{n=1}^{\infty} \frac{3}{10^n}$. This is a geometric series with $a_1 = \frac{3}{10}$ and $r = \frac{1}{10}$, so $S = \frac{3/10}{1 - \frac{1}{10}} = \frac{3/10}{9/10} = \frac{3}{9} = \frac{1}{3}$. The calculations proceed in the same manner as those in Sec. 6.5.

7.10.5 Tests for Convergence of Series Without Negative Terms

When working with infinite series, there are two separate but related objectives: determining whether the series converges or diverges, and finding the sum of a convergent series. Since a divergent series has no sum, the first-mentioned task must precede the second. We begin the discussion of tests for convergence with series lacking negative terms because of the following result about their associated sequences of partial sums.

Th. 7-16.
If a non-decreasing sequence is bounded from above, it converges.
(A sequence a_n is non-decreasing if, for all n and all $m > n$, $a_m \geq a_n$.)

Proof: Let A denote the set of values of the sequence. Since A is bounded from above (and nonempty), the Axiom of Completeness guarantees the existence of a least upper bound for A, which we represent by L. The claim is that L is also the limit of the sequence. Since L is the lub for A, given any $\epsilon > 0$ there is an element in A that exceeds $L - \epsilon$. This element, of course, is in the infinite sequence, so we denote it by a_n. Now, if m is bigger than n, then $a_m \geq a_n > L - \epsilon$ because the sequence is non-decreasing. Also, since L is the lub of A, it is also an upper bound, and for all $m > n$ (or less than it, for that matter), $a_m \leq L$. Hence, for any positive ϵ, there exists an n such that $m > n$ implies $L - \epsilon < a_m \leq L$ and hence $|a_m - L| < \epsilon$, Q. E. D.

This theorem not only provides a convergence test but also limits the ways in which a divergent infinite series with no negative terms can behave: Divergence occurs only if the limit of the sequence of partial sums is infinite – the series cannot diverge by "oscillation." This observation enables us to prove the following absolutely cardinal theorem.

Th. 7-17 (Comparison test for series with no negative terms).
Let a_n and b_n be two infinite sequences such that for all m bigger than a certain index n, $a_m \geq b_m \geq 0$. Then, if the infinite series $\sum a_n$ converges, so does $\sum b_n$, while if $\sum b_n$ diverges, then $\sum a_n$ diverges as well.

Proof: We ignore the finite sum of all terms with index less than n and concentrate on the *tails*, the infinite part of each series. Now, if $\sum_{m=1}^{\infty} a_m$ converges, then $\sum_{k=n}^{\infty} a_k$ also converges; therefore its sequence of partial sums is bounded from above. Since $b_k \leq a_k$ for the entire tail, the partial sums of $\sum_{k=n}^{\infty} b_k$ cannot exceed the corresponding sums for $\sum_{k=n}^{\infty} a_k$; therefore, they are bounded too. Since the b's are nonnegative, $\sum_{k=n}^{\infty} b_k$ converges by Th. 7-16; therefore $\sum_{k=1}^{\infty} b_k$ converges as well. For the other half, assume $\sum_{k=1}^{\infty} b_k$ divergent. Therefore, the tail $\sum_{k=n}^{\infty} b_k$ diverges as well, and, by the observation above, its partial sums must approach infinity. As before, but reversed, $a_m \geq b_m$; therefore, the partial sums of $\sum_{k=n}^{\infty} a_k$ cannot fall below those of $\sum_{k=n}^{\infty} b_k$. Since the latter approach infinity, the former must do likewise; therefore $\sum_{k=n}^{\infty} a_k$ diverges, as does $\sum_{k=1}^{\infty} a_k$.

This theorem provides us with several examples of both convergent and divergent series. For example, the harmonic series $\sum_{n=1}^{\infty} \frac{1}{n}$ diverges. We compare it to the divergent series from the preceding section: $\frac{1}{2} + \frac{1}{4} + \frac{1}{4} + \frac{1}{8} + \frac{1}{8} + \frac{1}{8} + \frac{1}{8} + \frac{1}{16} + \frac{1}{16} + \cdots$. Each term of $\sum \frac{1}{n}$ is greater than the corresponding term in that series, and all the terms in both series are positive; therefore the harmonic series diverges. On the other hand, consider $\sum_{n=1}^{\infty} \frac{1}{n!} = 1 + \frac{1}{2} + \frac{1}{6} +$

$\frac{1}{24} + \frac{1}{120} + \cdots$. After the second term, every term in this series is positive and less than the corresponding term in the convergent geometric series $\sum_{n=1}^{\infty} \frac{1}{2^{n-1}}$; therefore, $\sum \frac{1}{n!}$ converges. We can also say something about the sum of this series: it is less than the sum of the geometric series, which is 2. (In fact, this series sums to $e - 1 = 1.7183-$, as will be seen eventually.)

In addition to allowing us to show convergence or divergence of many series outright, the comparison test also stands behind the two most important tests for convergence: the ratio test and the integral test. These will be presented as the next two theorems.

Th. 7-18 (Ratio test).

Let $\sum a_n$ be a nonnegative infinite series and let $r = \lim_{n \to \infty} \frac{a_{n+1}}{a_n}$. Then:

$r > 1$ means $\sum a_n$ diverges;
$r < 1$ means $\sum a_n$ converges;
$r = 1$ gives no conclusive result.

Proof (abbreviated): If $r > 1$, then eventually the terms of the series increase, and the series diverges by the nth-term test. If $r < 1$, then let r' be any number between r and 1; eventually the terms of the sequence fall below the corresponding terms in the convergent $\sum (r')^n$, so $\sum a_n$ converges by the comparison test. Finally, both $\sum \frac{1}{n}$ and $\sum \frac{1}{n^2}$ produce $r = 1$, but the former diverges and the latter converges. (Indeed, $\sum \frac{1}{n^{1+\epsilon}}$ converges, as will be shown below.)

We note that in this theorem a nonexistence of r is not an inconclusive result. If r approaches infinity, then it exceeds 1, and the series diverges. If r oscillates, we check the values it oscillates between. The values of $\frac{a_{n+1}}{a_n}$ form a sequence themselves; if a tail of that sequence lies entirely between 0 and 1, the series converges, and a tail entirely above 1 signals a divergent series. If all tails have points above and below 1, an inconclusive result does occur.

Th. 7-19 (Integral test).

Let $f(n)$ be a nonnegative, non-increasing function defined for all real numbers ≥ 1. Then $\sum_{n=1}^{\infty} f(n)$ and $\int_1^{\infty} f(x)dx$ either both diverge or both converge; and, in the latter case, the sum of the infinite series lies between $\int_1^{\infty} f(x)dx$ and $f(1) + \int_1^{\infty} f(x)dx$.

(Notation: The symbol $\int_1^{\infty} f(x)dx$ means $\lim_{a \to \infty} \int_1^a f(x)dx$. The integral converges if and only if the limit exists.)

Proof: First, draw a graph of the situation. (See Fig. 7-22.) By Th. 7-10, $\int_1^a f(x)dx = \int_1^2 f(x)dx + \int_2^3 f(x)dx + \cdots + \int_{a-1}^a f(x)dx$. Letting a become infinite, the integral $\int_1^{\infty} f(x)dx$ becomes an infinite series. Since f is nonnegative, the infinite series has no negative terms. Now consider $\sum_{n=1}^{\infty} f(n)$. This is equal to the areas of the taller rectangles in Fig. 7-22. Since f is non-increasing, $f(n) \geq \int_n^{n+1} f(x)dx$. Hence, if $\int_1^{\infty} f(x)dx$ diverges, so does $\sum_{n=1}^{\infty} f(n)$. However, we can also consider $\sum_{n=1}^{\infty} f(n)$ as $f(1)$ plus the areas of the shorter rectangles in Fig. 7-22. Again, since f is non-increasing, $f(n+1) \leq \int_n^{n+1} f(x)dx$; therefore, if $\int_1^{\infty} f(x)dx$ converges, so does $\sum_{n=2}^{\infty} f(n)$ and therefore $\sum_{n=1}^{\infty} f(n)$. Once convergence is established, the inequality $\int_1^{\infty} f(x)dx \leq \sum_{n=1}^{\infty} f(n) \leq f(1) + \int_1^{\infty} f(x)dx$ becomes obvious as a result of the preceding discussion.

The integral test lends itself readily to the computation of the sum of a convergent series. This is apparent with the concluding inequality, and becomes even more so when

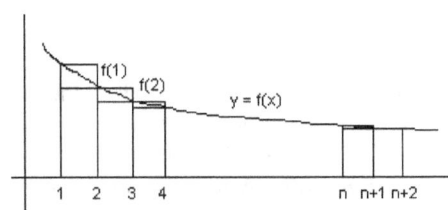

Fig. 7-22. The integral test on a nonincreasing function $f(x)$. The area under the curve to the right of the line $x = 1$ is greater than the areas of the shorter rectangles but smaller than the areas of the taller rectangles. Either of the sums of the rectangles' areas is an infinite series.

one realizes that any positive integer can be substituted for its 1's. Hence, we can estimate the sum of certain series by calculating the sum of the first several terms and then using the integral test to "fill in" the rest. For example, consider $\sum_{n=1}^{\infty}\frac{1}{n^3}$. This converges because the associated integral $\int_1^{\infty}\frac{dx}{x^3} = \lim_{a\to\infty}\int_1^a\frac{dx}{x^3} = \lim_{a\to\infty} -\frac{1}{2}\cdot\frac{1}{x^2}\Big|_1^a = \frac{1}{2}$ converges. The inequality tells us immediately that the sum of the series lies between ½ and 1½, but these limits are not very precise. To get a good estimate, we add the first ten terms of the series (total = 1.197531986), and then replace the 1's with 11's to get $\int_{11}^{\infty}\frac{dx}{x^3} \le \sum_{n=11}^{\infty}\frac{1}{n^3} \le \frac{1}{11^3} + \int_{11}^{\infty}\frac{dx}{x^3}$. The value of the integral is $\frac{1}{2}\cdot\frac{1}{11^2} = \frac{1}{242}$, and splitting the difference in the inequality by adding $\frac{1}{2\cdot11^3} = \frac{1}{2662}$ results in an estimate for $\sum_{n=1}^{\infty}\frac{1}{n^3}$ of 1.202039875. The actual sum of the series is 1.202056903, so the extra work involved in using the integral test increased the accuracy of the estimate from two decimal places to four. In other terms, the number of summands necessary to produce four-decimal-place accuracy was reduced from about 100 to twelve. (In fact, it requires 171 terms to reach a sum greater than 1.202039875.)

7.10.6 Series With Negative Terms: Alternating Series, Leibniz' Theorem, Absolute and Conditional Convergence

So far, all the series we have studied have contained only nonnegative terms. Most series, however, contain both positive and negative terms. Fortunately, the most important series that contain negative terms are *alternating*; that is, every other term is negative, and the ones in between are positive. Alternating series are covered by the easy-to-apply convergence theorem below.

Th. 7-20 (Leibniz' Theorem).
If $\sum a_n$ is a series that satisfies all of (1) – (3) below, it converges.
(1) $\sum a_n$ is alternating: the signs of any two consecutive terms are different.
(2) $|a_n| \ge |a_{n+1}|$ for all n bigger than some index M.
(3) $\lim_{n\to\infty} a_n = 0$.

Proof: For simplicity we assume $M = 0$. If an actual series requires $M > 0$, we combine the finite sum of all terms before a_M into a single term. We also assume that a_1 is positive. If a_1 is negative, we can start with a_2 which must be positive.
We write $\sum_{n=1}^{\infty} a_n$ as $a_1 - (|a_2| - |a_3|) - (|a_4| - |a_5|) - (|a_6| - |a_7|) - \cdots$. Because of (2), each of the parentheses encloses a positive or zero expression. Therefore, $s_1, s_3, s_5, \ldots, s_{2n+1}$ form a non-*increasing* sequence. We then rewrite $\sum_{n=1}^{\infty} a_n$ as $(|a_1| - |a_2|) + (|a_3| - |a_4|) + (|a_5| - |a_6|) + (|a_7| - |a_8|) + \cdots$. Again, because of (2), each parenthesized expression is nonnegative, so $s_2, s_4, s_6, \ldots, s_{2n}$ form a non-*decreasing* sequence. We see that $s_1 \ge s_2, s_3 \ge s_4 \ge s_2, s_5 \ge s_6 \ge s_4 \ge s_2$, etc., so that for all n, $s_{2n+1} \ge s_2$. Therefore, s_{2n+1} is a non-increasing sequence bounded from below, so it converges (Th. 7-16, with a minus sign inserted). Similarly, $s_2 \le s_3, s_4 \le s_5 \le s_3, s_6 \le s_7 \le s_5 \le s_3$, etc., so that s_{2n} is bounded from above and must converge. Finally, both s_{2n+1} and s_{2n} must converge to the *same* limit, as their difference at any point is a_{2n+1}, which approaches zero by (3). This implies that the merged sequence s_n, and therefore $\sum a_n$, must converge, Q. E. D.

This theorem not only makes it easy to determine the convergence of many series but also enables the quick estimation of the sum of a convergent alternating series. The argument above makes it clear that the sum S must lie between any two consecutive partial sums s_n and s_{n+1} of the series. Particularly, if the absolute values of the series's terms are decreasing slowly, we may "split the difference" between two consecutive partial sums to get an estimate for the total sum of the series.
Examples of series to which Leibniz' theorem does and does not apply include:
(a) The alternating harmonic series: $1 - \frac{1}{2} + \frac{1}{3} - \frac{1}{4} + \frac{1}{5} - \cdots = \sum_{n=1}^{\infty}\frac{(-1)^n}{n}$.
This series meets all three conditions in the theorem and therefore converges.
(b) The alternating reciprocals of factorials: $\frac{1}{2} - \frac{1}{6} + \frac{1}{24} - \frac{1}{120} + \cdots = \sum_{n=2}^{\infty}\frac{(-1)^n}{n!}$.
This series also meets all the conditions and converges.
(c) The series $1 - \frac{1}{2} - \frac{1}{3} + \frac{1}{4} - \frac{1}{5} - \frac{1}{6} + \frac{1}{7} - \frac{1}{8} - \cdots$.
This series is not an alternating series and therefore Leibniz' theorem does not apply.
(d) The series $2 - 1 + 1 - \frac{1}{2} + \frac{2}{3} - \frac{1}{3} + \frac{2}{4} - \frac{1}{4} + \frac{2}{5} - \frac{1}{5} + \cdots$.

This series, although alternating, does not have terms that decrease steadily in absolute value (e. g., $\frac{1}{3} < \frac{2}{4}$), so the theorem does not apply.

(e) The series $2 - \frac{3}{2} + \frac{4}{3} - \frac{5}{4} + \frac{6}{5} - \frac{7}{6} + \cdots = \sum_{n=1}^{\infty} (-1)^{n+1} \frac{n+1}{n}$.

This series is both alternating and decreasing in absolute value, but the terms do not approach zero. The sequences s_{2n} and s_{2n+1} both converge, but to different values. The series "diverges by oscillation," which could not happen if the series had only positive terms.

An example of using the theorem to compute a sum estimate uses the alternating harmonic series. We add (and subtract) the first ten terms to get 0.645635, then add half of the next, positive, term, 1/11, to get 0.691089. The exact sum of the alternating harmonic series is ln 2 = 0.693147, so we have a decent but not great value. The calculation in Sec. 7.5.2 produced a better result with about as much work; however, an improved series for ln 2 will shortly be developed that produces a more accurate value still.

Normally, we think of addition as commutative: 2 + 3 = 3 + 2 and 2 + (- 3) = (- 3) + 2. The same is true for sums of 3, 4, …, one billion, … terms, but with infinite series, some funny things can happen when the terms are scrambled. The outstanding example is the alternating harmonic series. As it is usually written, the sum is ln 2, but let's see what happens if we move the terms around. First, we separate the positives from the negatives: $1 + \frac{1}{3} + \frac{1}{5} + \frac{1}{7} + \cdots - \frac{1}{2} - \frac{1}{4} - \frac{1}{6} - \frac{1}{8} - \cdots$. Now, we have two divergent series instead of one convergent one. More intriguing is the result obtained when each odd-denominator term is paired with the term with twice its denominator: $\left(1 - \frac{1}{2}\right) + \left(\frac{1}{3} - \frac{1}{6}\right) + \left(\frac{1}{5} - \frac{1}{10}\right) + \cdots - \frac{1}{4} - \frac{1}{8} - \frac{1}{12} - \cdots$. The unpaired negative terms have denominators which are multiples of 4. Now, when the subtractions inside each parenthesis are carried out and the results rearranged in order of increasing denominators, we get $\frac{1}{2} - \frac{1}{4} + \frac{1}{6} - \frac{1}{8} + \frac{1}{10} - \frac{1}{12} + \cdots$. Each term in the original series has been replaced with a corresponding term exactly half as big. The sum of the series has been changed from ln 2 to ln $\sqrt{2}$ just by rearranging the terms! So, in the infinite case, addition is NOT always commutative.

If you wish to know what kinds of series may fall prey to this trick, I will tell you that it is the hallmark of series that are only *conditionally* convergent. The convergence of such a series is conditioned on the presence of negative terms in it; that is, if we replaced all the minus signs with plus signs, the series would become divergent. The alternating harmonic series is conditionally convergent because reversing its minus signs turns it into the divergent harmonic series. Not all series with negative terms are conditionally convergent: series (b) above remains convergent when its minus signs are replaced with plus signs. Since this operation amounts to taking the absolute values of all the terms, we say that series (b) is *absolutely* convergent. Series that are absolutely convergent are not susceptible to the rearranging of terms = changing the sum trick, so, for them, addition is still commutative. To "sum"-marize:

If $\sum a_n$ converges and $\begin{cases} \sum |a_n| \text{ converges, the series is absolutely convergent.} \\ \sum |a_n| \text{ diverges, the series is conditionally convergent.} \end{cases}$

Of course, for series with no negative terms, convergence is always absolute.

7.10.7 Power Series, Taylor's Theorem, and Computation

Given an infinite sequence a_n, the infinite series $\sum_{n=0}^{\infty} a_n x^n$ is called a *power series* in the variable x. Depending on the values of the a's, this series may converge for all values of x, some of them, or only when $x = 0$. More precisely, we have the following theorem:

Th. 7-21 (Convergence of power series).
A power series $\sum a_n x^n$ may behave in any one of these three ways (and only these three):
1. The series converges only when $x = 0$.
2. There exists a number r, called the *radius of convergence*, such that the series converges for all x satisfying $|x| < r$, and diverges whenever $|x| > r$.
3. The series converges for every real value of x.
In case 2, the series may either converge or diverge when $x = r$ or $- r$. In all cases except $x = r$ or $- r$, the convergence of a convergent power series is absolute; in those two instances, it may be either absolute or conditional.

The proof of the theorem hinges on the fact that, for positive numbers, $a > b$ implies $a^n > b^n$ for all positive integers n, as well as the fact that absolutely convergent series are convergent. (This latter fact is not as obvious as it seems, and may itself need a proof in a rigorous text.) With these two facts, proper application of the comparison test to series with $x = c$ and $x = d$, with $0 < c < d$, shows that the convergence of the latter implies that of the former, while the divergence of the former implies that of the latter. The property of absolute convergence is brought in to take care of situations where some of the a's, or c and d, are negative.

To find the radius of convergence of a power series, all we need do is determine $\lim_{n \to \infty} \left| \frac{a_{n+1}}{a_n} \right|$. The radius of convergence is the reciprocal of this limit. If the limit is zero, case 3 above applies, and if the limit is infinite, case 1 applies. In pathological situations, we may need to calculate $\left[\lim_{n \to \infty} \left| \frac{a_{n+k}}{a_n} \right| \right]^{1/k}$ for some k, or, if the a's are random (or seem to be), we may not have a power series at all, because an infinite sequence must be well-defined before it can be used for the a's in a power series. If the a's are bounded (only a finite number greater than M and only a finite number less that L, with $M > L > 0$), then the radius of convergence is 1.

Now, power series are important to mathematics because some of them serve as tools for calculating the values of transcendental functions such as sin, cos, exp, and ln (and other reasons as well). Toward that end, let us return to the example in Sec. 7.3.1: $P(x) = 8x^4 - 5x^3 + 7x - 9$. This function was found to satisfy $P(1) = 1$ and $P'(1) = 24$, so that when Δx was small, $P(1 + \Delta x)$ was close to $1 + 24\Delta x$. In this notation, we now replace 1 with the constant (for this example) x, and Δx with the variable a. This gives us $P(x + a) \approx P(x) + aP'(x)$, or $P(x + a) - P(x) - aP'(x) \approx 0$. Let us differentiate this equation with respect to a, remembering that x is a constant; we get $P'(x + a) - P'(x) \approx 0$; doing it again gives us $P''(x + a) \approx 0$ - but this second differentiation will give us an approximation that in general will not be good. In the example, we found $P(1.01) = 1.24332708$, so that $P(1.01) - P(1) - 0.01 \cdot P'(1) = 0.00332708$. Now, this number seems to closely match up with $P''(1) = 66$; in fact, it is very nearly $P''(1) \times \frac{(0.01)^2}{2}$. In the new notation, we have $P(x + a) - P(x) - aP'(x) \approx \frac{a^2}{2} P''(x)$, or $P(x + a) \approx P(x) + aP'(x) + \frac{a^2}{2} P''(x)$. The "residue" obtained from substituting our example into this notation is 0.00002708, which is nearly $P'''(1) \times \frac{(0.01)^3}{6}$. We note that $6 = 3!$ (and $2 = 2!$) for future reference. We now have $P(x + a) \approx P(x) + aP'(x) + \frac{a^2}{2!} P''(x) + \frac{a^3}{3!} P'''(x)$. The "residue" in the example is now down to 8×10^{-8} which is exactly $P^{(4)}(1) \times \frac{(0.01)^4}{24}$, and, of course, $24 = 4!$. The pattern should now be clear: the fourth-degree polynomial in $(x + a)$ is now rewritten as a fourth-degree polynomial in a alone: $P(x + a) = P(x) + aP'(x) + \frac{a^2}{2!} P''(x) + \frac{a^3}{3!} P^{(3)}(x) + \frac{a^4}{4!} P^{(4)}(x)$. (Remember that x is a constant, so that $P'(x)$, etc., are as well.) We note that we are building up a power series in the variable a, with the coefficient of a^n being $\frac{P^{(n)}(x)}{n!}$. This is true in general for polynomials, and it can be extended to all functions provided certain conditions are met. These conditions are stated below in:

Th. 7-22 (Taylor's theorem, Lagrange remainder).
If F is a function that has derivatives of orders up to $n + 1$ at x, and these derivatives are continuous on the closed interval $[x, x + a]$, then

$$F(x + a) = F(x) + aF'(x) + \frac{a^2}{2!} F''(x) + \frac{a^3}{3!} F^{(3)}(x) + \cdots + \frac{a^n}{n!} F^{(n)}(x) + \frac{a^{n+1}}{(n + 1)!} F^{(n+1)}(x + c)$$

for some c between 0 and a.

Actually, only F itself need be continuous on the closed interval; the derivatives need only be continuous on the open interval with the same endpoints.

This theorem, with the Lagrange remainder (the term involving c), is considered an extension of the Mean Value Theorem for Derivatives (Th. 7-7); indeed, if we let $n = 0$, Taylor's Theorem reduces to Th. 7-7. Th. 7-22 can be proven from that former result, using mathematical induction to ratchet up the value of n.

Unfortunately, this theorem in and of itself is not all that useful in computations because of the unspecified value of c in the remainder term. However, if we can guarantee that the remainder will approach zero (or at least get very small) as n increases, we can get a highly accurate result. Denoting the remainder term by $R_n(x, x + a)$, we determine the maximum possible value for it by taking the worst possible choice for c – the one that yields the largest value of $F^{(n)}(x + c)$. We call this biggest remainder $\max_c[R_n(x, x + a)]$. If x and a are held constant, then this maximum is a function of n alone. Finally, we have:

Th. 7-22A (Taylor series).

If F has derivatives of all orders in the closed interval $[x, x + a]$, and if $\lim_{n \to \infty} \max_c [R_n(x, x + a)] = 0$, then the infinite series $F(x) + aF'(x) + \frac{a^2}{2!}F''(x) + \frac{a^3}{3!}F^{(3)}(x) + \frac{a^4}{4!}F^{(4)}(x) + \cdots$ converges and has $F(x + a)$ as its sum.

The infinite series in this theorem is the Taylor series for the function F. It is frequently convenient to use zero for x in a Taylor series, in which case the series is named for another mathematician and is called a Maclaurin series instead. Maclaurin series are particularly useful for calculating e^x, $\cos x$, and $\sin x$. To illustrate, we write the Maclaurin series for e^x like this: Since e^x is its own derivative and is everywhere continuous, the first half of the conditions in Th. 7-22A is satisfied. As for the remainder, we have $R_n(0, a) = \frac{a^{n+1}}{(n+1)!}e^c$, where $0 < c < a$ and therefore $e^c < e^a$. Now, as n increases, $\frac{(n+1)!}{n!} = n + 1$ will eventually overwhelm $\frac{a^{n+1}}{a^n} = a$, no matter how big a is. Therefore, $\max_c [R_n(0, a)]$ will eventually start to decrease, at which point it must behave like x^n for $|x| < 1$, which approaches zero by result F in Sec. 7.10.1. This fulfills all the conditions of Th. 7-22A, so the series must converge for every a, and converge to e^a. Since $e^0 = 1$, the series is $1 + x + \frac{x^2}{2!} + \frac{x^3}{3!} + \frac{x^4}{4!} + \cdots$. Using the first four terms, the value of $e^{0.1}$ is calculated to be 1.10517. The first nine terms give e itself as 2.71828.

Similar reasoning shows that the Maclaurin series for $\sin x$ is $x - \frac{x^3}{3!} + \frac{x^5}{5!} - \frac{x^7}{7!} + \cdots$ and that for $\cos x$ is $1 - \frac{x^2}{2!} + \frac{x^4}{4!} - \frac{x^6}{6!} + \cdots$. Now watch as Taylor and Maclaurin series prove the trigonometric identity $\sin x + \cos x = \sqrt{2}\sin(x + \frac{\pi}{4})$. We determine the series for $\sin(x + \frac{\pi}{4})$ as $\sin\frac{\pi}{4} + x\cos\frac{\pi}{4} - \frac{x^2}{2!}\sin\frac{\pi}{4} - \frac{x^3}{3!}\cos\frac{\pi}{4} + \frac{x^4}{4!}\sin\frac{\pi}{4} + \frac{x^5}{5!}\cos\frac{\pi}{4} - \cdots$. Since $\sin\frac{\pi}{4} = \cos\frac{\pi}{4} = \frac{\sqrt{2}}{2}$, $\sin(x + \frac{\pi}{4}) = \frac{\sqrt{2}}{2} + x\frac{\sqrt{2}}{2} - \frac{x^2}{2!}\cdot\frac{\sqrt{2}}{2} - \frac{x^3}{3!}\cdot\frac{\sqrt{2}}{2} + \frac{x^4}{4!}\cdot\frac{\sqrt{2}}{2} + \frac{x^5}{5!}\cdot\frac{\sqrt{2}}{2} - \cdots$. When we multiply this series by $\sqrt{2}$ and separate odd and even powers of x, we find that we have recovered the sine and cosine Maclaurin series.

The series for the sine and cosine are both similar to the series for e^x. This is made even more clear by allowing x to be a complex number, say iz where z is real. Then the series for e^{iz} is $1 + iz + \frac{(iz)^2}{2!} + \frac{(iz)^3}{3!} + \frac{(iz)^4}{4!} + \cdots$. The laws of exponents hold for complex numbers, so $(iz)^n = i^n z^n$. Since $i^2 = -1$, $i^3 = i^2 i = -i$, $i^4 = (i^2)^2 = 1$, and therefore $i^{4n+k} = i^k$. Thus, we can rewrite the series for e^{iz} as $1 - \frac{z^2}{2!} + \frac{z^4}{4!} - \frac{z^6}{6!} + \cdots + i\left(z - \frac{z^3}{3!} + \frac{z^5}{5!} - \frac{z^7}{7!} + \cdots\right) = \cos z + i \sin z$. If we let $z = \pi$ so $\cos z = -1$ and $\sin z = 0$, we have $e^{\pi i} = -1$, or $e^{\pi i} + 1 = 0$. This equation, which ties together the five most important quantities in mathematics, is sometimes referred to as Euler's equation, although that title should go to $e^{iz} = \cos z + i \sin z$ instead. This equation unifies the exponential and trigonometric functions; despite their radically different behavior on the real numbers, they actually are quite similar on the complex numbers.

7.10.8 The Term-by-Term Integration and Differentiation Theorems

We begin this section by observing that the power series $\sum a_n x^n$ defines a function whose domain is the interval of convergence, regardless of the pattern displayed by the a's. The theorems of this section enable us to perform the operations of the calculus on these functions, which may give power series for other functions more difficult to compute.

Th. 7-23A (Term-by-term differentiation).

If $f(x) = \sum_{n=0}^{\infty} a_n x^n$ converges on the open interval $(-r, r)$, then $f'(x) = \sum_{n=1}^{\infty} n a_n x^{n-1}$ on that same interval.

The reason this is called term-by-term differentiation is that to find the derivative of an infinite-series function, we differentiate each of its infinitely many terms and add them up. Formula II for derivatives in Sec. 7.3.1 applies to any finite number of terms, and the step to the infinite is this theorem.

Th. 7-23B (Term-by-term integration).

If $f(x) = \sum_{n=0}^{\infty} a_n x^n$ converges on the open interval $-r < x < r$, then $\int f(x)dx = \sum_{n=0}^{\infty} \frac{a_n x^{n+1}}{n+1} + C$ on the same interval, and $\int_p^q f(x)dx = \sum_{n=0}^{\infty} \frac{a_n(q^{n+1}-p^{n+1})}{n+1}$ if both p and q are in this interval.

This theorem, except for the extra clause for definite integrals, is the exact analogue of Th. 7-23A.

The two theorems mention only the open interval of convergence because power series can either converge or diverge at the endpoints of that interval. Convergence at endpoints can be lost but not gained during differentiation, and gained but not lost during integration.

The main use of these theorems is, given a function with a known power series, to differentiate or integrate the series and thus quickly obtain the series for another function whose derivatives may be difficult to compute. For example, let $f(x) = \frac{1}{x^2+1} = 1 - x^2 + x^4 - x^6 + \cdots$ ($|x| < 1$, Sec. 7.10.4). Applying Th. 7-23B, we have $\int \frac{dx}{x^2+1} = \tan^{-1} x = x - \frac{x^3}{3} + \frac{x^5}{5} - \frac{x^7}{7} + \cdots$ ($|x| < 1$). Similarly, we have $\frac{1}{x+1} = 1 - x + x^2 - x^3 + \cdots$, and therefore $\ln(1+x) = x - \frac{x^2}{2} + \frac{x^3}{3} - \frac{x^4}{4} + \cdots$ for $|x| < 1$. In both of these cases, the series also converges when $x = +1$, and hence $\tan^{-1} 1 = \frac{\pi}{4} = 1 - \frac{1}{3} + \frac{1}{5} - \frac{1}{7} + \cdots$, while $\ln 2 = 1 - \frac{1}{2} + \frac{1}{3} - \frac{1}{4} + \cdots$. Neither of these series is particularly useful for computing π or $\ln 2$ because they converge with extreme slowness, but manipulations of these formulas can be used to facilitate practical computations. For example, in Sec. 5.6 we have the identity $\tan(A+B) = \frac{\tan A + \tan B}{1 - \tan A \tan B}$. As a statement about inverse tangents, this means that if $A = \tan^{-1} x$ and $B = \tan^{-1} y$, then $A + B = \tan^{-1} \frac{x+y}{1-xy}$ (not always a principal value). If we let $x = \frac{1}{2}$ and $y = \frac{1}{3}$ then $\frac{x+y}{1-xy} = 1$ and hence $\pi = 4\tan^{-1}\frac{1}{2} + 4\tan^{-1}\frac{1}{3}$. The two resulting series can be used for a practical calculation of pi to a few dozen digits. As for logarithms, we have $\ln(1+x) = x - \frac{x^2}{2} + \frac{x^3}{3} - \frac{x^4}{4} + \cdots$ and hence $\ln(1-x) = -x - \frac{x^2}{2} - \frac{x^3}{3} - \frac{x^4}{4} - \cdots$. Therefore, $\ln(\frac{1+x}{1-x}) = 2(x + \frac{x^3}{3} + \frac{x^5}{5} + \frac{x^7}{7} + \cdots)$. All these formulas hold only for $|x| < 1$, but as x ranges between -1 and 1, $\frac{1+x}{1-x}$ covers every positive real number. In fact, if $y = \frac{1+x}{1-x}$, $x = \frac{y-1}{y+1}$. Thus, to calculate $\ln 2$, one takes $y = 2$, $x = \frac{y-1}{y+1} = \frac{1}{3}$, and $\ln 2 = 2(\frac{1}{3} + \frac{1}{3\cdot 3^3} + \frac{1}{5\cdot 3^5} + \frac{1}{7\cdot 3^7} + \cdots)$. Only one term beyond those indicated produces a five-decimal-place value of $\ln 2$. The natural logarithms of larger integers can be found by letting $y = \frac{n+1}{n}$, then $x = \frac{1}{2n+1}$. Hence, $\ln \frac{3}{2} = 2(\frac{1}{5} + \frac{1}{3\cdot 5^3} + \frac{1}{5\cdot 5^5} + \frac{1}{7\cdot 5^7} + \cdots)$ and $\ln 3 = \ln 2 + \ln 3/2$. We calculate $\ln 4$ as $2 \ln 2$, $\ln 5$ as $\ln 4 + \ln \frac{5}{4}$, the latter being $2(\frac{1}{9} + \frac{1}{3\cdot 9^3} + \frac{1}{5\cdot 9^5} + \cdots)$, $\ln 6$ as $\ln 2 + \ln 3$, $\ln 7 = \ln 6 + \ln \frac{7}{6}$, $\ln 8 = 3 \ln 2$, $\ln 9 = 2 \ln 3$, etc., using series only for the logarithms of prime numbers.

7.10.9 Solutions of Certain Differential Equations by Power Series

A complicated-looking differential equation such as $x^2 \frac{d^2y}{dx^2} - 3x \frac{dy}{dx} + y = x^2$ can frequently be solved by assuming that a power-series solution $y = \sum a_n x^n$ exists and then substituting that assumed solution into the equation, applying Th. 7-23A, and equating coefficients of like powers of x. The result is usually a recurrence relationship among the a's, which can be used to specify an actual solution involving various constants of integration.

The simplest equation solvable by this process is $\frac{dy}{dx} = y$. Substituting the power-series solution, we have $\sum_{n=1}^{\infty} n a_n x^{n-1} = \sum_{n=0}^{\infty} a_n x^n$. We rewrite the summation on the left so that it features x^n instead of x^{n-1} to get $\sum_{n=0}^{\infty}(n+1)a_{n+1}x^n = \sum_{n=0}^{\infty} a_n x^n$. This tells us that $a_n = (n+1)a_{n+1}$, or $a_{n+1} = \frac{a_n}{n+1}$. The value of a_0 is arbitrary, so we denote it by c. Then $a_1 = \frac{a_0}{1} = c, a_2 = \frac{a_1}{2} = \frac{c}{2}, a_3 = \frac{a_2}{3} = \frac{c}{6}$, and $a_n = \frac{c}{n!}$, so the solution $\sum_{n=0}^{\infty} a_n x^n$ is actually $\sum_{n=0}^{\infty} \frac{c x^n}{n!} = c\sum_{n=0}^{\infty} \frac{x^n}{n!} = ce^x$.

Now consider $x\frac{dy}{dx} = 3y$. The power-series substitution produces $\sum_{n=1}^{\infty} n a_n x^n = \sum_{n=0}^{\infty} 3a_n x^n$. The left side has no constant term, so $a_0 = 0$. As for the others, $na_n = 3a_n$ implies either $n = 3$ or $a_n = 0$, so we conclude that $a_n = 0$ unless $n = 3$, and a_3 may have any value. Therefore, $y = cx^3$.

Second-order equations, of course, are more complicated. Taking the equation at the beginning of this section, we have $\sum_{n=2}^{\infty} n(n+1)a_n x^n - \sum_{n=1}^{\infty} 3na_n x^n + \sum_{n=0}^{\infty} a_n x^n = x^2$. Equating coefficients of like powers of x

gives $a_0 = 0, -3a_1 + a_1 = 0$ or $a_1 = 0$, $2a_2 - 6a_2 + a_2 = 1$ or $a_2 = -\frac{1}{3}$. Other coefficients of powers must be zero unless $n(n-1) - 3n + 1 = 0$, a quadratic equation solved by $n = 2 \pm \sqrt{3}$. Hence, the general solution to this equation is $y = c_1 x^{2+\sqrt{3}} + c_2 x^{2-\sqrt{3}} - \frac{x^2}{3}$.

Although this solution is not in a power-series form because of the irrational exponents, the reasoning is still valid. A solution of this type (perhaps with integer exponents on x, as in the previous example) is to be expected whenever, for each derivative $\frac{d^n y}{dx^n}$ in the equation, a term kx^n multiplies it. If we change the equation slightly to $(x^2 - 1)\frac{d^2 y}{dx^2} - 3x\frac{dy}{dx} + y = x^2$, the solution becomes a true power series. The solution has a_0 and a_1 as arbitrary constants, then $a_2 = \frac{a_0}{2}, a_3 = -\frac{a_1}{3}, a_4 = -\frac{3a_2+1}{12}, a_5 = -\frac{a_3}{10}, a_6 = \frac{a_4}{30}, a_7 = \frac{a_6}{7}, a_8 = \frac{13a_6}{56}, \ldots, a_{n+2} = a_n \frac{n^2-4n+1}{n^2+3n+2}$ for all $n > 2$. The interval of convergence of this power series is $-1 \le x \le 1$ regardless of the choice of a_0 and a_1.

The type of equation that can be solved by this method is any of the form $P_n(x)\frac{d^n y}{dx^n} + P_{n-1}(x)\frac{d^{n-1}y}{dx^{n-1}} + \cdots + P_2(x)\frac{d^2 y}{dx^2} + P_1(x)\frac{dy}{dx} + P_0(x)y = F(x)$, where $F(x)$ has a known power-series expansion and for each i, $P_i(x)$ is a polynomial in x of degree no higher than i. In particular, $P_0(x)$ must be a constant. This type is known as a linear, nth-order differential equation with polynomial coefficients. If F is zero, the equation is called *homogeneous*, otherwise it is non-homogeneous. In general, the solution to a non-homogeneous equation is the same as the corresponding homogeneous one, except for an extra expression that does not have an arbitrary constant in it.

7.10.10 The Series $\sum_{n=1}^{\infty} \frac{n^a}{x^n}$ and the "Christensen Coefficients"

Imagine that you are playing some game and need only a roll of 5 on one die to win. Your first roll in the situation, a 3, gets you no closer. Subsequent rolls of 4, 2, 6, 3, 4, ..., don't do any good either. You know you have a 1 in 6 chance of winning on any one roll, but how many rolls is it likely to take before you get the 5? It just might take forever. The mathematics of probability says that it is *almost* certain that you will eventually roll a 5, and, with a little calculus thrown in, tell you how long, on average, you will have to wait.

On each roll, you have a 1 in 6 chance of coming up with a 5, and therefore a 5 in 6 chance of *not* doing so. Hence, you have a $\frac{1}{6} \times \frac{5}{6}$ chance of rolling your first 5 on your second roll. Similarly, you have a $\frac{1}{6} \times \frac{5}{6} \times \frac{5}{6}$ chance of rolling your first 5 on the third try. In general, the probability of rolling the first 5 on your nth roll is $5^{n-1}/6^n$. To add the probabilities for all positive n, note that $\{5^{n-1}/6^n\}$ is a geometric progression with $a_1 = 1/6$ and $r = 5/6$. Hence (Sec. 7.10.4), $S = \frac{1/6}{1-\frac{5}{6}} = 1$. (Warning: A probability of 1 for an event does not mean it is certain when an infinite number of possibilities exist. For example, if you pick an integer at random, the probability that it is composite is 1, but prime numbers still exist.)

To find out how long, on average, you must wait to get a 5, you must multiply the nth term in the series by n and compute the sum. However, the series is no longer geometric. Instead of $\sum_{n=1}^{\infty} ar^{n-1}$, it is now $\sum_{n=1}^{\infty} nar^{n-1}$, and is known as the *expected* geometric series. The operation of multiplying the nth term in a series (finite or infinite) by n is an application of the *expectation operator*. This, of course, can be repeated, leading to the "a-fold expected geometric series," $\sum_{n=1}^{\infty} \frac{n^a}{x^n}$.

To sum the n-fold expected geometric series, we start with an ordinary geometric series. When $|x| < 1$, $1 + x + x^2 + x^3 + \cdots = \frac{1}{1-x}$. We multiply this equation by x: $x + x^2 + x^3 + x^4 + \cdots = \frac{x}{1-x}$ and then use Th. 7-23A: $1 + 2x + 3x^2 + 4x^3 + \cdots = \frac{1}{(1-x)^2}$, where the numerator on the right is derived from $(1-x)\cdot 1 - (x)\cdot(-1)$. Note that we have, in effect, applied the expectation operator to the left side.

Now, we repeat the process, first multiplying by x: $x + 2x^2 + 3x^3 + 4x^4 + \cdots = \frac{x}{(1-x)^2}$, and then differentiating: $1 + 4x + 9x^2 + 16x^3 + \cdots = \frac{(1-x)^2 \cdot 1 - x \cdot 2 \cdot (1-x) \cdot (-1)}{(1-x)^4} = \frac{x+1}{(1-x)^3}$. The expectation operator has now been applied twice to the left side. Repeating the process again results first in $x + 4x^2 + 9x^3 + 16x^4 + \cdots = \frac{x^2+x}{(1-x)^3}$ and then $1 + 8x + 27x^2 + 64x^3 + \cdots = \frac{x^2+4x+1}{(1-x)^4}$. The computation in the numerator is $(1-x)^3 \cdot (2x+1) - $

$(x^2 + x)(3)(1 - x)^2(-1) = [(1 - x)^2][(1 - x)(2x + 1) + 3x^2 + 3x] = [(1 - x)^2][2x + 1 - 2x^2 - x + 3x^2 + 3x]$. The second bracket simplifies to $x^2 + 4x + 1$, and the $(1 - x)^2$ reduces the denominator's exponent from 6 to 4.

Continuing to apply this process leads to higher powers of the integers on the left side, higher exponents on $(1 - x)$ in the right side's denominator, and larger polynomials in the right side's numerator. If we stop after the $(n + 1)$st multiplication by x, we will have something like $x + 2^n x^2 + 3^n x^3 + 4^n x^4 + \cdots = \frac{P_n(x)}{(1-x)^{n+1}}$, where the degree of $P_n(x)$ is n. We have on the left an n-fold expected geometric series, and on the right a closed-form sum for it (as long as $|x| < 1$). The coefficients of $P_n(x)$ are highly interesting; I immodestly call the the "Christensen coefficients." The coefficient of x^r in $P_n(x)$ is denoted $\left\langle {n \atop r} \right\rangle$. It can be shown that $\left\langle {n \atop r} \right\rangle = r \left\langle {n-1 \atop r} \right\rangle + (n - r + 1) \left\langle {n-1 \atop r-1} \right\rangle$, a recurrence formula similar to that for binomial coefficients. Also, $\left\langle {n \atop r} \right\rangle = \left\langle {n \atop n-r+1} \right\rangle$, similar to $\binom{n}{r} = \binom{n}{n-r}$. However, there are differences, as $\sum_{r=1}^{n} \left\langle {n \atop r} \right\rangle = n!$, but $\sum_{r=0}^{n} \binom{n}{r} = 2^n$. The first few rows of the Christensen coefficient table are shown below.

1						
1	1					
1	4	1				
1	11	11	1			
1	26	66	26	1		
1	57	302	302	57	1	
1	120	1191	2416	1191	120	1

To complete the derivation of the sum $\sum_{n=1}^{\infty} \frac{n^a}{x^n}$, we replace n with a in the equation $x + 2^n x^2 + 3^n x^3 + \cdots = \frac{P_n(x)}{(1-x)^{n+1}}$, then replace x with $1/x$ (so the equation now holds only if $|x| > 1$) and multiply numerator and denominator of the fraction on the right by x^{a+1}. Because of the law $\left\langle {a \atop r} \right\rangle = \left\langle {a \atop a-r+1} \right\rangle$, this does not affect $P_a(x)$. The final result is $\sum_{n=1}^{\infty} \frac{n^a}{x^n} = \frac{\sum_{r=1}^{a} \left\langle {a \atop r} \right\rangle x^r}{(x-1)^{a+1}}$, for a a positive integer and $|x| > 1$.

To use this formula to solve the problem at the beginning of this section, we let $x = 6/5$, the reciprocal of the ratio 5/6 of the original geometric series. We write the series as $\frac{1}{6} + \frac{2(\frac{1}{6})}{x} + \frac{3(\frac{1}{6})}{x^2} + \frac{4(\frac{1}{6})}{x^3} + \cdots$. To make the first term 1, we divide the entire series by 6 and multiply the terms by 6. We now have $\frac{1}{6}(1 + \frac{2}{x} + \frac{3}{x^2} + \frac{4}{x^3} + \cdots)$, and we must multiply each denominator by x to match the exponents up to their respective numerators. Therefore, we multiply the series by 6/5, changing the pre-factor to 1/5, to reach $\frac{1}{5}(\frac{1}{x} + \frac{2}{x^2} + \frac{3}{x^3} + \frac{4}{x^4} + \cdots)$. The series now has the form $\sum_{n=1}^{\infty} \frac{n^a}{x^n}$, where $a = 1$ and $x = 6/5$. Therefore, its sum is $\frac{6/5}{(\frac{6}{5}-1)^2} = 30$, and the pre-factor reduces the sum of the original series to 6. Intuition, in this case, is right. However, if our player is impatient and twice applies the expectation operator to get $\frac{1}{6} + \frac{4(\frac{1}{6})}{6/5} + \frac{9(\frac{1}{6})}{(\frac{6}{5})^2} + \cdots$, the formula says 66, not 36.

A closed-form expression for the coefficients $\left\langle {a \atop r} \right\rangle$ is $\left\langle {a \atop r} \right\rangle = r^a - \binom{a+1}{1}(r - 1)^a + \binom{a+1}{2}(r - 2)^a - \cdots$, where the $\binom{a+1}{1}$, etc. are binomial coefficients, and the terms alternate in sign until the finite sum ends at $\binom{a+1}{r}(r - r)^a$, which is zero. Thus, $\left\langle {6 \atop 4} \right\rangle = 4^6 - \binom{7}{1}3^6 + \binom{7}{2}2^6 - \binom{7}{3}1^6 = 4096 - 7 \cdot 729 + 21 \cdot 64 - 35 = 4096 + 1344 - (5103 + 35) = 5440 - 5138 = 302$, which matches the number given in the table above.

Chapter 8   Iterative Mathematics and Fractal Geometry

One of the hottest topics in contemporary mathematics is the study of iterative processes, which occur when the same function or transformation is applied repeatedly to something, and the related areas of chaos theory and fractal geometry. Literally millions of years of computer time have been spent generating the fantastic images produced by certain iterations, while chaos theory has provided insights into the limits of science's ability to predict the future. In particular, chaos appears to explain why long-term weather forecasts are notoriously unreliable. Fractal geometry, which is involved in the description of the images created by those iterations, also pervades nature. Many approximate fractals, produced by iterative processes, appear in your body, landscape features such as rivers (which are not straight), and coastlines (same observation), and trees (whose bark is not smooth and which have a lot of branches).

8.1 The Feedback Effect

The essence of iteration is feedback: the output of a function is re-used as the function's input. Taking the black-box metaphor from Sec. 7.1.1, we are attaching a loop to the box connecting output to input, and opening a window in the loop so we can watch the numbers going by. Symbolically, we are given a function f, and a value x to start the iteration with. We then calculate $f(x)$, $f(f(x))$, $f[f\{f(x)\}]$, etc. and watch these numbers for patterns. (Mathematics, after all, is largely the study of patterns.) A typical, rather boring, example, is $f(x) = 2x - 3$. If we start with $x = 1$, we get successively -1, -5, -13, -29, -61, etc., the numbers getting more negative with each succeeding iteration. Starting with 5 gives 7, 11, 19, 35, 67, etc., and the numbers increase without limit. Letting x begin at 3 produces something different: since $f(3) = 3$, the result after any number of iterations is 3. Such a number for which $f(x) = x$ is called a *fixed point* of the function. We try $x = 3 + \epsilon$, where $|\epsilon|$ is very small, and get $3 + 2\epsilon, 3 + 4\epsilon, ..., 3 + 2^n\epsilon$ after n iterations. No matter how small $|\epsilon|$ is, eventually n is so large that $|2^n\epsilon|$ is large. For this reason, 3 is called an *unstable* fixed point of $f(x) = 2x - 3$. If we use $f(x) = \frac{x+3}{2}$ instead, 3 is still the fixed point, but now it is stable: in fact, every real number has its iterates converge on 3. Therefore, 3 is called an *attractor* for $f(x) = \frac{x+3}{2}$.

Linear functions such as these can display only a few types of behavior under iteration; the two found so far are the most common. More complex functions exhibit much more interesting behavior. About all that is needed is that the function have a smooth peak (which linear functions never have).

8.1.1 $kx(1 - x)$

The function $kx(1 - x)$, where k is between 1 and 4, is a very simple function at first glance. It has a value of zero when $x = 0$ or 1, and a maximum (smooth peak) at $x = \frac{1}{2}$. If k is in the prescribed range, there is a fixed point other than zero, and all iterates are between 0 and 1 (if we start with such a number). As long as k is less than 3, the fixed point is an attractor for all starting values between 0 and 1, with values outside this range ending up at negative infinity, i. e., they become less than any preselected negative number after enough iterations. When k is exactly 3, the fixed point is only semi-stable. It is still an attractor, but is now very weak in that the ratio $\left|\frac{i_n - 2/3}{i_{n-1} - 2/3}\right|$ approaches 1 as a limit as $\left|i_n - \frac{2}{3}\right|$ approaches zero. (The fixed point is at 2/3 when $k = 3$, and i_{n-1} and i_n are successive iterates of any starting point between 0 and 1.) For all other k between 1 and 3, the ratio $\left|\frac{i_n - f}{i_{n-1} - f}\right|$ has a limit that is less than 1, so the fixed point f is a strong attractor.

The really interesting action comes as k increases above 3. Instead of a single attractor, we suddenly acquire an attractor consisting of two numbers p and q such that $f(p) = q$ and $f(q) = p$. The single fixed point is now unstable, and is sometimes called a *repeller*. The event of a single fixed-point attractor splitting into a two-point one is called a *bifurcation*. As k continues to increase, a series of bifurcations occurs, at shorter and shorter intervals, and the number of points in the attractor repeatedly doubles. As long as k remains below a critical value of about 3.57, however, all initial values' iterates still converge on the same cyclic sequence of values. Since the critical value of k is an irrational number, nobody knows exactly how iterates behave when k is at exactly this critical value, but I suspect that all iterates are drawn, very slowly, toward an infinite set of points that is totally disconnected like the Cantor dust (see below).

As k creeps above the critical value of approximately 3.57, a chaotic region begins. Iterates can now range over complete intervals. In a sort of reverse bifurcation, short intervals merge together as k continues to grow, with the final merger of two into one occurring when k reaches about 3.68. The chaos continues until $k = 4$, with the one interval to which the iterates are confined slowly widening until it reaches $0 \leq x \leq 1$ when k hits 4.

But the chaos does not continue uninterrupted. Intervals of order are sprinkled everywhere in the chaotic region. There is a window of order that starts out as an n-number cycle for each n greater than 2. The largest interval features a limit cycle of 3 values which serve as attractor for all starting points. It begins at $k = 3.82$ or so, and as k increases, bifurcations begin and eventually the iterates become chaotic again. However, the iterates are now confined to three short intervals. As k reaches about 3.84, the iterates are suddenly released into the much wider interval containing the three small ones. Between the beginning and ending of this order interval (including its semi-chaotic part), all of the structure in the entire $1 \leq k \leq 4$ interval is repeated at a smaller scale, including another interval of order featuring a cycle of three values (which appears as a 9-number cycle because there is one such interval for each of the three intervals in the larger cycle). The idea of a structure containing little copies of itself is one feature of fractal geometry.

8.1.2 General Iteration of Real-Valued Functions

The structure in the "bifurcation map" for $kx(1 - x)$ is practically independent of the function itself. The functions $k \sin x$ $(1 < k \leq \pi)$, x^{k-x} $(2 < k < 4)$, and $k(1 - (1 - 2x)^4)$ for $\frac{1}{4} < k \leq 1$ all share the same features, down to the order of the various windows of orderly behavior in the chaotic region and even the relative sizes of those windows (3 is always the largest, followed by 5). The absolute sizes of the windows differ from one function to the next. The transition from order to chaos follows the same pattern everywhere, and the gaps between one bifurcation and the next shrink in ratios that converge to the same number, called Feigenbaum's number, whose value is about 4.66. Above all, the bifurcation maps all share the property of containing small copies of themselves, and thus of being fractals.

8.2 Fractal Geometry

The word *fractal* was invented by the French-American mathematician Benoit Mandelbrot to describe a slowly growing collection of mathematical objects that had what seemed to ber mutually exclusive features, such as curves without tangent lines at any point, or that seemed to cover whole areas of the plane. Mandelbrot is the creator of fractal geometry, and has led in its application to an astonishingly wide array of scientific problems, ranging from how big to make dams on the Nile River to why the sky is dark at night, and beyond. Broadly speaking, a fractal is an object with a dimension conflict: measured one way, the object has a certain dimension, but looked at another way, it has a different dimension. More precisely, a fractal's Hausdorff dimension, D, must be greater than its topological dimension, D_T. While the Hausdorff dimension is too complicated for this book, the topological dimension is what we think of when we mention dimension casually; thus, a line segment has one dimension, the interior of a square two, and a baseball three. There are usually substitutes for the cumbersome Hausdorff dimension; one of them, the *similarity* dimension, will make a frequent appearance in this chapter.

8.2.1 Koch Curves: Initiators and Generators

The simplest kinds of fractals are the snowflake, or Koch, curves. These are named for the original curve designed by Helge von Koch in 1903. To create this curve, you start with an equilateral triangle (which Mandelbrot calls the *initiator*), and, on the middle third of each of its sides, draw another equilateral triangle, which, of course, will be one-third as large as the original. This creates a six-pointed star with 12 sides, each one-third as long as one of the initiator's sides. (Note that we have multiplied the perimeter of the figure by 4/3.) We now proceed to build another equilateral triangle on each of the middle thirds of the 12-sided figure's sides, creating a 48-sided figure and multiplying the perimeter by 4/3 again. Iterating this procedure ad infinitum produces the final Koch snowflake. Since the perimeter has been multiplied by 4/3 an infinite number of times, it has become infinite, despite the fact that the curve still fits in the same box that held the six-pointed star. Moreover, the final curve has a tangent at no point on it. Most striking of all, if you look at a small portion of the limit curve closely, you can see that it is made up of four copies of itself – each one-*third* the size of the original. The whole is less than the sum of its parts! These strange properties give a clue that the Koch snowflake is no ordinary curve. It is a fractal because the 4 copies at 1/3 scale give it a similarity dimension of log 4/log 3 or about 1.26, while its topological dimension is that of a curve: 1. (It does not matter what base the logarithms are taken with respect to here, because they are divided by each other. Recall that $\log_a x = \frac{\ln x}{\ln a}$ (Sec. 7.5.6); when we divide two logarithms to the same base by one another, the $\ln a$'s cancel out.)

The snowflake is only one of an infinite number of Koch-type curves that are all constructed by the same method: A line segment or regular polygon is used as the initiator, and each line segment in the curve produced by the last step is replaced by a series of shorter segments called the *generator*. Usually the generator consists of N segments, each $1/r$ as long as the single segment they replace. r is not necessarily an integer or even less than N, but the similarity dimension of the fractal that results after an infinite number of steps is always $\log N / \log r$. N must be less than r^2 in order to successfully carry out a fractal construction in the plane without overlap. In all such constructions (without overlap), the similarity dimension is identical to the Hausdorff dimension. All of the Koch-type curves lack tangents everywhere, have infinite length and zero area, and are fractals with $D_T = 1$ and $1 < D < 2$.

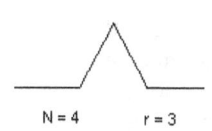

N = 4   r = 3

Fig. 8-1. The generator for the original Koch snowflake. The fractal dimension is about 1.26.

8.2.2 Cantor and Other Dusts

The now famous Cantor dust (named after the same mathematician who discovered the infinity of infinities in Sec. 6.5) is best conceived of by beginning with the line segment corresponding to the interval $0 \leq x \leq 1$ on the number line. We split this segment in two by removing its middle third, then do likewise to the resulting segments, and iterate this procedure ad infinitum, remembering to keep, not discard, the segment endpoints. A naïve mind might think that nothing will be left, and might justify this conclusion by noting that the amount left after n iterations is $\left(\frac{2}{3}\right)^n$ of the original segment; letting n approach infinity and using result F in Sec. 7.10.2, he deduces that the amount left over after an infinite number of iterations is zero. In so doing, he commits one of the most common fallacies about the infinite, as illustrated by the following two examples.

Relative to the infinite set of integers, the prime numbers are of measure zero, but primes do exist in infinite quantities. ("Measure," as used here, means roughly the chance of picking a number out of the larger set, here all integers, that happens to belong to the smaller set, here the primes.) Similarly, the rational numbers are of measure zero relative to the real numbers, but there are, nevertheless, an infinity of rational numbers.

The naïve mind's argument shows that the points that are left over after the infinity of middle-third removals are of measure zero relative to the whole line segment, but since the line segment has an infinite number of points, we can say nothing about the number of points left over based on just that measure. In fact, there are as many points left behind as there were removed, in spite of their measure of zero! The Cantor dust is precisely the set of all the points that are left behind.

The Cantor dust can also be described in two other ways: It is the largest bounded set which is mapped onto itself by the function $f(x) = 3\left(\frac{1}{2} - \left|x - \frac{1}{2}\right|\right)$, and it is the set of real numbers between 0 and 1 inclusive that have no 1's in their ternary (base-three) expansions. The function $f(x)$ is equal to $3x$ if $x \leq \frac{1}{2}$, and to $3 - 3x$ if $x \geq \frac{1}{2}$. If $x = \frac{1}{2}$, then $3x = 3 - 3x = 1\frac{1}{2}$. The ternary expansion of a number is like its decimal expansion, except that 3 is used in place of the decimal system's 10. Thus, 49 decimal appears as 1211 in ternary, and ¾ decimal (0.75) appears as 0.202020… ternary since $\frac{2}{3} + \frac{2}{27} + \frac{2}{243} + \cdots = \frac{2/3}{1 - \frac{1}{9}} = \frac{8/12}{8/9} = \frac{9}{12} = \frac{3}{4}$. The absence of ones in ¾'s ternary expansion shows that it is in the Cantor dust, which can also be verified by noting that $f(x)$ maps ¾ onto itself. That the two definitions are equivalent to one another can be seen by looking at the effect that f has on a number's ternary expansion. To apply f to a number in ternary, first exchange 0's and 2's if and only if the first 2 precedes the first zero in the expansion, then move the "ternary point" one place to the right. (This rule applies only to numbers between 0 and 1. Exchanging 0's and 2's subtracts the number from 1, and the first 2 precedes the first zero if and only if the number is bigger than ½.) It is easy to see that applying this rule to a number with no 1's will change it into another number with the same property, while a number containing a 1 will retain that digit. If f is iterated repeatedly, the 1 will eventually wind up to the left of the ternary point, making the number greater than one. Such numbers are sent to negative values by f, and once there they get more negative, since $f(x) = 3x$ for $x \leq \frac{1}{2}$.

Cantor dust is called a *dust* because it is totally disconnected; i. e., no complete interval of numbers, however short, is part of it. To prove this, take any two distinct numbers it it and their ternary expansions. Since the points are in the dust, their expansions have nothing but 0's and 2's in them. Also, since they are distinct, there must be a place where one expansion has a 0 opposite the other's 2. We find the first such spot and note that all numbers with a 1 in that spot and matching the two numbers in the dust to that spot's left lie between our two numbers but are not in the dust since they have a 1 in their ternary expansions.

The original construction of the Cantor dust is equivalent to a Koch-type construction where the initiator is a line segment and the generator is the same segment without its middle third. Therefore, the Cantor dust is made up of two copies of itself, each one-third the size of the original – another "mismatch." The Cantor dust's similarity dimension is therefore log 2/log 3 = 0.63, while its topological dimension is zero since it is totally disconnected. Since $D > D_T$, the Cantor dust is a fractal.

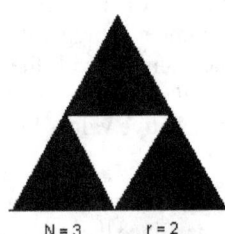

N = 2   r = 3

Fig. 8-2. The generator for the Cantor dust. The fractal dimension is about 0.63.

Now imagine the following construction: Beginning with the segment $0 \le x \le 1$, erase the middle half (¼ $< x <$ ¾). Then erase the middle quarter of the two remaining pieces, the middle eighth of each of the four shorter segments, and so on, with a relatively smaller portion taken out at each successive step. The result is still a totally disconnected dust with $D_T = 0$, but now a positive fraction of the original interval is left: $\frac{1}{2} \times \frac{3}{4} \times \frac{7}{8} \times \frac{15}{16} \times \frac{31}{32} \times ... = 0.288788$ of what we started with. The convergence of the infinite product causes the result to have a positive measure (the number just above), which means that the Hausdorff dimension is 1. (The similarity dimension does not apply here because the same process was not used at every step.) Since $D > D_T$, this dust is also a fractal despite the fact that its Hausdorff or fractal dimension is an integer. In fact, the set of irrational numbers between 0 and 1 is a dust of measure 1, the highest such possible. And this construction can be carried out in any number of dimensions; let us take the set in 4 dimensions of all points (w, x, y, z) where each letter is an irrational number between 0 and 1. This set is a dust with measure 1, $D = 4$, and $D_T = 0$, so it is very much a fractal. In short, there are ways to construct dusts with any desired measure or fractal dimension in any Euclidean space, with the obvious exception that D cannot exceed the dimension of the *embedding space*, the Euclidean space the dust is part of.

8.2.3 Ramified Curves and Fractal Lattices

The Koch curves introduced in Sec. 8.2.1 varied in wiggliness, but all of them had the property that a sufficiently small circle centered on any point on the curve had only two points in common with the curve. This section will cover curves that do not satisfy that property.

The number of points which all circles below a certain radius and a fixed center on a curve must have in common with that curve is called the curve's *order of ramification* for that point. "Normal" curves have this order equal to 2 at all but a few points on them. At the fork in a Y the order of ramification is 3; at the center of a plus sign it is 4. A ramified fractal has all but a few of its points with orders 3 or higher, and some have infinite orders.

Various Koch-like constructions are used to create these curves. Instead of a straight line, however, the initiator is usually a polygonal region, and the generator several subregions inside it. The construction procedure, as with Koch curves, is to replace each subregion in the current stage with a suitably reduced copy of the generator, and to repeat this process ad infinitum. The two curves which may be regarded as prototypical are the Sierpinski gasket and carpet. The gasket's order of ramification is 3 for most points and 4 for the rest, while the carpet's order is infinite. (See Figs. 8-3a and 8-3b.)

The Sierpinski gasket and Christensen lattice (see Fig. 8-3c) are fractal *lattices*: ramified curves that enclose an infinity of holes of various sizes, which touch each other at corners. In the gasket, the holes are triangles, while the lattice's holes are snowflake curves.

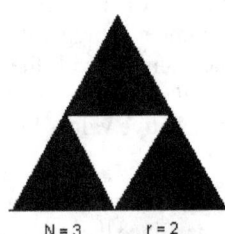

N = 3   r = 2

Fig. 8-3a. The generator for the Sierpinski gasket. The fractal dimension is about 1.58.

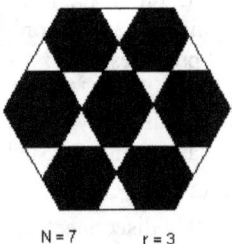

N = 7   r = 3

Fig. 8-3c. The generator for the Christensen lattice. The fractal dimension is about 1.77.

N = 8   r = 3

Fig. 8-3b. The generator for the Sierpinski carpet. The fractal dimension is about 1.89.

N = 3   r = 2

Fig. 8-4a. The stick generator for the Sierpinski gasket. In subsequent construction stages, this generator is placed on alternating sides of the result of the previous step, starting with the right side in the second stage.

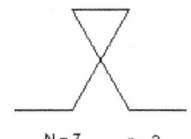

N = 7   r = 3

Fig. 8-4b. The stick generator for the Christensen lattice. Note how it includes the generator for the original Koch curve.

Both of these objects also have Koch constructions as well (see Figs. 8-4a and 8-4b); for the gasket, the initiator is a line segment, and the lattice's initiator is a regular hexagon – the generator is placed on the inside of each side of this hexagon and each of the many others that are formed in subsequent construction steps. The orders of ramification for the Christensen lattice are 6 for most points and 4 for nearly all the rest, with a very few having the "normal" order of 2.

8.2.4 Natural Fractals

All fractals, strictly speaking, are only mathematical constructs. But many natural objects – indeed, if you look closely enough, almost *every* natural object – are irregular, bumpy, twisted, or otherwise not conveniently summarized by ordinary mathematical objects such as straight lines and circles. It was, in fact, largely on account of the irregularity of natural objects that Mandelbrot invented fractal geometry: to study, as he writes, "the morphology of the amorphous." The classic example of a natural fractal is a coastline. In 1921, L. F. Richardson noticed that the more closely you looked at a typical coastline, the longer it seemed to be. Quantitatively, reducing the length of the segments used to approximate a coastline by a factor of r increased the number of segments needed not by a factor of r, but rather r to a power of about 1.2. Now Richardson knew nothing about Koch curves, which date from 1903, so this curious fact remained isolated until the well-read Mandelbrot appeared. When Mandelbrot heard of Richardson's result, already knowing about Koch curves and the Hausdorff dimension, he immediately saw that the 1.2 was a Hausdorff dimension and that Koch curves with similar dimensions could be used to model real coastlines. The word "fractal" first appeared in 1975. Since then, Mandelbrot and others have added a seemingly endless variety of natural objects to the list of approximate fractals. (I say "approximate" because all natural fractals carry within themselves an "inner cutoff" because atoms are not fractal.) The branching pattern of blood vessels in the body, bronchioles in the lungs, and nephrons in the kidneys are all modeled by the ramified (literally, "branching") curves of the previous section. Other fractals include the distribution of galaxies in the universe, the mountains on Earth, the craters on the moon, and the surface of tree bark. The Cantor dust becomes a model for transmission errors in a phone line. Even stock price movements have a fractal model – one which, by the way, allows no room for a system to "beat the market."

In essentially all these situations, Mandelbrot found that introducing a random element into the mathematical fractal greatly improved the resemblance between it and what it was supposed to represent. The Koch snowflake does not closely resemble a real coastline despite the fact that its dimension falls within the typical dimension range for coastlines. The reason is that its irregularities are, paradoxically, too regular! Introducing several random choices into the Koch construction, then adding in the influence of plate tectonics, produces coastlines and mountain ranges that are indistinguishable from the real thing.

8.3 Complex Iteration: Julia Sets and the Mandelbrot Set

The iteration results in Sec. 8.1, while interesting enough, are only a sliver of what can be realized. Allowing the variable x to be a complex number makes all the difference between a row of colored dots and a color photograph, or between a photograph of an object and the object itself.

In a typical example, select an arbitrary complex number c and another equally arbitrary one denoted by z. We then iterate the function $f(z) = z^2 + c$ repeatedly and see whether the iterates "blow up" by getting very large or stay reasonably close to zero. Repeating this process for numerous values of z, preferably with the aid of a computer, and keeping track of which z's lead to iterates that stay bounded forever (in practice, as long as we tell the computer to wait) produces what is called the *Julia set* for the value of c. The set is named after Gaston Julia, who first showed that these sets are not ordinary pieces of the complex plane. As Julia had no computer available to him, he had no idea how extraordinary his sets are. Human language comes up short in describing them, but Mandelbrot, who had computers, suspected they were fractals. When $c = 0$, the Julia set is just a circle, but all other Julia sets have been shown to be fractal.

Julia sets come in three basic types: scattered dusts, which are usually clumpy; ramified fractal curves; and solid areas whose boundaries are fractals (or the circle when $c = 0$). This distinction led Mandelbrot to define another set: the set of all c's whose Julia sets are solid areas. Using just this definition, it would take forever (even with a computer) to draw this set, but it was shown that all you needed to know was whether zero, when used as z, led to iterates that stayed bounded. With this improvement, it was finally possible to get a good look at Mandelbrot's set (now called the *Mandelbrot set*). What was revealed will blow your mind.

From a distant vantage point, the Mandelbrot set is a *cardioid* (heart-shaped curve) with a fringe of circles attached, some large, but most small. As you get closer, you see that between each pair of circles – which aren't quite circular – a whole series of smaller circles is attached. The "circles" themselves have smaller circles attached to them, and, like the snowflake's triangles, so on ad infinitum. And that's not all. Each "circle" has an infinity of filaments, which are themselves fractal, emanating from it, and scattered along each filament are "islands" whose shape is that of – the Mandelbrot set! The closer you look at the edge of the Mandelbrot set, the wilder it appears!

On a more serious note, the Mandelbrot set has a definite mathematical structure. Let X be a point on the far positive real axis in the complex plane (which is just a Cartesian coordinate plane in which the point (a, b) represents the complex number $a + bi$), C be the cardioid's cusp, whose coordinates are $(1/4, 0)$, and P a point on the cardioid. If the positive angle XCP, when measured in fractions of a complete revolution, is a rational number m/n in lowest terms, then a "circle," which I call a *bud*, is attached to the cardioid at P. The number m/n serves as the "address" of the bud. All points within the bud have iterates that are attracted to a limit cycle containing precisely n complex numbers, and all filaments coming from this bud have points at which precisely n filaments come together. These filaments can be grouped by length into m subsets. Points within the main cardioid have iterates that are attracted to a fixed point (a different point for each starting point), and points within islands show limit cycles with various numbers of points, usually a multiple of the n for the bud that the island's filament comes from. All islands show the way back home: their cusps always are in the direction that leads back to the main set. The Mandelbrot set itself is not a fractal as it has a well-defined area, but its boundary is a fractal curve with a dimension of exactly 2, meaning that it nearly fills a region of the complex plane.

As with the function $kx(1 - x)$, which, when iterated, displays all the structure of more complicated functions, the Mandelbrot set generating function $z^2 + c$ shows almost all the complexities of more advanced functions like $c \sin z$, $z^n + c$ for $n \neq 2$, $c \exp z$, and $c(z + \frac{1}{z})$. Many of these functions, in fact, show islands shaped like the Mandelbrot set, but not all of them have the closer-you-look, wilder-it-looks property of the Mandelbrot set itself.

8.4 Newton's Method, And When It Doesn't Work

One of the applications of differential calculus that did not receive coverage in the last chapter is a method for obtaining approximations to solutions of equations that cannot be solved analytically, such as $\ln x = x - 3$ or $x = \cos x$. The technique, called *Newton's method*, first puts the equation into the form $F(x) = 0$, and then picks a value x_0 such that $F(x_0)$ is small. Then, the difficult-to-analyze graph of F is replaced with the easy-to-analyze (enter the differential calculus) tangent line to that graph at the point $(x_0, F(x_0))$. This line, of course, has slope $F'(x_0)$, and therefore its equation is $y - F(x_0) = F'(x_0)[x - x_0]$ (Sec. 5.7.2). To find out where the line crosses the x-axis, we replace y with 0 in this equation and solve for x, to get $x_1 = x_0 - \frac{F(x_0)}{F'(x_0)}$.

The tangent line is a good approximation to the graph of F, but it is not the same as the graph. Therefore, this x_1 will, in general, not satisfy $F(x_1) = 0$. However, we can feed it back into Newton's method to get a better approximation x_2, and feed that number in to get x_3, and so on, iterating the method until we have a number x_n such that $F(x_n)$ is as close to zero as we can get with the accuracy of our calculations.

At least in theory, that is. With some functions, the successive approximations will oscillate back and forth or even get farther away from the root instead of closer to it. Most functions have a "no-man's land," a set of points that, when used as the starting point for Newton's method, yield chaotic behavior in the iterates.

The situation is worse (or, from an artistic viewpoint, better) in complex numbers. The equations usually have several solutions, each one of which has a certain set of points which, when used as starting point, produce iterates that converge on the solution. But on the border between two such sets – watch out! The rules of the game dictate that whenever two sets meet, all of the others must be there as well. Therefore, the boundary is a ramified fractal curve. Furthermore, there may be islands of points (not just a few scattered ones, as in the real case) that, when used as starting point, lead to cyclic behavior or chaos. These islands are especially prominent when the equation is of the form $(z^2 + c)^2 + c = \epsilon$ for certain ranges of ϵ and any c. In this case, the islands have the shape of Julia sets corresponding to values of the c used there that are between -2 and 1/4. (Note the resemblance of the equation here to the equation that generated the Julia sets in the first place!)

The possibility of abnormal behavior in Newton's method is of more than aesthetic importance. Engineers and mathematicians seeking solutions to complicated equations need a reliable method for solving them, and we have seen that Newton's method has potential pitfalls. The best way to ensure a quick solution via Newton's method or other methods is to start with a guess that is close to the solution, but if the solution is unknown, how can you tell whether you are close? One way out is to calculate $F(x)$, the quantity which is to be made equal to zero, for a grid of points in the complex plane (or equally spaced numbers if F is a function of a real number), and then to start Newton's method with the point that gives $F(x)$ its smallest absolute value. This point is almost always a safe starting point.

Chapter 9   Combinatorial Theory

This chapter takes mathematics back to its most fundamental activity: counting. What makes combinatorial theory advanced mathematics is the nature of what is being counted. Instead of rocks or beads on an abacus, this branch of math counts things like the number of ways 10 children can be lined up in a row (3,628,800) and the number of ways 100 can be broken down into smaller integers such as $90 + 8 + 2$ or $50 + 23 + 9 + 9 + 3 + 3 + 1 + 1 + 1$. (Exactly 190,569,292 such ways exist.)

9.1 One-to-One Correspondence and Counting

One of the most important techniques of combinatorial theory is establishing a one-to-one correspondence between two sets where we already know the number of objects in one of the sets. A one-to-one correspondence between two sets A and B is a function from A to B such that (1) Each element of A appears exactly once as the first member of an ordered pair in the function; and (2) Each element of B appears exactly once as the second member of an ordered pair in the function. An invertible function is a one-to-one correspondence. The precise definition of an infinite set is one that can be placed into one-to-one correspondence with a *proper* subset of itself (i. e., a subset that leaves out at least one member of the larger set).

Ordinary counting is a form of establishing a one-to-one correspondence. The "known" set here is a subset of the positive integers with the property that if x is in the subset and $x > 1$, then $x - 1$ is also in the subset. This restricts the subsets allowed to $\{1\}$, $\{1, 2\}$, $\{1, 2, 3\}$, etc. When a child counts the monkeys in a picture, he will point to each in turn and recite a positive integer, thus establishing a one-to-one correspondence between the monkeys and the integers he recites. The child makes sure that the set of integers has the aforementioned property by reciting them in serial order. The largest (and last) integer used is then assigned to the set of monkeys.

A typical use of one-to-one correspondence is involved in the following problem: How many ways can eight rooks be placed on a normal chessboard without there being any possible captures? Certainly we must have one rook on each row and one rook in each column to avoid having to place two rooks in the same row or column, where they could capture each other. Let us write down the number of the column in which the rook on the first row stands (1 is at the far left, 8 is at the far right), and after it the column number of the rook on the second row, and so on. We will have eight numbers, each will be from 1 to 8, and of course no number will be used twice and all numbers from 1 to 8 will be used. In other words, we will have a *permutation*, or rearrangement, of the numbers from 1 to 8. We will shortly find out that there are $8 \times 7 \times 6 \times 5 \times 4 \times 3 \times 2 \times 1 = 8! = 40,320$ of these. Therefore, there are 40,320 ways to place the rooks.

9.2 Binomial Coefficients

Combinatorial theory could not advance very far without a serious study of the binomial coefficients, which are associated with powers of $(x + y)$ and were introduced in Chapter 4. In order to show their applicability to the theory, we must start with two basic facts:

1. If two independent events E and F can happen in x and y ways respectively, then both events can occur simultaneously in xy ways.

2. If two mutually exclusive events G and H can happen in u and v ways respectively, then either event can occur in $u + v$ ways.

(Events are independent if the outcome of neither affects the other, and they are mutually exclusive if they cannot both happen at the same time.)

The reason that 1. is true is that we can assign a positive integer to each possible outcome of E, and similarly for F. Then, given any two outcomes, we can form an ordered pair with the first member corresponding to the outcome of E and the second to the outcome of F. Given any of the x possible first members, we may have any of the y second members, since E and F are independent, and therefore we have xy ordered pairs, each of which corresponds to exactly one simultaneous outcome of E and F.

The proof of 2. is even simpler: We concatenate the list of possible outcomes of G with that of H. Since G and H are mutually exclusive, the same outcome cannot be listed twice. Therefore, by the definition of addition, the combined list has $u + v$ outcomes on it.

For example, consider the list of numbers 0, 1, 2, ..., 98, 99. Each of these corresponds to a selection of a tens digit and a ones digit independently from the set of digits, which has ten members. Therefore, there are $10 \times 10 = 100$ numbers from 0 to 99 inclusive. To see how many of these are either multiples of eight or end in the digit 3, we note that the former are 0 x 8, 1 x 8, 2 x 8, ..., 12 x 8 = 96, but not 13 x 8 since that is greater than 99. So there are thirteen multiples of eight (one for each number between 0 and 12 inclusive), and ten whose ones digit is three, since we have ten choices, including zero, for the other digit. Since 3 is odd while 8 is even, no number can end in 3 and be a multiple of 8 at the same time, and our "events" are mutually exclusive. Therefore, there are 13 + 10 = 23 numbers that meet the description. They are: 0, 3, 8, 13, 16, 23, 24, 32, 33, 40, 43, 48, 53, 56, 63, 64, 72, 73, 80, 83, 88, 93, and 96.

We can now prove several theorems which involve binomial coefficients.

Theorem 9-1. The number of ways of arranging n objects, all different, in a row is $n!$.

Proof: We may take any object of our n for the first in the row. Once the first has been selected, we may choose any of the other $n-1$ to be the second object. After the first two have been selected, we have $n-2$ left to choose among to be third. We continue in this manner until we have only one object left, which we "choose" to be the last object in the row. We denote the selections by $E_1, ..., E_n$ and observe that they can happen in $n, ..., 1$ ways. By fact 1, they can all happen together in $n(n-1)(n-2)...(2)(1) = n!$ ways.

Th. 9-2. The number of ways of arranging r objects, taken from a larger set of n different objects, in a row is $n!/(n-r)! = n(n-1)...(n-r+1)$.

Proof: As above, but we stop with the rth object in the row, for which we have $n-(r-1)=(n-r+1)$ choices as we have already selected $r-1$ out of the n objects to get to that point.

Th. 9-3 (Combinations). The number of ways of selecting r objects out of a set of n different objects is $\frac{n!}{(n-r)!r!} = \binom{n}{r}$.

Proof: We approach the problem of Th. 9-2 in a different way. First we select the set of r objects (let's say this is done in w ways), and then we arrange the r objects in a row. By Th. 9-1, the latter can be done in $r!$ ways. Therefore, the number of ways to arrange r out of n in a row is $r!w$. But we already know that it is $n!/(n-r)!$ (Th. 9-2). Therefore, $r!w = n!/(n-r)!$ and $w = \frac{n!}{(n-r)!r!}$. Since w is the number of ways to pick r out of n, the theorem is proved.

In order to justify the use of the notation in Chapter 4, we must prove that those binomial coefficients are the same numbers as those defined by Th. 9-3. First we must show that the symbol $\binom{n}{r}$ as defined above satisfies $\binom{n}{0} = \binom{n}{n} = 1$, which is obvious from the factorial form of the definition and the definition $0! = 1$. Then, we must prove the recurrence relation in Chapter 4.

Th. 9-4. $\binom{n}{r} = \binom{n-1}{r} + \binom{n-1}{r-1}$.

Proof: Let us assume that one of our n objects is called a. With regard to the r objects we are selecting and a, we have two choices: leave a out or put it in - we cannot do both. If we leave a out, we select the r objects from the $n-1$ that are not a in $\binom{n-1}{r}$ ways. If we do put a in, the other $r-1$ must come from the $n-1$ non-a's, in $\binom{n-1}{r-1}$ ways. Since we must either leave a out or put it in, but not both, the total number of ways to select r out of n is $\binom{n-1}{r} + \binom{n-1}{r-1}$. (Fact 2 applies to this situation.)

Th. 9-5. $\binom{n}{r} = \binom{n}{n-r}$.

This theorem can be proven by observing that in selecting r objects out of n to be in the set, we are also selecting $(n-r)$ out of n to be in the complement – the "rejects." Since each set has a unique complement, the number of sets must equal the number of complements.

We can also prove this theorem algebraically by noting that $\binom{n}{n-r} = \frac{n!}{(n-r)!(n-(n-r))!} = \frac{n!}{(n-r)!r!} = \binom{n}{r}$.

Th. 9-6. $\binom{n}{r+1} = \frac{n-r}{r+1}\binom{n}{r}$.

First, the algebraic proof: $\binom{n}{r+1} = \frac{n!}{(n-(r+1))!(r+1)!} = \frac{n(n-1)(n-2)...(n-r)}{(1)\,(2)\,(3)...(r+1)} = \frac{n-r}{r+1} \times \frac{n(n-1)(n-2)...(n-r+1)}{(1)\,(2)\,(3)...(r)} = \frac{n-r}{r+1} \times \binom{n}{r}$.

Now, the proof by counting. In how many ways can we select $r+1$ objects out of n, given that we have already picked r of them? Obviously, the answer is $n-r$, since that is how many we have left to choose from. If we repeat this choice with each of the $\binom{n}{r}$ possible r-member subsets, we will have $(n-r)\binom{n}{r}$ subsets with $r+1$ members. However, each of these subsets will show up several times. For example, the subset (a, b, c, d) will be generated from $(a, b, c), (a, b, d), (a, c, d)$, and (b, c, d). Any of its four elements could have been the "last" one, so this set shows up 4 times. In general, each $(r+1)$-element subset will show up $r+1$ times; in order to count them once each we must divide $(n-r)\binom{n}{r}$ by $r+1$. When we do this, we obtain the expression in the theorem.

Example: $\binom{6}{2} = \frac{6-1}{1+1}\binom{6}{1} = \frac{5}{2} \cdot 6 = 15$, and $\binom{6}{3} = \frac{6-2}{2+1}\binom{6}{2} = \frac{4}{3} \cdot 15 = 20$.

One consequence of this theorem is that a row of binomial coefficients starts off small, gets bigger, reaches a peak in the middle, and then shrinks as r approaches n. If a set of coefficients with the same n and varying r's is plotted on a graph, the resulting curve will have a large hump in the middle. In Fig. 9-1, the coefficients with 10 as their top are plotted. The curve, when smoothed out a little, closely resembles the famous "bell curve" – in fact, as the number on top of the graphed binomial coefficients increases indefinitely, the curve converges to the bell curve.

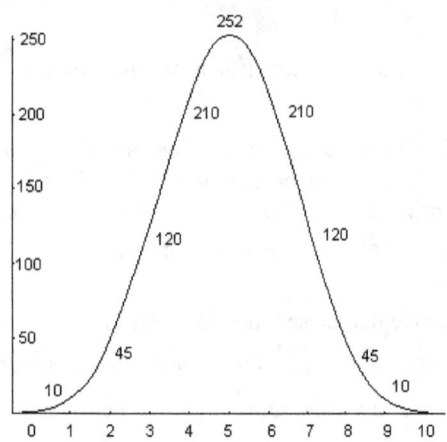

Fig. 9-1. The binomial coefficients with 10 on top are plotted. This curve looks much like the "bell curve."

The bell curve is the graph of the *normal probability distribution*, which describes the behavior of a wide variety of measurements taken on people. Height and I.Q. are among the normally distributed quantities pertaining to people. Temperature and rainfall records also show the normal distribution. Binomial coefficients mimic the normal distribution because both illustrate random phenomena. Imagine that ten pennies are tossed repeatedly and a record is kept of how many heads show up at each toss. With each toss, you are in effect randomly selecting a subset of the pennies, namely the ones which show up heads. Since there are 252 subsets with five members versus only one subset with ten, five coins will show heads that many times as often as there will be ten heads showing. In fact, after 1,024 tosses, the graph derived from the record of head counts will look like the chart of binomial coefficients in Fig. 9-1. (In practice, the two graphs will seldom match exactly, because, in selecting the subsets randomly, it is very unlikely that you will select all 1,024 once each.)

9.3 Poker and Bridge Odds

Did you ever want to know why a flush beats a straight in poker? Or have you gotten several bridge hands without any aces or face cards in them and would like to know how unlucky you were? The answers can be found in binomial coefficients.

A flush beats a straight for the simple reason that it is harder to make a flush. In order to have a flush, all five of your cards must come from one 13-card suit. In addition, there are four suits, in each of which a flush can be had. Therefore, there are $4 \times \binom{13}{5} = 5{,}148$ ways to make a flush. As for the straight, there are ten possible runs of five ranks, from A-2-3-4-5 to 10-J-Q-K-A, and, in each of them, the cards can be of any of the four suits. Therefore, there are $10 \times 4^5 = 10{,}240$ possible straights. So a flush is almost twice as rare as a straight. To find the odds against either, one determines the number of different poker hands. Each hand is a five-card subset of the 52-card deck, so there are $\binom{52}{5} = 2{,}598{,}960$ of them. Dividing this figure by the possible straights and flushes results in 1 chance in 254 (rounded) for a straight and one chance in 505 for a flush.

Bridge hands have 13 cards instead of 5, so the numbers get bigger. There are $\binom{52}{13} = 635{,}013{,}559{,}600$ possible bridge hands. In order to have, say, all four aces, we select the remaining nine cards from the 48 non-aces in $\binom{48}{9} = 1{,}677{,}106{,}640$ ways. Since we have no choices (or, technically, one choice) on how to pick the aces, this number is the total of ways to have a four-ace hand. The chance of a four-ace hand is thus 1 in 379, more likely than a flush but less likely than a straight.

As for the honorless hand, in order to have it, all 13 cards must come from the 36 that are tens or below. Therefore, there are $\binom{36}{13} = 2{,}310{,}789{,}600$ such hands. Surprisingly, you are 38 percent more likely to be dealt a hand without any points at all than to be dealt all the aces.

9.4 Probability

We can calculate the odds on other types of hands as well. All that is necessary is to be able to count how many hands fit the description. Once this is done, we divide by the total number of bridge or poker hands to calculate the odds. This illustrates the general law of probability: The chance of event E happening is equal to the number of ways E can happen divided by the number of ways E can happen or not happen. The total we divide by is called the *size of the sample space*. So, Probability of $E = \frac{\text{\# of ways } E \text{ can happen}}{\text{Size of the sample space}}$. An example from craps will serve to illustrate: What is the probability of throwing a natural (7 or 11)? When two dice are thrown there are 36 elements in the sample space since each die has six sides and the dice are independent of each other (Fact 1, 6 x 6 = 36). We can count the ways to throw a seven as 1 and 6, 2 and 5, 3 and 4, 4 and 3, 5 and 2, as well as 6 and 1. (The two dice are different from each other, so "1 and 6" is not the same as "6 and 1.") In order to throw an 11 we need either 5 and 6 or 6 and 5. Thus, there are eight ways to throw a natural (we cannot get both 7 and 11 at the same time, so Fact 2 applies), so the probability of throwing a natural is $\frac{8}{36}$ or $\frac{2}{9}$. Similarly, the probability of throwing craps (2, 3, or 12) can be calculated as $\frac{1}{9}$. So you are only twice as likely to win on the first roll as you are to lose. If you do not win or lose on the first roll, you are considerably more likely to lose than to win on a later roll, so the overall probability that you will win at craps is about 48%. This makes craps a little better than roulette, but not quite as good as most casinos' slot machines.

9.5 Multinomial Coefficients

We now introduce a generalization of the binomial coefficients called *multinomial* coefficients. We define the multinomial coefficient $\binom{n}{a_1, a_2, \dots, a_k}$ where $a_1 + a_2 + \dots + a_k = n$ to be $\frac{n!}{a_1! \times a_2! \times \dots \times a_k!}$ and note that this is the number of ways of arranging n objects in a row where we consider a_1 of them to be indistinguishable from one another, an additional a_2 of them to be indistinguishable from one another but different from the a_1, and so on up to the a_k. Thus, the number of ways to arrange the letters in the word *Mississippi* is $\binom{11}{1, 4, 4, 2} = \frac{39{,}916{,}800}{24 \times 24 \times 2} = 34{,}650$.

The reason these numbers are called multinomial coefficients is that they are related to the expansion of $(x_1 + x_2 + \dots + x_k)^n$ in the same way that binomial coefficients are related to $(x + y)^n$. We note that multinomial coefficients with just two numbers on the bottom are just binomial coefficients: $\binom{n}{r, n-r} = \binom{n}{r}$.

Theorem 9-7 (Multinomial Theorem)

$(x_1 + x_2 + \dots + x_k)^n = \Sigma \binom{n}{a_1, a_2, \dots, a_k} x_1^{a_1} x_2^{a_2} \dots x_k^{a_k}$, where the sum extends over all solutions to $a_1 + a_2 + \dots + a_k = n$ and any of the a's can be zero.

Proof: How many times does the term $x_1^{a_1} x_2^{a_2} \dots x_k^{a_k}$ arise when $(x_1 + x_2 + \dots + x_k)^n$ is multiplied out? We must select one x from each of the n factors. To arrive at a term $x_1^{a_1} x_2^{a_2} \dots x_k^{a_k}$, we must select a_1 x_1's, a_2 x_2's, and so on. This gives us n x's, a_1 of which are all identical (they are all x_1), a_2 of which are identical but different from the a_1, and so on. Therefore, the number of ways of arriving at $x_1^{a_1} x_2^{a_2} \dots x_k^{a_k}$ is $\binom{n}{a_1, a_2, \dots, a_k}$.

Th. 9-8. The sum of all the multinomial coefficients with n on top and whose bottom has k numbers is k^n.

We prove this theorem directly from the last one by letting all the x's equal 1: $(x_1 + x_2 + \cdots + x_k)^n = (1 + 1 + \cdots + 1)^n = k^n$, and on the right side of Th. 9-7's equation we have all the multinomial coefficients whose top is n and whose bottom has k numbers, each multiplied by 1.

We note in passing that $\binom{4}{3,1,0} = \binom{4}{0,3,1} = \binom{4}{1,0,3} = \cdots$, but each of these multinomial coefficients appears separately in Th. 9-7 and Th. 9-8.

Now I can give you the proof of Fermat's Little Theorem that I promised you in Sec. 6.4.3. Recall that this theorem states that if p is any prime, then $n^p - n$ is a multiple of p, where n may be any integer.

We start by writing down all the multinomial coefficients whose top is p and whose bottom has n numbers. By Th. 9-8 their sum is n^p. Now, n of these coefficients are 1 (those whose bottom is 0, 0, …, 0, p, 0, …, 0), so the others must add up to $n^p - n$. These "others" all have at least two non-zero numbers on their bottom and therefore all the numbers on their bottom are less than p. Since p is prime, the Fundamental Theorem of Arithmetic guarantees that the product of the factorials of these bottom numbers will not be a multiple of p. Since $p!$ is obviously a multiple of p, each of the "other" multinomial coefficients is a multiple of p. Therefore, their sum, $n^p - n$, is a multiple of p as well.

Here is another proof of the same theorem that does not use multinomial coefficients.

Imagine a circle of p colored beads, each of which can be any of n colors. We can rotate this circle to any of p positions, so some of the possible arrangements of beads, such as those in Fig. 9-2, are actually identical. We distinguish between two types of arrangements: one type where all the beads are the same color, and another where the beads are different colors. Since we have n colors for each of p beads, there are n of the first kind and $n^p - n$ of the second type. Now, if we rotate this circle of beads into p different positions, each arrangement where the beads are different colors becomes p arrangements, and we wish to consider all these arrangements to be the same. To count them once each, we must divide $n^p - n$ by p, and, since we must obtain an integer, $n^p - n$ must be a multiple of p. Where do we use the fact that p is a prime? Well, if p is not a prime, some arrangements with more than one color of bead, such

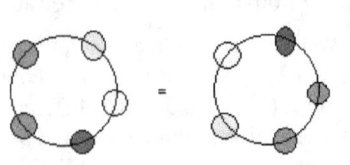

Fig. 9-2. These two arrangements of colored beads in a ring can be turned into each other by a rotation through two places counterclockwise.

as that in Fig. 9-3, can be turned into themselves by some rotations. The arrangement in Fig. 9-3 is turned into itself by a rotation through 2 or 4 positions. However, if the number of beads is prime, this cannot occur. To see why, imagine that we have an arrangement of p beads in positions numbered going clockwise zero to $p - 1$, and that this arrangement is turned into itself by a rotation through k positions ($0 < k < p$). This rotation carries the bead in position 0 into position k, so both of these beads must be the same color. It carries bead k into $2k$ mod p, so these beads must be the same color too. By the time we get back to bead 0 we have used up all the beads; since p is prime the numbers 0, k, $2k$ mod p, $3k$ mod p, …, $(p - 1)k$ modulo p exhaust the numbers 0, 1, 2, …, $p - 1$ (Sec. 6.6), so all the beads must be the same color.

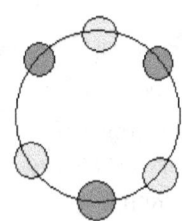

Fig. 9-3. An arrangement of beads that can be turned into itself by a rotation through two or four places. Note that the number of beads is not a prime.

9.6 Stirling Numbers

We close this mercifully short chapter with a section on Stirling numbers, which are used to convert binomial coefficients into polynomials and vice versa. The Stirling numbers of the first kind are defined by the equation $x(x - 1)(x - 2) \ldots (x - n + 1) = \sum_{r=1}^{n} s_{n,r} x^r$, where the left side is equal to $n! \binom{x}{n}$ and the Stirling numbers are the $s_{n,r}$. The expression on the left is often called the nth *factorial power* of x, and is symbolized by $x^{(n)}$. Given this definition, we can compute some of the Stirling numbers: $s_{1,1} = 1$, $s_{2,1} = -1$, $s_{2,2} = 1$, $s_{3,1} = 2$, $s_{3,2} = -3$, and $s_{3,3} = 1$. In general, $s_{n,n} = 1$ and $s_{n,1} = (-1)^{n-1}(n - 1)!$ The general recurrence formula for Stirling numbers of the first kind is $s_{n,r} = s_{n-1,r-1} - (n - 1)s_{n-1,r}$, which can be derived as follows: We start with the definition for $x^{(n-1)} = x(x - 1)(x - 2) \ldots (x - n + 2) = \sum_{r=1}^{n-1} s_{n-1,r} x^r$ and then multiply by $(x - n + 1)$ to get $x^{(n)}$ on the left and $(x - (n - 1)) \sum_{r=1}^{n-1} s_{n-1,r} x^r$ on the right. We now apply the distributive property to the right to get $x \sum_{r=1}^{n-1} s_{n-1,r} x^r - (n - 1) \sum_{r=1}^{n-1} s_{n-1,r} x^r = \sum_{r=1}^{n-1} s_{n-1,r} x^{r+1} - \sum_{r=1}^{n-1} (n - 1)s_{n-1,r} x^r$. Replacing r with $r - 1$ in the first summation results in $x^{(r)} = \sum_{r=2}^{n} s_{n-1,r-1} x^r - \sum_{r=1}^{n-1} (n - 1)s_{n-1,r} x^r$. Since $s_{n,0}$ is always zero, we may write the equation again as $x^{(n)} = \sum_{r=1}^{n} s_{n-1,r-1} x^r - \sum_{r=1}^{n-1} (n - 1)s_{n-1,r} x^r$. Since, by definition, $x^{(n)} = \sum_{r=1}^{n} s_{n,r} x^r$, we have $s_{n,r} = s_{n-1,r-1} - (n - 1)s_{n-1,r}$.

These Stirling numbers have properties that are similar to, but not identical with, those of binomial coefficients. If we let $x = 1$ in the definition, we find $\sum_{r=1}^{n} s_{n,r} = 0$, and if we let $x = -1$, we find that $\sum_{r=1}^{n} |s_{n,r}| = n!$. Also, we have $s_{n,n-1} = -\binom{n}{2}$. Reading across any row in a table of Stirling numbers, starting at $s_{n,\,n}$ and working towards $s_{n,\,1}$, we find small numbers at first, then large ones, then smaller ones again, just like with binomial coefficients. However, there is no symmetry relation $s_{n,\,n-r} = s_{n,\,r}$ as there is with binomial coefficients.

Stirling numbers of the second kind are used to convert ordinary polynomials into sums of factorial powers or binomial coefficients. The definition is $x^n = \sum_{r=1}^{n} t_{n,r} x^{(r)}$, where, as before, $x^{(r)} = x(x-1)(x-2)\ldots(x-r+1)$ and the $t_{n,\,r}$ are the Stirling numbers of the second kind. The recurrence relation for these numbers is $t_{n,r} = t_{n-1,r-1} + r t_{n-1,r}$. It is much more difficult to derive this relation from the definition. Because of it, we have $t_{n,1} = t_{n,n} = 1$ and $t_{n,2} = 2^{n-1} - 1$. The definition of Stirling numbers of the second kind has the following consequence: When the one-term polynomial x^n is differenced to arrive at a constant, as in Sec. 4.6, the number at the beginning of the rth row of differences is $r!\, t_{n,r}$. This can be proved from an equation near, but not right at, the end of Sec. 4.6.

Another use for Stirling numbers of the second kind is to count the number of ways a set with n members can be divided up into exactly r subsets, none overlapping and none empty. To divide a set with, say, 5 members into 3 subsets we have two possibilities: either the 5th element, e, is one of the subsets by itself, or else e is part of a larger subset. For e to be a subset by itself, the other 4 members must be in the other two subsets, and there are $t_{4,2}$ ways to do that. If e is part of a larger subset we can delete it and get a partition of the other four elements into three subsets. When we add e back, we have three choices on where to put it, so there are $3t_{4,3}$ ways that e can be part of a larger subset. Hence, $t_{5,3} = t_{4,2} + 3t_{4,3}$. Like Stirling numbers of the first kind, those of the second kind start out small, get bigger in the middle, and then shrink again as we move across a row of them. There is no symmetry relation for these numbers either.

Chapter 10   Linear Algebra

This chapter will cover objects called *vectors*, *matrices* (singular: *matrix*), and *linear transformations*. The characteristic feature of these objects is that they have (or can be considered to have) several parts, each of which is a number. There are ways to add, subtract, and sometimes multiply these objects, but as a rule not all the properties of ordinary addition, subtraction, and (especially) multiplication are preserved. This is the reason that I called vectors and similar objects merely "number-like" at the end of Chapter 3.

10.1 Vectors in R^n

The simplest objects on which linear algebra operates are vectors in R^n. A vector in R^n is an ordered set of n real numbers which are operated on as a unit. For example, (3, 2) is a vector in R^2, and (7, -6, 5, π, -18/5) is a vector in R^5. (Recall that the set of real numbers is denoted by R.) I stress that (7, -6, 5, π, -18/5) is completely different from (-18/5, 5, π, -6, 7) even though the numbers in both vectors are the same because they appear in different orders in the two vectors.

Two vectors can be added or subtracted if they have the same number of components, or parts. We simply add or subtract the corresponding parts: (3, 2) + (4, 5) = (3 + 4, 2 + 5) = (7, 7) and (3, 2) − (4, 5) = (3 − 4, 2 − 5) = (-1, -3). With this definition, it should be obvious that the vector (0, 0, ..., 0) in R^n is an identity element for vector addition because adding it to any vector leaves that vector unchanged. This vector is called the *zero vector* because all of its components are zeros. There is also an inverse element to any vector; the additive inverse of $(v_1, v_2, ..., v_n)$ is $(-v_1, -v_2, ..., -v_n)$ because adding these two vectors results in the zero vector. Adding vectors is commutative because adding real numbers is, and adding vectors is clearly associative as long as we keep careful track of which is the first, second, etc. component of each vector. We have just proved:

Theorem 10-1. The vectors in R^n form an Abelian group relative to vector addition.

A vector in R^n is best thought of in geometric terms, as a displacement in an n-dimensional coordinate space. (See end of Sec. 5.8.) The sum of two vectors is a displacement equal to the sum of the displacements corresponding to the vectors, by the so-called *parallelogram rule*. The two vectors are drawn starting from the same point, which can be taken to be the origin of the coordinate space. We then draw the parallelogram two of whose sides are the vectors to be added, and their sum is the diagonal of the parallelogram that starts at the origin. (See Fig. 10-1.) Note that we can draw v_2 as starting from the point where v_1 ends and arrive at the same sum. By the way, the other diagonal of the parallelogram gives us $v_1 − v_2$ or $v_2 − v_1$ depending on which direction we traverse it in. If we move from the point where v_1 ends to the point where v_2 ends, we travel the displacement corresponding to $v_2 − v_1$, and if we traverse the diagonal in the opposite direction, we travel $v_1 − v_2$.

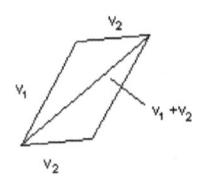

Fig. 10-1. The parallelogram law of vector addition.

A vector in R^n can also be multiplied by a real number. To do this, we simply multiply each component of the vector by the number. Hence, 3 × (9, -4, 7) = (27, -12, 21). Multiplying a vector by 1 clearly doesn't change anything, and multiplying any vector in R^n by zero results in the zero vector in R^n. Multiplying a vector by -1 results in the additive inverse of that vector. In terms of displacements, r times a vector is a vector $|r|$ times as long and in precisely the same direction if r is positive, and precisely the opposite direction if r is negative. It should be obvious that for any vectors u and v in R^n and real numbers x and y that $x(u + v) = xu + xv$, $(x + y)u = xu + yu$, and $(x \cdot y)u = x(yu)$ because these rules are simply the distributive and associative properties of multiplication. These rules are important because they are some of the axioms of a *vector space*, an abstract algebraic system that almost got mentioned in Chapter 6.

10.2 Vector Spaces over a Field

The definition of a vector space involves an arbitrary field F. F can be the real field, the rational field, the complex field, or even a finite field such as Z_7. There is also a nonempty set V, the vectors, a binary operation on V called *vector addition* and symbolized by +, and a function, called *scalar multiplication*, from the Cartesian product of F and V into V. This is not technically a binary operation because the sets F and V are distinct. The scalar product of x and v is denoted xv or $x(v)$ as with ordinary numbers. In any vector space, the following axioms must hold for all u and v in V and all x and y in F.

A_0. Relative to vector addition, V is an Abelian group in which the identity element is denoted 0_V and the inverse of a vector v is denoted $-v$.

M_1. $x(u + v) = xu + xv$.
M_2. $(x + y)u = xu + yu$.
M_3. $(x \cdot y)u = x(yu)$.
M_4. For the multiplicative identity e in F, $eu = u$.

The vectors in R^n form a vector space over the real field R. We have just seen how the vector space axioms A_0 and $M_1 - M_4$ reduce to already known properties of real numbers.

Surprisingly, it is not necessary to stipulate that vector addition be commutative. This property can be derived from the other vector space axioms. We do this by taking any two vectors u and v and considering the additive inverse of $(u + v)$. This inverse is clearly $-v + -u$ since $(u + v) + (-v + (-u))$:

$= u + (v + (-v)) + (-u)$ as vector addition is associative
$= u + 0_V + (-u)$
$= u + (-u) = 0_V$. At the same time, $-(u + v)$:
$= -e(u + v)$ since $-e + e = 0$ in the field F
$= -eu + (-ev)$ by axiom M_1
$= -u + (-v)$ by axiom M_4.

Therefore, $-v + (-u) = -u + (-v)$. Since u and v were any two vectors, $-u$ and $-v$ can be any two vectors as well, so vector addition must be commutative.

10.3 Dimension of a Vector Space

Each vector space has associated with it an integer called its *dimension*. To define this integer we must proceed slowly.

10.3.1 Linear Span and Linear Combinations

For any set $S\{v_1, v_2, \dots, v_n\}$ of vectors in V, we have a much larger set called the *linear span* of S, denoted $L(S)$. This is the set of all vectors that can be written in the form $a_1v_1 + a_2v_2 + \dots + a_nv_n$, where the a's come from the vector space's field F. By definition, if S is empty, $L(S) = 0_V$. As an example, take the vectors $(2, 5, 0)$ and $(4, 3, 0)$ in R^3. The linear span of these two vectors contains all vectors of the form $(x, y, 0)$ in R^3 because the simultaneous equations $2a + 4b = x$ and $5a + 3b = y$ always have a solution (a, b) regardless of the values of x and y. We note that 0_V is always in the linear span of a set of vectors, since the field F always has a zero element that "annihilates" any vector in scalar multiplication. A sum of the form $a_1v_1 + a_2v_2 + \dots + a_nv_n$ is called a *linear combination* of v_1 through v_n. Thus, the linear span of a set S is the set of all linear combinations of the vectors in S.

10.3.2. Linear Independence

A set $S\{v_1, v_2, \dots, v_n\}$ of vectors is called *linearly independent* if the equation $a_1v_1 + a_2v_2 + \dots + a_nv_n = 0_V$ is satisfied only by $a_1 = a_2 = \dots = a_n = 0$. Equivalently, S is linearly independent if every vector in $L(S)$ can be written <u>uniquely</u> as a linear combination of vectors from S (i. e., it can be done in only one way). The set with $(2, 5, 0)$ and $(4, 3, 0)$ is linearly independent because the equations $2a + 4b = x$ and $5a + 3b = y$ always have a unique solution, viz., $a = \frac{4y - 3x}{14}$ and $b = \frac{5x - 2y}{14}$. However, the set containing $(2, 4, 0)$ and $(3, 6, 0)$ is not linearly independent because $2a + 3b = 0$ and $4a + 6b = 0$ can be solved with $a = 3$ and $b = -2$, as well as an infinite number of other solutions.

We may also define linear independence as the condition that none of the vectors in S be expressible as a linear combination of the others. For, if we have $v_n = a_1v_1 + a_2v_2 + \dots + a_{n-1}v_{n-1}$, we then have $0_V = a_1v_1 + a_2v_2 + \dots + a_{n-1}v_{n-1} + (-e)v_n$, where, as before, e is the multiplicative identity in F. Our second example set is linearly dependent because its second vector is $\frac{3}{2}$ times its first. Obviously, any set of vectors that includes the zero vector is linearly dependent.

10.3.3 Bases and Dimensions

A *basis* of a vector space V is a linearly independent set of vectors $b_1, b_2, ..., b_n$ whose linear span is all of V, i. e., every vector in V is expressible, uniquely, as a linear combination of the b's. If a vector space has a basis with n elements, we say that it is of *dimension n*. Some vector spaces cannot be spanned by any finite collection of vectors; we say that these vector spaces are of infinite dimension.

We note that if B is a basis for V, then adding any vector in V to B produces a linearly dependent set. For any vector in V can be written as a linear combination of vectors from B, and no vector in a linearly independent set can be written as a linear combination of the others. Using this fact, we can prove that all bases of a vector space have the same number of elements, and therefore that the dimension of a vector space is well-defined.

The vector space R^n has dimension n. We define the *standard basis* of R^n as the set of vectors $(1, 0, 0, ..., 0)$, $(0, 1, 0, 0, ..., 0)$, ..., $(0, 0, ..., 0, 1)$. Clearly, any vector in R^n can be written uniquely as a linear combination of these vectors. The coincidence of the vector space dimension of R^n and the geometric dimension of the Euclidean space that we imagine vectors in R^n to inhabit is the reason that the term "dimension" is used in the study of vector spaces.

Now to prove that the dimension of a vector space is well-defined.

Theorem 10-2. The dimension of a vector space is unique.
Proof: Imagine that V has two bases, $v_1, v_2, ..., v_m$ and $w_1, w_2, ..., w_n$ with $n > m$. We now delete v_1 from the basis of v's and replace it with a w that is independent of the other v's. There will always be such a w, for, since every vector in V is dependent on the w's, if every w is dependent on $v_2, ..., v_n$, every vector in V would be dependent on v_2, ..., v_n including v_1 which was part of the basis. Our new set will still be linearly independent since the set of v's is and we chose the w to be independent of the other v's. Now, let is write our w as a linear combination of all the v's, including v_1: $w = a_1 v_1 + a_2 v_2 + \cdots + a_m v_m$, and consider the combination $\frac{e}{a_1} w + \frac{-a_2}{a_1} v_2 + \frac{-a_3}{a_1} v_3 + \cdots + \frac{-a_m}{a_1} v_m$, where the divisions $\frac{e}{a_1}, \frac{-a_2}{a_1}$, etc. mean $(a_1)^{-1}$, $(-a_2)(a_1)^{-1}$, etc. in the field F. We note that a_1 cannot be zero because if it were then w would be dependent on the v's other than v_1. By inspection, we can see that this combination is precisely equal to v_1. We now take an arbitrary vector x in V and write it as a linear combination of all the v's: $x = p_1 v_1 + p_2 v_2 + \cdots + p_m v_m$. Substituting the linear combination that we just had for v_1 in this combination gives $x = \frac{p_1}{a_1} w + (\frac{-a_2 p_1}{a_1} + p_2) v_2 + (\frac{-a_3 p_1}{a_1} + p_3) v_3 + \cdots + (\frac{-a_m p_1}{a_1} + p_m) v_m$. This gives us x expressed as a linear combination of w and $v_2, ..., v_m$, showing that this set spans V and is therefore a basis. We now repeat the entire process: we throw out v_2 and replace it with a w independent of our first w and $v_3, ..., v_m$. As before, the new set is still a basis. We can continue this process until all the v's have been thrown out and we have a basis consisting exclusively of w's. But there are only m of them, so n must be equal to m.

10.4 Linear Transformations

Given two vector spaces V and W over the same field F, we define a *linear transformation* from V to W as a function $T: V \rightarrow W$ that satisfies $T(ax + by) = aT(x) + bT(y)$ for all a and b in F and all x and y in V. One example of a linear transformation is the derivative function on the vector space containing all the polynomials in the variable t. This function maps t^3 into $3t^2$, and, by formulas II and III in Sec. 7.3.1, it satisfies the defining condition. By the way, this polynomial space is of infinite dimension. A basis for it is the set $\{1, t, t^2, t^3, ...\}$ which has an infinite number of elements.

10.4.1 Image and Kernel

The *image* of a transformation T, abbreviated im T, is the set of all vectors in W that the function T has as the second member of its ordered pairs. The *kernel* of a transformation, abbreviated ker T, is the set of all vectors in V that are transformed into the zero vector in W. The *rank* of a transformation is defined as the dimension of its image, and the *nullity* of a transformation is the dimension of its kernel. The zero vector in V is always part of the kernel, and if x and y are in ker T, then so is $ax + by$, since $T(ax + by) = aT(x) + bT(y) = a0_W + b0_W = 0_W$. Similarly, if x and y are in im T, then so is $ax + by$. To see this, let u be a vector in V whose image under T is x (there is always such a u since x is in im T), and v a vector in V whose image is y. Then, $T(au + bv) = aT(u) + bT(v) = ax + by$. This completes the proof that the image and kernel of a linear transformation are vector spaces and therefore have dimensions.

10.4.2 Injective, Surjective, and Bijective Transformations

A transformation T is called *injective* or *one-to-one* if the equation $T(x) = T(y)$ implies $x = y$, or, equivalently, its kernel contains nothing but the zero vector and its nullity is therefore zero. T is called *surjective* or *onto* if im T is all of W, or, equivalently, its rank is equal to the dimension of W. T is *bijective* if it is both injective and surjective. A bijective transformation has an inverse transformation $T^{-1}: W \rightarrow V$, so bijective transformations are also called *invertible*. We note that given any nontrivial (i. e., its image is more that the zero vector in W) transformation, we can restrict the spaces V and W so that the transformation is bijective, provided that the vector spaces have finite dimension.

10.4.3 The Dimension Theorem

We now can state and prove the most important theorem in elementary linear algebra.

Theorem 10-3. (Dimension Theorem)
For any linear transformation $T: V \rightarrow W$, rank(T) + nullity(T) = dimension(V).
To prove this theorem, we start by selecting a basis for ker T, which will have nullity(T) elements. We call this basis $k_1, k_2, \ldots, k_n$, where $n = $ nullity(T). We then extend this basis into a basis for all of V by selecting dimension(V) – nullity(T) additional vectors from V. We call these vectors $v_1, v_2, \ldots, v_r$, where $r = \dim V - n$. I now claim that $T(v_1), T(v_2), \ldots, T(v_r)$ form a basis for im T. For any vector x in V can be written as a linear combination of the k's and v's since they from a basis for V: $x = a_1 k_1 + a_2 k_2 + \cdots + a_n k_n + b_1 v_1 + b_2 v_2 + \cdots + b_r v_r$. Then $T(x) = a_1 T(k_1) + a_2 T(k_2) + \cdots + b_1 T(v_1) + \cdots + b_r T(v_r)$ since T is a linear transformation. But $T(k_1) = T(k_2) = \cdots = 0_W$ since the k's are in ker T. Therefore, $T(x) = b_1 T(v_1) + b_2 T(v_2) + \cdots + b_r T(v_r)$, and we have $T(x)$ as a linear combination of the $T(v_i)$. Since the k's and v's form a basis for V, the b's must be unique for any particular x. Also, since x was any vector in V, $T(x)$ can be any vector in im T. Therefore, the $T(v_i)$ form a basis for im T, meaning dimension(im T) = rank(T) = r = dim(V) – n. Hence, rank(T) + nullity(T) = dimension(V).

With this theorem, we can prove that, in order for $T: V \rightarrow W$ to be injective, we must have dim $V \leq$ dim W, and in order for T to be surjective, we must have dim $V \geq$ dim W. Therefore, if T is bijective (invertible), dim V must equal dim W. Of special interest in this regard are linear transformations from a vector space into itself. Since $V = W$ in this case, the transformation may be invertible. We call a transformation from V into itself *singular* if ker T contains more than the zero vector; a transformation from V into itself must be either singular or invertible.

10.5 The Dot Product

We now return to the somewhat less abstract world of vectors in R^n. Given two vectors $(a_1, a_2, \ldots, a_n)$ and $(b_1, b_2, \ldots, b_n)$ we define their *dot product*, $a \cdot b$, to equal $a_1 b_1 + a_2 b_2 + \cdots + a_n b_n$. The dot product satisfies $a \cdot b = b \cdot a$, $a \cdot (b_1 + b_2) = a \cdot b_1 + a \cdot b_2$, and $(ca) \cdot b = c(a \cdot b)$. We say that a and b are *orthogonal* if $a \cdot b = 0$; two vectors are orthogonal if and only if their displacements in the n-dimensional coordinate space of Sec. 10.1 are perpendicular. The length of a vector v in R^n is defined to be $\sqrt{v \cdot v}$; the length of every vector in R^n is at least zero, and only the zero vector has length zero. The length of a vector is also called its *norm*, and the symbol $\|v\|$ is used to denote the length of v. If the length of v is 1, we say that v is a *unit* vector. A basis $v_1, v_2, \ldots, v_n$ of R^n is said to be *orthonormal* if $v_i \cdot v_j = \delta_{ij}$ for all i and j between 1 and n, where δ_{ij} is the *Kronecker delta*, defined as zero if $i \neq j$ and one if $i = j$. In an orthonormal basis, all the vectors are unit vectors and they are all orthogonal to each other. The standard basis of R^n is an orthonormal basis. If we are given any basis of R^n, we can transform it into an orthonormal basis via a process called the *Gram-Schmidt orthonormalization process*, but this is very complicated.

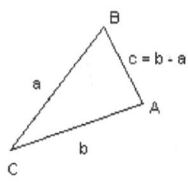

Fig. 10-2. The dot product of two vectors is equal to the product of their lengths and the cosine of the angle between them.

The dot product of two vectors is equal to the product of their lengths and the cosine of the angle between them. The angle between (3, 4) and (4, 3) is therefore $\cos^{-1} 0.96 \approx 16$ degrees. We can prove this relation by applying the law of cosines to the triangle in Fig. 10-2. The law of cosines says $\cos C = \frac{a^2 + b^2 - c^2}{2ab}$. In terms of dot products, this is $\cos C = \frac{a \cdot a + b \cdot b - c \cdot c}{2\|a\|\|b\|}$. But $c = b - a$, so $c \cdot c = (b-a) \cdot (b-a) = b \cdot b - b \cdot a - a \cdot b + a \cdot a$ and $\cos C = \frac{a \cdot b + b \cdot a}{2\|a\|\|b\|} = \frac{a \cdot b}{\|a\|\|b\|}$, and therefore $a \cdot b = \|a\|\|b\| \cos C$. If the vectors are parallel, the angle between them is zero, so their dot product is the product of their lengths.

10.6 The Cross Product

We now restrict attention to the vector space R^3. In R^3, the standard basis is abbreviated to i, j, and k, and each vector is written as the sum of real numbers times these three unit vectors. Instead of $(4, 6, -3)$ we write $4i + 6j - 3k$. In R^3 we can define the *cross* product of two vectors u and v, written $u \times v$. This is another vector whose length is equal to the product of the lengths of u and v and the sine (not cosine) of the angle between them, and whose direction is perpendicular to both u and v and such that an ordinary screw rotated in the direction that carries u into v advances in the direction of $u \times v$. With this definition, we see that $u \times v = -(v \times u)$. The cross product acts like the dot product in other ways: $u \times (v + w) = (u \times v) + (u \times w)$ and $(cu) \times v = c(u \times v)$. Instead of the rule involving the screw, it may be easier to remember the cyclic permutation rules: $i \times j = k$, $j \times k = i$, and $k \times i = j$, while $k \times j = -i$, $j \times i = -k$, and $i \times k = -j$. If u and v are parallel, then $u \times v = 0_{R^3}$. The length of $u \times v$ is equal to the area of the parallelogram whoise adjacent sides are u and v.

We can develop the formula for the cross product of two general vectors in R^3 from the cyclic permutation rules and the linearity relationships. The cross product of $a_1 i + a_2 j + a_3 k$ and $b_1 i + b_2 j + b_3 k$ is $(a_2 b_3 - a_3 b_2)i + (a_3 b_1 - a_1 b_3)j + (a_1 b_2 - a_2 b_1)k$. This can be written as an order-3 determinant, as will be seen in the next section.

10.7 Matrices and Determinants

Like a vector, a matrix is composed of several numbers, but they are arranged in a rectangular pattern, not a simple row. The matrix $\begin{bmatrix} 1 & 2 & 3 \\ 4 & 5 & 6 \end{bmatrix}$ has two rows and three columns, so we say that it is a 2 x 3 matrix. Changing the order of the rows or columns changes the matrix: $\begin{bmatrix} 1 & 2 & 3 \\ 4 & 5 & 6 \end{bmatrix}$ is different from $\begin{bmatrix} 4 & 5 & 6 \\ 1 & 2 & 3 \end{bmatrix}$ or $\begin{bmatrix} 1 & 3 & 2 \\ 4 & 6 & 5 \end{bmatrix}$. A matrix with only one row is called a *row vector*, and a matrix with only one column is called a *column vector*. These act much like ordinary vectors. The size of a matrix is also called its *order*.

10.7.1 Adding and Subtracting Matrices

Two matrices can be added or subtracted if they are of the same order. The corresponding components are simply added or subtracted. To add $\begin{bmatrix} 1 & 2 & 3 \\ 4 & 5 & 6 \\ 7 & 8 & 9 \end{bmatrix}$ and $\begin{bmatrix} 1 & 4 & 7 \\ 2 & 5 & 8 \\ 3 & 6 & 9 \end{bmatrix}$ we add the corrsponding components and get $\begin{bmatrix} 2 & 6 & 10 \\ 6 & 10 & 14 \\ 10 & 14 & 18 \end{bmatrix}$. When we subtract the second matrix from the first we get $\begin{bmatrix} 0 & -2 & -4 \\ 2 & 0 & -2 \\ 4 & 2 & 0 \end{bmatrix}$. It should be obvious that if A and B are any two matrices of the same size, then $A + B = B + A$ and $A - B = -(B - A)$.

10.7.2 Multiplying Matrices

Multiplying matrices is much more complicated. Two matrices A and B can be multiplied if and only if the number of columns in A is the same as the number of rows in B. If A is an m x n matrix and B an n x p matrix, then their product AB is an m x p matrix. The element in the ith row and jth column of AB is the dot product of the ith row of A and the jth column of B. Thus, $\begin{bmatrix} 1 & 4 \\ 2 & 3 \end{bmatrix} \times \begin{bmatrix} 4 & 5 \\ 3 & -2 \end{bmatrix} = \begin{bmatrix} 1 \times 4 + 4 \times 3 & 1 \times 5 + 4 \times (-2) \\ 2 \times 4 + 3 \times 3 & 2 \times 5 + 3 \times (-2) \end{bmatrix} = \begin{bmatrix} 16 & -3 \\ 17 & 4 \end{bmatrix}$. To multiply in the other direction, we calculate $\begin{bmatrix} 4 & 5 \\ 3 & -2 \end{bmatrix} \times \begin{bmatrix} 1 & 4 \\ 2 & 3 \end{bmatrix} = \begin{bmatrix} 4 \times 1 + 5 \times 2 & 4 \times 4 + 5 \times 3 \\ 3 \times 1 + (-2) \times 2 & 3 \times 4 + (-2) \times 3 \end{bmatrix} = \begin{bmatrix} 14 & 31 \\ -1 & 6 \end{bmatrix}$. It is not even close; AB is not the same as BA. The square matrices with elements from R form only a ring, not a field. (It can be shown that matrix multiplication is associative and distributive over matrix addition, but the proofs are tedious.)

10.7.3 Rank of a Matrix

Each matrix, rectangular or square, has a *rank*. The rank of a matrix is defined as the dimension of the linear span of its rows. We note that the rank of a matrix cannot be greater than the number of its rows or columns. Thus a 4 x 3 matrix cannot have a rank greater than 3. The rank of a matrix with all its elements equal to zero, called a *null* matrix, is zero. We can also define the rank of a matrix to be the dimension of the linear span of its columns; it can be shown that these definitions are equivalent, but again the proof is tedious. A matrix is said to be *singular* if its rank is strictly less than the smaller of the number of its rows and columns. Hence, $\begin{bmatrix} 1 & 2 & 3 \\ 4 & 5 & 6 \\ 7 & 8 & 9 \end{bmatrix}$ is singular; a square matrix is singular if and only if one of its rows is a linear combination of the others, and the third row of this matrix is twice the second row minus the first.

10.7.4 Determinants of Square Matrices

The *determinant* of the square matrix $\begin{bmatrix} a & b \\ c & d \end{bmatrix}$ is defined to be $ad - bc$. For larger matrices, we take all products of one element from each row and one from each column, and multiply each by $(-1)^n$, where n is the number of exchanges necessary to get the order of the columns back into the natural order, when the rows start in natural order. We then add all of these signed products. There will be $r!$ of them for the determinant of an r x r matrix. The determinant of $\begin{bmatrix} a_{11} & a_{12} & a_{13} \\ a_{21} & a_{22} & a_{23} \\ a_{31} & a_{32} & a_{33} \end{bmatrix}$ is equal to $a_{11}a_{22}a_{33} + a_{12}a_{23}a_{31} + a_{13}a_{21}a_{32} - a_{13}a_{22}a_{31} - a_{12}a_{21}a_{33} -$ $a_{11}a_{23}a_{32}$. Under this definition, it can be seen that exchanging two rows or two columns of a matrix changes only the determinant's sign. Therefore, the determinant of a matrix with two identical rows or columns is zero. The determinant can be viewed as a function from all ordered sets of n elements, each of which is an ordered set of n real numbers, to the real numbers; that is, as a function from $(R^n)^n$ into R. A function of this type that has the property stated immediately above is called *alternating*. The determinant is also linear: $\det(xa + yb, c, d, \ldots) = x \det(a, c, d, \ldots) + y \det(b, c, d, \ldots)$, and so for every component. It can be seen by inspection that $\det(e_1, e_2, \ldots, e_n)$, where the e's are the standard basis for R^n and the ith component of e_i is 1, is equal to 1. The determinant function is the only function from $(R^n)^n$ to R that is alternating, linear in each component, and sends the standard basis for R^n to 1.

Going back to matrices, we have the important fact that the determinant of a square *singular* matrix is always equal to 0. The determinant of the product of two matrices is the product of their determinants. Hence if A is singular, then AB is singular for any matrix B. The cross product of $a_1 i + a_2 j + a_3 k$ and $b_1 i + b_2 j + b_3 k$ can be written $\det \begin{bmatrix} i & j & k \\ a_1 & a_2 & a_3 \\ b_1 & b_2 & b_3 \end{bmatrix}$.

10.7.5 The Inverse of a Matrix

The matrix $\begin{bmatrix} 1 & 0 & \cdots & 0 \\ 0 & 1 & \cdots & 0 \\ \vdots & \vdots & \ddots & \vdots \\ 0 & 0 & \cdots & 1 \end{bmatrix}$ is called the *identity* matrix. If I is the n x n identity matrix, then $IA = AI = A$ for all n x n matrices A. The determinant of the identity matrix is 1. If A is non-singular (its determinant is not zero), we can always find a matrix B such that $AB = BA = I$. This B is called the *inverse* of A, denoted A^{-1}. We have $\det(A^{-1}) = 1/\det A$ and $(AB)^{-1} = B^{-1}A^{-1}$. Calculating the inverse of a matrix can be very time-consuming. The fastest way is to use the so-called *augmented matrix* method. In this, we write our given matrix to the left of the identity matrix of the same size. We then apply *row operations* to the augmented matrix until the left side is the identity matrix. The right side is then the inverse matrix. Row operations are of three types: Interchanging two rows, multiplying a row by a nonzero number, and adding or subtracting a multiple of one row to or from another row. We can determine the inverse of the matrix $\begin{bmatrix} a & b \\ c & d \end{bmatrix}$ as follows: Starting with $\begin{bmatrix} a & b \\ c & d \end{bmatrix}\begin{bmatrix} 1 & 0 \\ 0 & 1 \end{bmatrix}$, we first multiply row 2 by a: $\begin{bmatrix} a & b \\ ac & ad \end{bmatrix}\begin{bmatrix} 1 & 0 \\ 0 & a \end{bmatrix}$. Now we subtract c times row 1 from row 2: $\begin{bmatrix} a & b \\ 0 & ad - bc \end{bmatrix}\begin{bmatrix} 1 & 0 \\ -c & a \end{bmatrix}$. We now divide row 2 by $(ad - bc)$, getting

$\begin{bmatrix} a & b \\ 0 & 1 \end{bmatrix} \begin{bmatrix} 1 & 0 \\ \frac{-c}{ad-bc} & \frac{a}{ad-bc} \end{bmatrix}$. Then, subtract b times row 2 from row 1: $\begin{bmatrix} a & 0 \\ 0 & 1 \end{bmatrix} \begin{bmatrix} 1+\frac{bc}{ad-bc} & \frac{-ab}{ad-bc} \\ \frac{-c}{ad-bc} & \frac{a}{ad-bc} \end{bmatrix}$. In the upper left of the

matrix on the right, $1+\dfrac{bc}{ad-bc} = \dfrac{ad-bc}{ad-bc} + \dfrac{bc}{ad-bc} = \dfrac{ad-bc+bc}{ad-bc} = \dfrac{ad}{ad-bc}$. We now divide row 1 by a to get

$\begin{bmatrix} 1 & 0 \\ 0 & 1 \end{bmatrix} \begin{bmatrix} \frac{d}{ad-bc} & \frac{-b}{ad-bc} \\ \frac{-c}{ad-bc} & \frac{a}{ad-bc} \end{bmatrix}$. The identity matrix is on the left, so the matrix on the right is the inverse of our original

matrix, $\begin{bmatrix} a & b \\ c & d \end{bmatrix}$. Note that the determinant of this matrix appears in the denominator of all four of the entries in the inverse matrix. This will be true of larger matrices as well, with more complicated expressions in the numerators.

If by chance we have a singular matrix to start with, at some point during the computation we will get, on the left side, a row with all zeros. This serves as a signal to stop: the identity matrix has no row with all zeros. For example, the singular matrix from Sec. 10.7.3: $\begin{bmatrix} 1 & 2 & 3 \\ 4 & 5 & 6 \\ 7 & 8 & 9 \end{bmatrix}$. We start with $\begin{bmatrix} 1 & 2 & 3 \\ 4 & 5 & 6 \\ 7 & 8 & 9 \end{bmatrix}\begin{bmatrix} 1 & 0 & 0 \\ 0 & 1 & 0 \\ 0 & 0 & 1 \end{bmatrix}$ and subtract 4 times row 1

from row 2: $\begin{bmatrix} 1 & 2 & 3 \\ 0 & -3 & -6 \\ 7 & 8 & 9 \end{bmatrix}\begin{bmatrix} 1 & 0 & 0 \\ -4 & 1 & 0 \\ 0 & 0 & 1 \end{bmatrix}$. Now we subtract 7 times row 1 from row 3: $\begin{bmatrix} 1 & 2 & 3 \\ 0 & -3 & -6 \\ 0 & -6 & -12 \end{bmatrix}\begin{bmatrix} 1 & 0 & 0 \\ -4 & 1 & 0 \\ -7 & 0 & 1 \end{bmatrix}$.

And then we subtract 2 times row 2 from row 3: $\begin{bmatrix} 1 & 2 & 3 \\ 0 & -3 & -6 \\ 0 & 0 & 0 \end{bmatrix}\begin{bmatrix} 1 & 0 & 0 \\ -4 & 1 & 0 \\ 1 & -2 & 1 \end{bmatrix}$. The third row on the left is all zeros, so the matrix we started out with must be singular.

10.8 Linear Transformations as Matrices

A typical linear transformation is $T(x, y, z) = (4x - 3y + 7z, 2x + 4y - 3z)$. This transformation from R^3 into R^2 can be represented by the 3 x 2 matrix $\begin{bmatrix} 4 & 2 \\ -3 & 4 \\ 7 & -3 \end{bmatrix}$. Then, applying this transformation to any vector in R^3 multiplies that vector by this matrix. In general, a linear transformation from R^m to R^n will be represented by a matrix with m rows and n columns. A transformation from R^n into itself will be represented by a square matrix. The composition of two transformations is represented by the product of their matrices: $T_2(T_1(x))$ is represented by AB if T_1 is represented by A and T_2 by B. Since matrix multiplication is not commutative, composition of transformations is not commutative. If a transformation from R^n to itself is invertible, it will be represented by a matrix that is invertible, and if the transformation is singular, it will be represented by a matrix that is singular.

We may select a nonstandard basis for R^n and write our vectors as linear combinations of this nonstandard basis. Then the matrix representing T will change in a regular way. If T is represented by M with respect to the standard basis, it will be represented by BMB^{-1} with respect to the nonstandard basis, where B is the matrix that represents the nonstandard basis as linear combinations of the standard basis. The matrix B is called a *transition* matrix, and BMB^{-1} is called a *conjugate* of M.

If T is a transformation from one vector space into a different one, the fact that we can select a basis for each independently allows us to represent T by a very simple matrix. We start by selecting a basis $k_1, ..., k_n$ for ker T and extending it to a basis for V by adding $v_1, ..., v_r$ as in the proof of the Dimension Theorem in Sec. 10.4.3. Then $T(v_1)$, ..., $T(v_r)$ form a basis for im T and we extend this to a basis of W by adding $w_1, ..., w_p$. With these bases, T is represented by $\begin{bmatrix} I & N \\ N & N \end{bmatrix}$, where I is the r x r identity matrix and the N's are null matrices with all their elements equal to zero.

10.9 Simultaneous Linear Equations and Cramer's Rule

A system of linear equations such as $\begin{cases} a_{11}x_1 + a_{12}x_2 + a_{13}x_3 = b_1 \\ a_{21}x_1 + a_{22}x_2 + a_{23}x_3 = b_2 \\ a_{31}x_1 + a_{32}x_2 + a_{33}x_3 = b_3 \end{cases}$ can be represented in matrix form as AX = B, where X and B are column vectors. A system of linear equations is solved by a set of values for the x's that satisfies all the equations simultaneously. For any set of equations, there may be a unique solution, no solutions, or infinitely many solutions. The number of equations need not equal the number of unknowns, so the matrix A need not be square.

The necessary and sufficient condition for a system of linear equations to have a (not necessarily unique) solution is that the rank of the coefficient matrix A equal the rank of the augmented matrix $A|B$. If A is square and non-singular, this condition will always be met, and, as a bonus, the solution will be unique.

The fastest way to solve a system of linear equations is called *Gaussian elimination*. In this, we write the augmented matrix $A|B$ and perform row operations on it to eliminate (make vanish) the off-diagonal elements. Given the system above, we start by multiplying row 2 by a_{11}, and then subtracting a_{21} times row 1 from row 2. This eliminates x_1 from the second equation. We now multiply row 3 by a_{11} and then subtract a_{31} times row 1 from row 3. That eliminates x_1 from the third equation. x_2 can be eliminated from the third equation in similar fashion using the second equation. We now have an augmented matrix that looks like this: $\begin{bmatrix} a_{11} & a_{12} & a_{13} & b_1 \\ 0 & c_{22} & c_{23} & b'_2 \\ 0 & 0 & c_{33} & b'_3 \end{bmatrix}$. This matrix is called an *upper triangular* matrix since only the elements on or above the principal diagonal are non-zero. If c_{33} and c_{22} are not zero, the matrix is not singular and the system has a unique solution. To find this solution, we may proceed in either of two ways: eliminate c_{23}, a_{13}, and a_{12} from the matrix by more row operations like those used to eliminate a_{21}, a_{31}, and a_{32}; or *back-substitution*: solving the almost-trivial third equation for x_3 by dividing it by c_{33}, substituting that solution into the second equation and solving it for x_2, and finally substituting x_3 and x_2 into the first equation and solving it for x_1.

If A is a square and non-singular matrix, $AX = B$ has a unique solution given by $X = A^{-1}B$. However, this result is of theoretical importance only, since calculating A^{-1} usually takes much more work than Gaussian elimination. A slightly better method is to use what is called *Cramer's rule*: the solution $\begin{bmatrix} x_1 \\ x_2 \\ \vdots \\ x_n \end{bmatrix}$ of $AX = B$ is given by $x_i = \frac{\det(A_i)}{\det(A)}$, where each A_i is formed from A by replacing the ith column of A with B. Note that if A is non-singular, its determinant will be non-zero.

Unfortunately, calculating the determinant of a large matrix directly from the definition is impossibly time-consuming. Instead, we use what is known as *Laplace development*. To understand this, we must define a *minor* of a square matrix. The minor of a_{ij}, the element of A in row i and column j, is simply A with its ith row and jth column crossed out. Thus, the minor of 6 in $\begin{bmatrix} 1 & 2 & 3 \\ 4 & 5 & 6 \\ 7 & 8 & 9 \end{bmatrix}$ is $\begin{bmatrix} 1 & 2 \\ 7 & 8 \end{bmatrix}$. The minor of an element in an n x n matrix is an $(n-1)$ x $(n-1)$ matrix. We then attach a sign to the minor as follows: If $i + j$ is even, we use a plus sign, and if $i + j$ is odd, we use a minus sign. The determinants of these signed minors are called *cofactors*. Now, we use any row or column j of the matrix and calculate $a_{1j}\text{cof}_{1j} + a_{2j}\text{cof}_{2j} + \ldots + a_{nj}\text{cof}_{nj}$ – and this is equal to $\det(A)$. This reduces the job of evaluating an order-n determinant to that of evaluating n determinants of order $n-1$. As the method stands, it does not save a lot of time. However, we can choose a row or column that is mostly zeros to save some time. Better yet, we can perform row operations on A to create a column with a lot of zeros. Interchanging two rows of A changes the determinant's sign (Sec. 10.7.4). Multiplying a row of A by m multiplies the determinant by m. And adding or subtracting a multiple of one row to or from another does not change the determinant. The best way of all is to use row operations to reduce A to an upper triangular matrix; the determinant of an upper triangular matrix is the product of the elements on the principal diagonal.

Chapter 11   Mathematics and Computers

Curiously, mathematicians don't use computers all that much. Computers are very good at arithmetic, and arithmetic is a branch of mathematics. However, nobody has found a way to make computers creative, and mathematics is at its heart a creative discipline. This chapter will focus on ways that computers can be used to do some serious mathematics.

In 1977, computers were used to prove the famous four-color theorem of topology, a branch of mathematics related to geometry. The theorem states that any map drawn on a plane can be colored with four colors no matter how complicated the nations on the map get, as long as each nation is limited to one continuous piece of land. The problem had occupied some of the best mathematicians in the world until it was finally solved by two people named Haken and Appel, with a big assist from computers. Because of this, there is still some lingering doubt about the theorem. Its falsity is unlikely, but it would take centuries for a human to verify the proof, and people don't live that long.

Another area where computers are indispensable to mathematics is the calculation, to millions or even billions of decimal places, of key mathematical constants like pi and e. Pi is now known to at least 1.24 *trillion* decimal places; before the advent of the computer, it was known to "only" 527. Why calculate pi to such precision when a mere forty decimal places will serve for any conceivable practical application? Part of the answer is similar to the typical reason for climbing Mount Everest: because it's there. A more serious reason is to scrutinize the digits carefully for any sort of pattern or regularity. It is believed that the digits in pi are completely random, but nobody knows for sure. Unfortunately, no finite number of calculated digits can prove that pi is random; there may be no pattern in the first ten trillion digits, but a pattern starting with digit ten trillion and one. Carl Sagan (not a mathematician, but an astronomer) explored this possibility at the end of his novel "Contact."

11.1 BASIC Computer Programming

If you have a computer at home (and if you bought this book, you probably do), you can use it to do some mathematics. However, you must know a high-level programming language in order to accomplish anything. Therefore, the next few pages will contain a tutorial on the BASIC (which stands for Beginner's All-purpose Symbolic Instruction Code) language. There are many different versions of this language; what is presented here is Microsoft's "Qbasic" as of 1997. (See, I just dated myself.) Microsoft has made BASIC increasingly difficult to find in later versions of its Windows operating system, and it may no longer be included at all. However, there are versions of the language available for purchase at a modest price in computer stores such as Fry's.

11.1.1 Variables: Simple and Subscripted

The data manipulated by computers running BASIC comes in pieces called *variables*. A variable may be *simple*, consisting of only one piece of data, or *subscripted*, containing many pieces. Before you can use a subscripted variable, you must tell the computer how many pieces of data it contains. You do this with a DIM (for *dimension*) statement. DIM A(100) reserves 100 pieces of data under the name A. To refer to one of those pieces, you use the subscript: A(69) is the 69th piece of data.

Variables also come in three flavors, according to the type of data they contain. Real variables are most often used. Also called *floating-point* numbers, these variables simulate real numbers. Because computers are finite, a real variable cannot represent any real number, but only a limited range and with a limited precision. In Microsoft Qbasic, real variables have about seven significant digits of precision, and can range up to about 10^{35}. Real variables have names consisting of letters and numbers; the first character must be a letter. They are distinguished from other variables by not having a *type suffix* at the end of their names.

Integer variables, as their name implies, represent integers. Again, only a finite range of values can be accommodated, but in this case precision is perfect. The range of values allowed is -32,768 to +32,767. Integer variables have names like real ones, but they have a type suffix, %, added. Subscripted integer variables are a handy way to represent large numbers beyond the range of real variables. (See Sec. 11.1.8 below.)

The third type of variable is the *string* variable. In most versions of BASIC, the type suffix for a string variable is $. String variables represent characters that can be printed or displayed on a monitor. Usually, before output is performed, numeric data is converted into strings and manipulated with the string functions such as LEFT$, RIGHT$, and STR$. These functions will be discussed later in the tutorial.

11.1.2 The Assignment Statement

The heart of all BASIC programs is the *assignment*, or LET, statement. It takes the form LET *variable = expression*, and causes the variable on the left to be assigned the value of the expression on the right. In most versions of BASIC, the keyword LET is optional. A statement such as $N = N + 1$ makes no sense in algebra, but in BASIC, it causes the value of N, no matter what it might be, to increase by one. This is usually referred to as an *increment*.

11.1.3 Arithmetic

All BASIC versions support five fundamental arithmetic operations: addition, subtraction, multiplication, division, and exponentiation. The symbols for these are, respectively, +, -, *, /, and ^. The order of operations follows the conventions in Sec. 3.1.2: operations inside parentheses are performed first, then exponentiation from left to right, multiplication and division from left to right, and finally addition and subtraction from left to right. Therefore, $4 + 6 * 7 ^\wedge 2 = 298$. The * is available on most keyboards as a shifted 8, while the ^ is a shifted 6.

11.1.4 Mathematical Functions

Most versions of BASIC support a range of mathematical functions. The exponential function, for example, is obtained with EXP(X), which produces e^X. The natural logarithm is obtained by using LOG(X). SIN(X), COS(X), and TAN(X) yield the sine, cosine, and tangent respectively of the angle whose *radian* measure is X. If you want another trigonometric function, you must use the reciprocal relations in Sec. 5.6; to obtain sec X, for example, you write 1/COS(X). The BASIC function ATN(X) returns the radian measure of the angle whose tangent is X. Unfortunately, this is the only inverse trigonometric function provided in most BASICs. To obtain other inverse trigonometric functions, you must play a little with ATN. For example, to obtain $\sin^{-1} X$, write ATN(X/SQR(1 – X*X)). The SQR function produces the square root of a number. X*X usually executes a little faster than X^2, which most BASICs evaluate as EXP(2*LOG(X)).

11.1.5 String Functions

In BASIC, string functions are distinguished by having the type suffix $ attached to their names. The function STR$(X) returns the string that would be printed if a PRINT X statement were to be executed. This string can then be manipulated. The function LEFT$(X$, N) returns the first N characters of X$; the function RIGHT$(X$, N) returns the last N characters of X$; and the MID$(X$, N, T) function returns T consecutive characters of X$, starting with the Nth. The function LEN(X$) returns the number of characters in X$. The following section of code converts an integer variable, X%, whose value is between 0 and 9999, into a four-character string:

```
X$ = STR$(X%)
L = LEN(X$)
X$ = RIGHT$(X$, L-1)
X$ = ”000” + X$
X$ = RIGHT$(X$, 4)
```

The third statement strips off the left character of X$, which is usually a space. The + in the fourth statement performs string concatenation, not addition. If A$ = "DOG" and B$ = "CAT" then A$ + B$ = "DOGCAT" and B$ + A$ = "CATDOG". Note that to assign a value to a string variable, you must use quotation marks; if you say X$ = CAT, you will get an error message.

11.1.6 The FOR-NEXT Loop

When programming a computer, we frequently wish to perform the same set of instructions over and over, usually a fixed number of times. The BASIC statements that do this are the FOR and NEXT statements.

The BASIC statement FOR J = K TO L STEP M initializes the value of J, a real variable, to the value of the expression K. J is called a *loop control variable*. The following statements are then executed until a NEXT J statement is encountered. When the NEXT is executed, the expression M is added to J's value and the new value is compared to

L. If J now exceeds L (assuming M is positive), the FOR-NEXT loop terminates and execution continues with the statement after the NEXT. Otherwise, the computer returns to the statement immediately after the FOR and executes the loop again with the new value of J. The STEP M part of the FOR statement is optional; if you do not specify a STEP, the computer uses 1 as the step size. The FOR-NEXT loop is a powerful way to work with subscripted variables, also known as *arrays*. To add 1 to every element in an array with 100 elements, you use the following piece of code:

```
FOR I = 1 TO 100
A(I) = A(I) + 1
NEXT I
```

FOR-NEXT loops can be *nested*; that is, one loop contains another, which may contain still others. The following section of code calculates the product of two matrices A and B and stores it in the matrix C:

```
FOR I = 1 TO M
FOR J = 1 TO P
S = 0
FOR K = 1 TO N
S = S + A(I, K) * B(K, J)
NEXT K
C(I, J) = S
NEXT J
NEXT I
```

The matrix A is an M x N matrix, B is an N x P matrix, and C is an M x P matrix. To dimension an M x P matrix, you write DIM C(M, P) – but BASIC syntax rules require you to use actual numbers, not variables, in any DIM statement. Note that the order of the loop control variables in the NEXT statements is the reverse of the order in the FOR statements. Failure to observe this rule will cause an error.

11.1.7 Conditional Branches: The IF Statement

Sometimes, we wish the computer to take an action only if a particular condition holds. The BASIC statement used for this purpose is the IF statement, also called a *conditional branch*. It takes the form IF *condition* THEN *statement*, where the *condition* usually involves a *comparison operator* such as = (equals), <> (does not equal), <= (equals or is less than), or >= (equals or is greater than). The instruction does exactly what you think it does: it executes the statement after THEN if and only if the condition before THEN is true. If the condition is false, the statement after THEN is skipped over and execution continues with the next line after the IF statement. Frequently, the satatement after THEN is a GOTO statement (sometimes called an *unconditional branch*, although this is an oxymoron). The GOTO statement takes the form GOTO *line-number* and causes the program to jump to the statement labeled with the specified line number. The line number must be a numeral; it cannot be a variable or an expression. GOTO can be used by itself as well as part of an IF statement. For example, the following piece of code prints out all the powers of two that are less than 1000:

```
  X = 1
1 X = X * 2
  PRINT X
  IF X < 500 THEN GOTO 1
```

In this fragment, the 1 in 1 X = X * 2 is a line number. Some versions of BASIC require that each statement have a unique line number; Microsoft Qbasic does not. And, in case you've forgotten, X = X * 2 multiplies the value of X by 2. Note that the IF statement uses 500, not 1000, as the comparison value. This is because by the time the IF is executed, the current value of X has already been printed. We could have written the code with the IF before the PRINT, in which case the comparison value would have been 1000.

11.1.8 Handling REALLY Big Numbers

You now know enough BASIC to use the language for what is formally known as *multi-precision arithmetic* and is informally known as dealing with big numbers. In this, a subscripted integer variable is used to store a large number a little bit at a time. Usually, each element in the array represents four digits of the number; this may be reduced to three in some instances. The following section of code adds the numbers stored in the arrays A% and B% and stores their sum in the array C%:

```
      FOR I = 1 TO N
      C%(I) = A%(I) + B%(I)
      IF C%(I) < 10000 THEN GOTO 1
      C%(I) = C%(I) – 10000
      C%(I – 1) = C%(I – 1) + 1
 1    NEXT I
```

Here, N is the number of elements in the arrays A%, B%, and C%. It is one-fourth the maximum number of digits these subscripted variables can represent. The leftmost (or most significant) digits in the numbers are held in the first elements of the arrays, and the rightmost (least significant) digits are in the Nth elements. The two statements after the IF handle the "carry" in the addition from one element to the one before (to the left of) it.

To multiply a large number by a small one takes just a little more work. The INT(X) function produces the integer part of X: INT(4.3712) = 4. If X is an integer, INT(X) = X. The following section of code multiplies the number in A% by M and stores the result in B%:

```
      W = 0
      FOR I = N TO 1 STEP – 1
      K = A%(I) * M + W
      W = INT(K/10000)
      B%(I) = K – 10000 * W
      NEXT I
```

In this piece, W acts as a carry from one element of B% to the next. The STEP – 1 cause the calculation to proceed from the least significant digits to the most significant, just as in the standard algorithm for multiplication. In order for this calculation to proceed, the variable K must not exceed the precision limit for real variables provided in BASIC. This limits the value of M to less than one thousand.

To divide a large number by a small one, we basically perform the above calculation in reverse. As with long division, we proceed from left to right, or A%(1) to A%(N):

```
      W = 0
      FOR I = 1 TO N
      K = A%(I) + 10000 * W
      B%(I) = INT(K/D)
      W = K – D * B%(I)
      NEXT I
```

Here, we are dividing A% by D and storing the result in B%. The final value of W is the remainder in the division. D must be less than one thousand, just like M above.

Handling numbers with lots of decimal places is not really different from handling large numbers. The most significant digits are stored in A%(1) and the least significant in A%(N). The following program calculates e to 100 decimal places:

```
      DIM A%(26), B%(26)
      A%(1) = 10000
      FOR J = 2 to 72
      W = 0
      FOR I = 1 to 26
      K = A%(I) + 10000 * W
      A%(I) = INT(K/J)
      W = K – J * A%(I)
      NEXT I
      FOR I = 1 to 26
      B%(I) = B%(I) + A%(I)
      IF B%(I) < 10000 THEN GOTO 1
```

```
B%(I) = B%(I) – 10000
B%(I – 1) = B%(I – 1) + 1
1   NEXT I
NEXT J
```

When the outer (J) loop terminates, the numbers in B% will represent the non-integer part of *e*. The actual value, of course, is 2 plus this non-integer part. There are several things to note about this program. The first is that we have essentially placed the addition block of code a page above after the division block of code just above. The second is that we have two FOR-NEXT loops with the same loop control variable, I. This is OK as long as the first loop is terminated before the second one starts; having two current loops with the same variable will cause an error. And the third thing is that A% and B% are both 26 numbers, thus representing 104 digits. This is because of the truncation error at the end of each division. We cannot account for the digits beyond 100 without keeping track of at least 104. The value of J is limited to 72 because 73! is the first factorial that exceeds 10^{104}, so that 1/72! is the last factorial reciprocal that can affect the 104th decimal place of *e*.

To actually get the computer to print the value of *e*, we would tack onto this program a piece of code similar to that in Sec. 11.1.5, and follow it with a PRINT X\$; instruction. The semicolon at the end of the PRINT statement causes the printer or monitor to remain on the same line so that the next 4 digits are printed immediately to the right of the last four.

We can also multiply one large number by another, although it is a good deal trickier. The following section of code multiplies A% by B%, storing the result in C%. Note that A% and B% have N elements each and C% has 2N. Also note that each element represents only three digits of a large number, not four. This is to keep the size of S within the precision limit for real variables.

```
M = 1000
FOR I = 2 * N TO N + 1 STEP – 1
FOR J = I – N TO N
S = A%(J) * B%(I – J)
S1 = INT(S/M)
S2 = S – M * S1
C%(I) = C%(I) +S2
C%(I – 1) = C%(I – 1) + S1
NEXT J
S3 = INT(C%(I)/M)
S4 = C%(I) – M * S3
C%(I) = S4
C%(I – 1) = C%(I – 1) + S3
S5 = INT(C%(I – 1)/M)
S6 = C%(I – 1) – M * S5
C%(I – 1) = S6
C%(I – 2) = C%(I – 2) + S5
NEXT I
FOR I = N TO 3 STEP – 1
FOR J = 1 TO I – 1
S = A%(J) * B%(I – J)
S1 = INT(S/M)
S2 = S – M * S1
C%(I) = C%(I) + S2
C%(I – 1) = C%(I – 1) + S1
NEXT J
S3 = INT(C%(I)/M)
S4 = C%(I) – M * S3
C%(I)= S4
C%(I – 1) = C%(I – 1) + S3
S5 = INT(C%(I – 1)/M)
S6 = C%(I – 1) – M * S5
C%(I – 1) = S6
C%(I – 2) = C%(I – 2) + S5
```

```
NEXT I
S = A%(1) * B%(1)
S1 = INT(S/M)
S2 = S – M * S1
C%(2) = C%(2) + S2
C%(1) = C%(1) + S1
IF C%(2) < M THEN END
C%(2) = C%(2) – M
C%(1) = C%(1) + 1
```

As before, we work from right to left, from C%(2 * N) to C%(1). The END statement near the end of the program simply tells the computer to stop. It should be noted that we have two essentially identical sections of code within the two FOR I loops. If A% and B% both have 2N elements, of which the first N are zeros, these two loops could be combined into one, although that would cause the program to run slower.

This program will only work if the numbers in A% and B% have less that about 100 digits. If you wish to multiply larger numbers, you could let M be 100 and work with only two digits at a time. This will make the program take longer to run, but will increase the size of the numbers you can manipulate to more than 600 digits.

11.2 Equation Solving with Spreadsheet Programs

Fortunately, there are several software programs that enable you to work with numbers (although not large ones) without having to learn a programming language. However, learning to use Microsoft Excel is almost as difficult as learning BASIC. This section will show you how to use Excel or a similar program to solve equations like $e^x = 3x$ and $x = \cos x$ that cannot be solved analytically. The latest versions of these programs come with routines that automate the process of solving any equation, but there is more mathematics to be learned by using the primitive way illustrated below.

11.2.1 Circular References

If you read any book on spreadsheets, you will find a section on circular references. It will probably say that circular references are to be avoided at all costs, rather like the plague. The type of circular reference typically found in a spreadsheet is hazardous to its health, but carefully chosen circular references can be used to solve non-analytic equations.

So what is a circular reference? It is simply a situation where the formula in a cell of the spreadsheet refers, usually indirectly, to the very cell in which it sits. A common example is the formula =SUM(A1.A10) entered into cell A10. The formula adds the contents of cells A1 through A10 and places the result into cell A10, thus changing the sum. This type of circular reference is to be avoided, because it is "unsolvable," that is, there is no way to satisfy the formula. Here, it is probable that the formula in cell A10 should be =SUM(A1.A9) instead, which does not create a circular reference when it is placed into cell A10. More subtly, cell J10 can refer to cells J1 through J9, J5 refer to A5 through I5, and C5 refer to J10. This is known as a "daisy chain" or indirect circular reference. Depending on the nature of the formulas involved, this circular reference may be unsolvable or solvable. It is the solvable circular references that we wish to use to get numerical solutions to non-analytic equations.

11.2.2 Using Circular References to Solve Equations

The first step in solving equations via the circular reference is to transform the equation you are solving into one of the form $x = F(x)$. Usually, there are several ways this can be done. For example, with $e^x = 3x$, we can divide by three to obtain $x = \dfrac{e^x}{3}$ or take natural logarithms to obtain $x = \ln 3x$. We must choose $F(x)$ so that $\left|\dfrac{dF}{dx}\right| < 1$ in the neighborhood of the solution. For $e^x = 3x$, there are two solutions, one less than 1 and one more than 1. To find the solution that is less than one, we must use $x = \dfrac{e^x}{3}$, and to find the one that is more than one, we must use $x = \ln 3x$.

The second step is to translate $F(x)$ onto the language of the particular spreadsheet program you are using. If you are using Excel (and most people now are), $\dfrac{e^x}{3}$ becomes =EXP(X)/3 and $\ln 3x$ becomes =LN(3*X). You then seed the process by entering a guess as to the solution into cell A1 (and A2 in this case). One is a good first guess for this problem, as both solutions are not very far from 1. Then you enter the translated $F(x)$ into cell B1 (and B2 here), replacing x wherever it occurs with the cell reference A1 (and A2 for the formula entered into B2). So, to solve $e^x = 3x$, enter =EXP(A1)/3 into B1 and =LN(A2*3) into B2. You will see an improved guess to each solution in cells B1 and B2. Thus far, we have not created a circular reference. But now we change cell A1 to =B1 and A2 to =B2. This creates a circular reference, as A1 refers to B1 and B1 to A1. In the latest version of Excel, cell A1 (and A2 in this example) will display 0, and the program will give you a warning stating that the worksheet contains a circular reference. To proceed, click on the Office button (the circle with four colors at the extreme upper left of the screen), then click on Excel Options in the lower right of the window that pops up, then on Formulas in the upper left of the next window. Then check the box labeled "Enable iterated calculations" and set "Maximum iterations" and "Maximum change." When you do this, it is a good idea to set the number of iterations to a moderately-sized prime number like 11 or 13, and to set "Maximum change" to the desired accuracy of your solution. This minimizes the chance that the process ends up in a limit cycle, instead of a limit point, without your realizing it. After a few presses of F9, I obtained solutions of 0.619061287 and 1.512134552 to $e^x = 3x$.

11.2.3 Why the Method Works

This technique is a fairly straightforward application of iterative mathematics, as described in Sec. 8.1. We seek a fixed point of the function $F(x)$, and find it by repeatedly feeding back F's output into itself. This is what the circular reference does: it forces the computer to iterate F. The restriction $\left|\frac{dF}{dx}\right| < 1$ is there to make sure that the fixed point is an attractor rather than a repeller. To see why, imagine that z is the actual answer to $x = F(x)$ and that our current guess is $z + \epsilon$, where $|\epsilon|$ is very small. Now, since $z = F(z)$, $F(z + \epsilon)$ will be approximately $z + \epsilon \left(\frac{dF}{dz}\right)$. It is now clear that if $\left|\frac{dF}{dz}\right|$ is greater than 1, the iterate of $z + \epsilon$ is farther from z than $z + \epsilon$ itself. If $-1 < \frac{dF}{dz} < 0$, the iterate of $z + \epsilon$ will be on the opposite side of z from $z + \epsilon$ itself, and the successive iterates will oscillate back and forth around z, closing in on it all the while. If $0 < \frac{dF}{dz} < 1$, the iterate of $z + \epsilon$ is on the same side of z, and the successive iterates will monotonically approach z. If $\left|\frac{dF}{dz}\right| = 1$, then the iterates will converge or diverge with agonizing slowness, depending on how $\frac{dF}{dz}$ behaves near z. The fixed point is here only semi-stable. An example is the iterates of $kx(1-x)$ when $k = 3$. When $k = 3$, the fixed point is $z = \frac{2}{3}$ and $\frac{dF}{dz}$ is exactly -1.

11.3 Mathematics and the Internet

When I started writing this book in 1993, few people had even heard of the Internet. Now, in 2010, everybody knows at least something about it, and millions of Americans use it every day. Yes, there is plenty of mathematics on the Web, and your computer can be used to do some serious mathematics.

One area in which you can help is the hunt for ever-larger Mersenne primes (see Sec. 4.4.3). To join the search, just point your Web browser to v5www.mersenne.org and click on Downloads in the page that appears. You will be presented with several links to the actual program code; select the one that matches your operating system. Programs are available for Windows, Macintosh, UNIX, and a few other less common operating systems. The download will take several seconds with a DSL or cable modem connection, and a few minutes with a dial-up connection. (The program is about two megabytes in size.) The program, once downloaded, will ask you questions about your computer and find out other information on its own, then give you a set of choices on what work you would like your computer to do for GIMPS (the Great Internet Mersenne Prime Search). Most people will want to run first-time Lucas-Lehmer tests (described in Sec. 4.4.3), but trial factoring, double-checking Lucas-Lehmer tests, and other options are available. Once you choose, the program will start doing the work you have selected, using the Internet to get assignments from the GIMPS server and report back to it when it is done. Depending on the size of the numbers it gets and your processor's speed, it will take several days to a few months to test a number which may have over ten million digits. The program runs in the background, so you can use your computer for other things while the testing is going on, but doing certain computationally intensive things, like playing certain video games, will slow the Mersenne-testing program. (A warning: The program will try to get another assignment after the first one is finished. If you want to drop out of GIMPS, you must select the "Quit GIMPS" option under the Advanced menu at the top of the program's window.)

As of May 2010, GIMPS has found 13 prime numbers, ranging in size from 420,000 to almost 13 million digits, in 13 and a half years of existence. This is a significant acceleration in the rate of Mersenne prime discoveries, as one had been discovered about every three years since computers were first turned on the problem in 1947. It is due partly to the use of thousands of computers connected to the Internet and part of GIMPS, and partly to an increase in the frequency of Mersenne primes. There are three Mersenne primes with exponents between 20 and 26 million, and three more with exponents between 37 and 44 million, the tightest cluster of Mersenne primes, as measured by the logarithms of their exponents, yet found. There are now 47 known Mersenne primes, but whether there are any more is for more computation to find out. It is likely that there are an infinity of them, but the mathematician with sufficient brains and time to prove that hasn't been born yet.

I would have liked to include in this chapter a section on using your computer and the Internet to calculate pi, but the pursuit of more digits in this number has gotten beyond the power of a typical home computer. In 2002, two Japanese researchers used 660 hours on a supercomputer to obtain 1,241,100,000,000 (that's almost 1¼ *trillion!*) digits of pi. And, in the years since 2002, there have probably been even more digits calculated. Of course, you can still use your computer to calculate pi; it just won't be breaking any new ground, as it would with GIMPS. To do this, use the relation $\pi = 4(\tan^{-1}\frac{1}{2} + \tan^{-1}\frac{1}{3})$, or the improved $\pi = 16\tan^{-1}\frac{1}{5} - 4\tan^{-1}\frac{1}{239}$, and apply the formula for $\tan^{-1}x$ in Sec. 7.10.8 and the programming techniques in this chapter.

Chapter 12   Conclusion

We have come to the end of this book. After slogging through eleven chapters covering most of the major branches of mathematics (the chapter on calculus alone covered several dozen pages and about a year's worth of college), you have reached the conclusion. A certain portion of the reading audience is bound to say "So what?" I can only respond that if you say "So what?" in response to this book, you should not buy it. You are not genuinely interested in mathematics. On the other hand, if you are intrigued by the content here and wish to find out more, I can tell you that in every chapter of this book, except maybe the first two, there is a lot of math that I did not include, mostly because I don't know it myself or because it is too technical for my intended audience. I have tried to include only those topics that I felt I could explain to the average adult. As a result, some good math that I would have liked to include had to be left out. For example, in Chapter 4 I had to leave out the theorem and proof that if p is a prime greater than 5, either F_{p-1} or F_{p+1} is a multiple of p, and F_p itself is either one more than or one less than a multiple of p. In Chapter 5, I could only state, not prove, the reflection properties of conic sections, and in Chapter 7 I did not venture into the tricky area of multivariable calculus.

I have included examples wherever possible; however, Chapter 10 is rather lacking in this area. Examples are important, as they serve to concrete-ize many very abstract propositions. Even in the first two chapters, there are a lot of examples, though the material is not very abstract. I hope that the examples I did put in serve to illustrate the concepts well.

12.1 Practical Applications of mathematics

Mostly, people study mathematics for the same reason they pay $80 million for a Van Gogh painting: because it is beautiful. As G. H. Hardy once said, "There is no permanent place in this world for ugly mathematics." But math would not have been invented if it did not have practical applications, however ugly they might be. (The more abstract branches invented in the last few decades may not have applications now, but will probably acquire some in the future.) Anyone who wishes to become a scientist simply must learn a lot of mathematics. Math is the universal language of science. In physics, for example, there is Newton's equation, $F = \frac{d(mv)}{dt}$, briefly mentioned in Sec. 7.4.3. This is a differential equation stating that Force = rate of change of Momentum. It is also a vector equation, because both force, F, and velocity, v, are vectors, with direction as well as magnitude. Thus, both Chapters 7 and 10 are used to study it.

Sections 7.4 and 7.8 are completely about applications of calculus. That is the branch of math with the most concrete applications. In order to study calculus, however, you must know algebra, geometry, and trigonometry, the branches of mathematics that are commonly lumped together as "pre-calculus" math. Many of the applications of mathematics are to other branches of math. One goal of modern mathematics is to find connections between seemingly unrelated branches of math. The Fundamental Theorem of Calculus is a connection between the seemingly unrelated areas of integral and differential calculus.

The branch of mathematics with the second-most applications is probably geometry. Surveyors apply geometry, especially trigonometry, to real-world situations on a regular basis. Architects use geometry to make sure the buildings they design do not tip over even in a strong wind. Even pool players need to know geometry in order to knock balls into the pockets. Artists like M. C. Escher use geometry in their pictures.

The part of math with the fewest applications is number theory, which was Hardy's area of specialization. Knowing that $2^{20,996,011} - 1$ is prime can serve no practical purpose. But even this area has acquired an application. The theory of numbers is now used in cryptography, and the security of data in the United States rests on the perceived difficulty of factoring large numbers as opposed to the relative ease of testing large numbers for primality.

Here is how the RSA encryption system, named for its inventors, R. Rivest, A. Shamir, and L. Adleman, works. First, the person or organization who wishes to send and receive encrypted messages finds two large prime numbers p and q. These numbers are then multiplied to produce N, one part of the "public" key. The other part of the public key, a, is a prime number less than $(p-1)(q-1)$. The "private" key, b, is a number that, when multiplied by a, is one more than a multiple of $(p-1)(q-1)$. Such a number will always exist and be unique among numbers less than $(p-1)(q-1)$ provided that a is prime. The organization then publishes N and a while keeping b secret. Someone who wants to send an encrypted message then first digitizes it by following a simple scheme such as A = 01, B = 02, etc. Then he breaks it into chunks that are comparable in size to N but a little smaller, and raises each chunk m to the power of a, modulo N. To do this does not require the computation of the enormous number m^a itself. (See Sec. 4.5 for a discussion of modular arithmetic.) The encrypted message is this modular remainder.

The recipient of the message then raises it to the power b, modulo N. This step "magically" uncovers the original message, m. The mathematics behind this is as follows: $(m^a \bmod N)^b \bmod N = m^{ab} \bmod N = m^{(p-1)(q-1)+1} \bmod N = (m^{p-1})^{q-1} \cdot m \bmod N = m \bmod N = m$ because m is less than N. The last step is an application of Fermat's Little Theorem (Sec. 4.4.3) in the form that, if p is a prime number, then $m^{p-1} = 1 \bmod p$. (Remember that $N = p \times q$.) In order to decrypt messages, you must know the private key, b, and in order to find b, you must know p and q, the prime factors of N. This is where the fun comes in, as nobody is really sure that factoring is intractable. It's just that factoring appears to be much more difficult than primality testing, and the size of the numbers that can be prime-tested in a reasonable amount of computer time is much larger than the size of the composite numbers that can be factored in a reasonable amount of computer time. In theory, any encryption system can be broken, given enough computational resources, but this system is as secure as any that has been tried and is widely used.

So math *does* have practical applications, and in the most unlikely areas to boot. It is extremely unlikely that you will go through your adult life without being touched by math at least once. Many types of job use math regularly, not just scientific careers. Bankers, for example, use algebra to determine the monthly payments necessary to pay off a loan, and use a little arithmetic to ascertain whether the prospective borrower can afford them. Economists use calculus to determine the optimum price at which a commodity should be sold and the amount of it that will be produced. Artists use geometry to produce the correct perspective effects in their paintings and sketches. Even ditch-diggers use mathematics, although not in a rigorous way, when they decide how much dirt they will try to move with the next shovelful.

12.2 Suggestions for Further Reading

During my life and the writing of this book, I have read many books on mathematics and studied the subject in college. I can recommend several books for the reader who wishes to learn more about math than I can provide here. Douglas Hofstadter's "Godel, Escher, Bach: An Eternal Golden Braid" contains much material on formal languages, including TNT, the language from which I abstracted "TNT Jr." in Chapter 6. He talks about Godel's Incompleteness Theorems, which prove that there will always be undecidable statements in mathematics. The last part of Hofstadter's book talks about artificial intelligence and the implications of Godel's work for the field. A warning: this book was written in 1979, and computer capabilities have increased enormously since then. However, there has not been a huge degree of progress in AI, which serves to underscore Hofstadter's points.

Another book I can recommend is Philip Davis and Reuben Hersh's "The Mathematical Experience." This book covers many aspects of mathematics, including unsavory ones such as its use in war and gematria, the assigning of numerical values to letters of the alphabet and manipulation of these numbers to produce a desired result. This book, too, mentions Godel's theorems and discusses their implications for the rest of mathematics, in particular, the loss of certainty that all mathematicians of the previous hundred years had sought to achieve. In 1900, it was thought that it is possible to prove, beyond the shadow of a doubt, that what we think we know in mathematics really is true, and that every question in mathematics has a definite answer. However, Godel's 1931 paper demolished that hope, and, ever since, mathematicians have had to live with nagging doubts. I know of two non-trivial undecidable statements in math, Euclid's parallel postulate (Sec. 5.2) and Cantor's continuum hypothesis (Sec. 6.5); there surely are many others that just haven't yet been proven to be undecidable. "The Mathematical Experience" also deals with mathematical intuition, the famous "Riemann hypothesis," and many other topics.

Also on the recommended list is the admirably non-technical "How to Solve It" by George Polya. This classic deals at length with a fairly easy problem and expands on the various strategies that can be brought to bear on it. Some of the advice is valid for non-mathematical problems as well as problems in math, so it can appeal to non-mathematicians as well as those who are interested in math.

Finally, there is "Infinity and the Mind" by Rudy Rucker. It deals with infinity in all its forms, and infinity's role in the foundations of mathematics. At times, this book gets highly technical, but its non-technical parts raise interesting questions and deal with several paradoxes that arise from the study of the infinite. The book deals with Godel's work and its implications for computers, but stresses that the human mind is a computer, too: Godel's theorems place limits on what *people* can do, as well as machines. Of note in this book is its treatment of the liar or Epimenides paradox: the problems that result from trying to assign a truth value to the sentence "This sentence is not true." Rucker reaches the conclusion that mathematical truth is undefinable, at least in a formal-language way, but still claims that he knows mathematical truth when he sees it.

12.3 Final Thoughts

If this book stimulates even one person (besides myself) to learn and enjoy mathematics, it will have been worth the effort. What started out as a way to solve the mystery of how a young boy could multiply 103 by 107 in his head has become a 100,000-word book which took me 12 years (off and on) to write. While I was discovering (not inventing) the math in this book, I experienced numerous moments of "Zahlvergnugen" which resembled the "runner's high" experienced by marathoners. Finding out something that very few other people know makes me happy, and sharing that thing with others makes me even happier. I hope that discovering these things for yourself makes you happy, too.

Index to Mathematicians

Section(s)

Table 1   Square Triangular Numbers and Quasi-Isosceles Pythagorean Triples

In the top table, the square triangular number in the position under "Position" is in the same row under "Number," and its square root and triangular root (defined in Section 4.3) are under those two columns respectively. Thus, the fourth square triangular number is 41616, its square root is 204, and its triangular root is 288.

In the second table, the solutions to $A^2 + B^2 = C^2$ which also satisfy $B = A + 1$ are given under the A, B, and C columns. Thus, the fourth such solution is A=119, B=120, and C=169.

It should be noted that there are an infinite number of both square triangular numbers and quasi-isosceles Pythagorean triples.

Position	Number	Square Root	Triangular Root
1	1	1	1
2	36	6	8
3	1225	35	49
4	41616	204	288
5	1413721	1189	1681
6	48024900	6930	9800
7	1631432881	40391	57121
8	55420693056	235416	332928
9	1882672131025	1372105	1940449
10	63955431761796	7997214	11309768

Position	A	B	C
1	0	1	1
2	3	4	5
3	20	21	29
4	119	120	169
5	696	697	985
6	4059	4060	5741
7	23660	23661	33461
8	137903	137904	195025
9	803760	803761	1136689
10	4684659	4684660	6625109

Table 2   Fibonacci Numbers and their Factors

This table contains the first 25 Fibonacci numbers and their prime factors. For n a prime greater than 5, it also has the factors of $F_n + 1$ or $F_n - 1$, whichever is a multiple of n. For example $F_{13} = 233$, a prime number and one less than $2 \times 3^2 \times 13$.

n	F_n	Factors of F_n
1	1	-
2	1	-
3	2	Prime
4	3	Prime
5	5	Prime
6	8	2^3
7	13	Prime ($2 \times 7 - 1$)
8	21	3×7
9	34	2×17
10	55	5×11
11	89	Prime ($2^3 \times 11 + 1$)
12	144	$2^4 \times 3^2$
13	233	Prime ($2 \times 3^2 \times 13 - 1$)
14	377	13×29
15	610	$2 \times 5 \times 61$
16	987	$3 \times 7 \times 47$
17	1597	Prime ($2 \times 17 \times 47 - 1$)
18	2584	$2^3 \times 17 \times 19$
19	4181	37×113 ($2^2 \times 5 \times 11 \times 19 + 1$)
20	6765	$3 \times 5 \times 11 \times 41$
21	10946	$2 \times 13 \times 421$
22	17711	89×199
23	28657	Prime ($2 \times 7 \times 23 \times 89 - 1$)
24	46368	$2^5 \times 3^2 \times 7 \times 23$
25	75025	$5^2 \times 3001$

Table 3   Modular Reciprocals

This table contains the reciprocals, modulo the primes P from 17 to 97, for each positive integer from 1 to P-1. Reciprocals for P = 5, 7, 11, and 13 can be found in Fig. 4-6.

To use the table, first find the block for the prime P you are using as modulus. Then find the tens digit of your number at the far left, and the ones digit in the row at top. Where the row and column intersect is your modular reciprocal. Thus, the reciprocal of 14 modulo 23 is 5, found in the block headed "P=23" in the intersection of the "1" row and "4" column. The reciprocals of single-digit numbers are found in the "0" row. Thus, the reciprocal of 7 modulo 19 is 11.

P = 17

	0	1	2	3	4	5	6	7	8	9
0	-	1	9	6	13	7	3	5	15	2
1	12	14	10	4	11	8	16			

P = 19

	0	1	2	3	4	5	6	7	8	9
0	-	1	10	13	5	4	16	11	12	17
1	2	7	8	3	15	14	6	9	18	

P = 23

	0	1	2	3	4	5	6	7	8	9
0	-	1	12	8	6	14	4	10	3	18
1	7	21	2	16	5	20	13	19	9	17
2	15	11	22							

P = 29

	0	1	2	3	4	5	6	7	8	9
0	-	1	15	10	22	6	5	25	11	13
1	3	8	17	9	27	2	20	12	21	26
2	16	18	4	24	23	7	19	14	28	

P = 31

	0	1	2	3	4	5	6	7	8	9
0	-	1	16	21	8	25	26	9	4	7
1	28	17	13	12	20	29	2	11	19	18
2	14	3	24	27	22	5	6	23	10	15
3	30									

P = 37

	0	1	2	3	4	5	6	7	8	9
0	-	1	19	25	28	15	31	16	14	33
1	26	27	34	20	8	5	7	24	35	2
2	13	30	32	29	17	3	10	11	4	23
3	21	6	22	9	12	18	36			

P = 41

	0	1	2	3	4	5	6	7	8	9
0	-	1	21	14	31	33	7	6	36	32
1	37	15	24	19	3	11	18	29	16	13
2	39	2	28	25	12	23	30	38	22	17
3	26	4	9	5	35	34	8	10	27	20
4	40									

Table 3 (cont.)

P = 43

	0	1	2	3	4	5	6	7	8	9
0	-	1	22	29	11	26	36	37	27	24
1	13	4	18	10	40	23	35	38	12	34
2	28	41	2	15	9	31	5	8	20	3
3	33	25	39	30	19	16	6	7	17	32
4	14	21	42							

P = 47

	0	1	2	3	4	5	6	7	8	9
0	-	1	24	16	12	19	8	27	6	21
1	33	30	4	29	37	22	3	36	34	5
2	40	9	15	45	2	32	38	7	42	13
3	11	44	25	10	18	43	17	14	26	41
4	20	39	28	35	31	23	46			

P = 53

	0	1	2	3	4	5	6	7	8	9
0	-	1	27	18	40	32	9	38	20	6
1	16	29	31	49	19	46	10	25	3	14
2	8	48	41	30	42	17	51	2	36	11
3	23	12	5	45	39	50	28	43	7	34
4	4	22	24	37	47	33	15	44	21	13
5	35	26	52							

P = 59

	0	1	2	3	4	5	6	7	8	9
0	-	1	30	20	15	12	10	17	37	46
1	6	43	5	50	38	4	48	7	23	28
2	3	45	51	18	32	26	25	35	19	57
3	2	40	24	34	33	27	41	8	14	56
4	31	36	52	11	55	21	9	54	16	53
5	13	22	42	49	47	44	39	29	58	

P = 61

	0	1	2	3	4	5	6	7	8	9
0	-	1	31	41	46	49	51	35	23	34
1	55	50	56	47	48	57	42	18	17	45
2	58	32	25	8	28	22	54	52	24	40
3	59	2	21	37	9	7	39	33	53	36
4	29	3	16	44	43	19	4	13	14	5
5	11	6	27	38	26	10	12	15	20	30
6	60									

Table 3 (cont.)

P = 67

	0	1	2	3	4	5	6	7	8	9
0	-	1	34	45	17	27	56	48	42	15
1	47	61	28	31	24	9	21	4	41	60
2	57	16	64	35	14	59	49	5	12	37
3	38	13	44	65	2	23	54	29	30	55
4	62	18	8	53	32	3	51	10	7	26
5	63	46	58	43	36	39	6	20	52	25
6	19	11	40	50	22	33	66			

P = 71

	0	1	2	3	4	5	6	7	8	9
0	-	1	36	24	18	57	12	61	9	8
1	64	13	6	11	66	19	40	46	4	15
2	32	44	42	34	3	54	41	50	33	49
3	45	55	20	28	23	69	2	48	43	51
4	16	26	22	38	21	30	17	68	37	29
5	27	39	56	67	25	31	52	5	60	65
6	58	7	63	62	10	59	14	53	47	35
7	70									

P = 73

	0	1	2	3	4	5	6	7	8	9
0	-	1	37	49	55	44	61	21	64	65
1	22	20	67	45	47	39	32	43	69	50
2	11	7	10	54	70	38	59	46	60	68
3	56	33	16	31	58	48	71	2	25	15
4	42	57	40	17	5	13	27	14	35	3
5	19	63	66	62	23	4	30	41	34	26
6	28	6	53	51	8	9	52	12	29	18
7	24	36	72							

P = 79

	0	1	2	3	4	5	6	7	8	9
0	-	1	40	53	20	16	66	34	10	44
1	8	36	33	73	17	58	5	14	22	25
2	4	64	18	55	56	19	76	41	48	30
3	29	51	42	12	7	70	11	47	52	77
4	2	27	32	68	9	72	67	37	28	50
5	49	31	38	3	60	23	24	61	15	75
6	54	57	65	74	21	62	6	46	43	71
7	35	69	45	13	63	59	26	39	78	

Table 3 (cont.)

P = 83

	0	1	2	3	4	5	6	7	8	9
0	-	1	42	28	21	50	14	12	52	37
1	25	68	7	32	6	72	26	44	60	35
2	54	4	34	65	45	10	16	40	3	63
3	36	75	13	78	22	19	30	9	59	66
4	27	81	2	56	17	24	74	53	64	61
5	5	70	8	47	20	80	43	67	73	38
6	18	49	79	29	48	23	39	57	11	77
7	51	76	15	58	46	31	71	69	33	62
8	55	41	82							

P = 89

	0	1	2	3	4	5	6	7	8	9
0	-	1	45	30	67	18	15	51	78	10
1	9	81	52	48	70	6	39	21	5	75
2	49	17	85	31	26	57	24	33	35	43
3	3	23	64	27	55	28	47	77	82	16
4	69	76	53	29	87	2	60	36	13	20
5	73	7	12	42	61	34	62	25	66	86
6	46	54	56	65	32	63	58	4	72	40
7	14	84	68	50	83	19	41	37	8	80
8	79	11	38	74	71	22	59	44	88	

P = 97

	0	1	2	3	4	5	6	7	8	9
0	-	1	49	65	73	39	81	14	85	54
1	68	53	89	15	7	13	91	40	27	46
2	34	37	75	38	93	66	56	18	52	87
3	55	72	94	50	20	61	62	21	23	5
4	17	71	67	88	86	69	19	64	95	2
5	33	78	28	11	9	30	26	80	92	74
6	76	35	36	77	47	3	25	42	10	45
7	79	41	31	4	59	22	60	63	51	70
8	57	6	84	90	82	8	44	29	43	12
9	83	16	58	24	32	48	96			

Table 4 Squares, Cubes, Square Roots and Cube Roots of Numbers 1 to 100

To use this table, simply find the column containing the data you want, and look opposite the number you are investigating. At the intersection of that row and column is the data you want. Hence, the cube root of 43 is (approximately) 3.5034. Linear interpolation in the square root column should be accurate to three decimal places for $n > 25$, and in the cube root column for $n > 20$.

n	n^2	n^3	$\sqrt{n}$	$\sqrt[3]{n}$	n	n^2	n^3	$\sqrt{n}$	$\sqrt[3]{n}$
1	1	1	1.0000	1.0000	41	1,681	68,921	6.4031	3.4482
2	4	8	1.4142	1.2599	42	1,764	74,088	6.4807	3.4760
3	9	27	1.7321	1.4422	43	1,849	79,507	6.5574	3.5034
4	16	64	2.0000	1.5874	44	1,936	85,184	6.6332	3.5303
5	25	125	2.2361	1.7100	45	2,025	91,125	6.7082	3.5569
6	36	216	2.4495	1.8171	46	2,116	97,336	6.7823	3.5830
7	49	343	2.6457	1.9129	47	2,209	103,823	6.8557	3.6088
8	64	512	2.8284	2.0000	48	2,304	110,592	6.9282	3.6342
9	81	729	3.0000	2.0801	49	2,401	117,649	7.0000	3.6593
10	100	1,000	3.1623	2.1544	50	2,500	125,000	7.0711	3.6840
11	121	1,331	3.3166	2.2240	51	2,601	132,651	7.1414	3.7084
12	144	1,728	3.4641	2.2894	52	2,704	140,608	7.2111	3.7325
13	169	2,197	3.6056	2.3513	53	2,809	148,877	7.2801	3.7563
14	196	2,744	3.7417	2.4101	54	2,916	157,464	7.3484	3.7798
15	225	3,375	3.8730	2.4662	55	3,025	166,375	7.4162	3.8030
16	256	4.096	4.0000	2.5198	56	3,136	175,616	7.4833	3.8259
17	289	4,913	4.1231	2.5713	57	3,249	185,193	7.5498	3.8485
18	324	5,832	4.2426	2.6207	58	3,364	195,112	7.6158	3.8709
19	361	6.859	4.3589	2.6684	59	3,481	205,379	7.6811	3.8930
20	400	8,000	4.4721	2.7144	60	3,600	216,000	7.7460	3.9149
21	441	9,261	4.5826	2.7589	61	3,721	226,981	7.8103	3.9365
22	484	10,648	4.6904	2.8020	62	3,844	238,328	7.8740	3.9579
23	529	12,167	4.7958	2.8439	63	3,969	250,147	7.9373	3.9791
24	576	13,824	4.8990	2.8845	64	4,096	262,144	8.0000	4.0000
25	625	15,625	5.0000	2.9240	65	4,225	274,625	8.0623	4.0207
26	676	17,576	5.0990	2.9625	66	4,356	287,496	8.1240	4.0412
27	729	19,683	5.1962	3.0000	67	4,489	300,763	8.1854	4.0615
28	784	21,952	5.2915	3.0366	68	4,624	314,432	8.2462	4.0817
29	841	24,389	5.3852	3.0723	69	4,761	328,509	8.3066	4.1016
30	900	27,000	5.4772	3.1072	70	4,900	343,000	8.3667	4.1213
31	961	29,791	5.5678	3.1414	71	5,041	357,911	8.4261	4.1408
32	1,024	32,768	5.6569	3.1748	72	5,184	373,248	8.4853	4.1602
33	1,089	35,937	5.7446	3.2075	73	5,329	389,017	8.5440	4.1793
34	1,156	39,304	5.8310	3.2396	74	5,476	405,234	8.6023	4.1983
35	1,225	42,875	5.9161	3.2711	75	5,625	421,875	8.6603	4.2172
36	1,296	46,656	6.0000	3.3019	76	5,776	438,976	8.7178	4.2358
37	1,369	50,653	6.0828	3.3322	77	5,929	456,533	8.7750	4.2543
38	1,444	54,872	6.1644	3.3620	78	6,084	474,552	8.8318	4.2727
39	1,521	59,319	6.2450	3.3912	79	6,241	493,039	8.8882	4.2908
40	1,600	64,000	6.3246	3.4200	80	6,400	512,000	8.9443	4.3089

Table 4 (cont.)

n	n^2	n^3	$\sqrt{n}$	$\sqrt[3]{n}$
81	6,561	531,441	9.0000	4.3267
82	6,724	551,368	9.0554	4.3445
83	6,889	571,787	9.1104	4.3621
84	7,056	592,704	9.1652	4.3795
85	7,225	614,125	9.2195	4.3968
86	7,396	636,058	9.2736	4.4140
87	7,569	658,503	9.3274	4.4310
88	7,744	681,472	9.3808	4.4480
89	7,921	704,969	9.4340	4.4648
90	8,100	729,000	9.4868	4.4814
91	8,281	753,571	9.5394	4.4979
92	8,464	778,688	9.5917	4.5144
93	8,649	804,357	9.6437	4.5307
94	8,836	830,584	9.6954	4.5468
95	9,025	857,375	9.7468	4.5629
96	9,216	884,736	9.7980	4.5789
97	9,409	912,673	9.8489	4.5947
98	9,604	941,192	9.8995	4.6104
99	9,801	970,299	9.9499	4.6261
100	10,000	1,000,000	10.0000	4.6416

Table 5   Factor Table to 1000

This table contains the smallest prime factor of each number ending in 1, 3, 7, or 9 between 3 and 1000. Numbers ending in other digits are multiples of 2, 5, or both. To use the table, find all but the rightmost digit of your number in the columns under N, and the last digit in the row at the top. A P in that location means that your number is prime; otherwise, the smallest factor is listed. To complete the factorization, divide your number by this factor, and repeat the process with the quotient. Thus, 189 is found, in the row starting with 18 and under the 9 column, to have 3 as a factor. 189 = 3 x 63; 63 is also found to have 3 as a factor, and 63 = 3 x 21; 21 has 3 as a factor, and 21 = 3 x 7. Since 7 is shown as a prime, the process ends with 189 having the prime factors of 3^3 x 7.

N	1	3	7	9	N	1	3	7	9	N	1	3	7	9	N	1	3	7	9
0	-	P	P	3	25	P	11	P	7	50	3	P	3	P	75	P	3	P	3
1	P	P	P	P	26	3	P	3	P	51	7	3	11	3	76	P	7	13	P
2	3	P	3	P	27	P	3	P	3	52	P	P	17	23	77	3	P	3	19
3	P	3	P	3	28	P	P	7	17	53	3	13	3	7	78	11	3	P	3
4	P	P	P	7	29	3	P	3	13	54	P	3	P	3	79	7	13	P	17
5	3	P	3	P	30	7	3	P	3	55	19	7	P	13	80	3	11	3	P
6	P	3	P	3	31	P	P	P	11	56	3	P	3	P	81	P	3	19	3
7	P	P	7	P	32	3	17	3	7	57	P	3	P	3	82	P	P	P	P
8	3	P	3	P	33	P	3	P	3	58	7	11	P	19	83	3	7	3	P
9	7	3	P	3	34	11	7	P	P	59	3	P	3	P	84	29	3	7	3
10	P	P	P	P	35	3	P	3	P	60	P	3	P	3	85	23	P	P	P
11	3	P	3	7	36	19	3	P	3	61	13	P	P	P	86	3	P	3	11
12	11	3	P	3	37	7	P	13	P	62	3	7	3	17	87	13	3	P	3
13	P	7	P	P	38	3	P	3	P	63	P	3	7	3	88	P	P	P	7
14	3	11	3	P	39	17	3	P	3	64	P	P	P	11	89	3	19	3	29
15	P	3	P	3	40	P	13	11	P	65	3	P	3	P	90	17	3	P	3
16	7	P	P	13	41	3	7	3	P	66	P	3	23	3	91	P	11	7	P
17	3	P	3	P	42	P	3	7	3	67	11	P	P	7	92	3	13	3	P
18	P	3	11	3	43	P	P	19	P	68	3	P	3	13	93	7	3	P	3
19	P	P	P	P	44	3	P	3	P	69	P	3	17	3	94	P	23	P	13
20	3	7	3	11	45	11	3	P	3	70	P	19	7	P	95	3	P	3	7
21	P	3	7	3	46	P	P	P	7	71	3	23	3	P	96	31	3	P	3
22	13	P	P	P	47	3	11	3	P	72	7	3	P	3	97	P	7	P	11
23	3	P	3	P	48	13	3	P	3	73	17	P	11	P	98	3	P	3	23
24	P	3	13	3	49	P	17	7	P	74	3	P	3	7	99	P	3	P	3

Table 6   Common and Natural Logarithms

Linear interpolation should be accurate to three decimal places for $n > 20$ in the $\log_{10}$ (common) column and for $n > 30$ in the $\ln$ (natural) column.

x	$\log_{10} x$	$\ln x$	x	$\log_{10} x$	$\ln x$	x	$\log_{10} x$	$\ln x$
1	0.0000	0.0000	36	1.5563	3.5835	71	1.8513	4.2627
2	0.3010	0.6931	37	1.5682	3.6109	72	1.8573	4.2767
3	0.4771	1.0986	38	1.5798	3.6376	73	1.8633	4.2905
4	0.6021	1.3863	39	1.5911	3.6636	74	1.8692	4.3041
5	0.6990	1.6094	40	1.6021	3.6889	75	1.8751	4.3175
6	0.7782	1.7918	41	1.6128	3.7136	76	1.8808	4.3307
7	0.8451	1.9459	42	1.6232	3.7377	77	1.8865	4.3438
8	0.9031	2.0794	43	1.6335	3.7612	78	1.8921	4.3567
9	0.9542	2.1972	44	1.6435	3.7842	79	1.8976	4.3695
10	1.0000	2.3026	45	1.6532	3.8067	80	1.9031	4.3820
11	1.0414	2.3979	46	1.6628	3.8286	81	1.9085	4.3944
12	1.0792	2.4849	47	1.6721	3.8501	82	1.9138	4.4067
13	1.1139	2.5650	48	1.6812	3.8712	83	1.9191	4.4188
14	1.1461	2.6391	49	1.6902	3.8918	84	1.9243	4.4308
15	1.1761	2.7080	50	1.6990	3.9120	85	1.9294	4.4427
16	1.2041	2.7726	51	1.7076	3.9318	86	1.9345	4.4543
17	1.2304	2.8332	52	1.7160	3.9512	87	1.9395	4.4659
18	1.2553	2.8904	53	1.7243	3.9703	88	1.9445	4.4773
19	1.2788	2.9444	54	1.7324	3.9890	89	1.9494	4.4886
20	1.3010	2.9957	55	1.7404	4.0073	90	1.9542	4.4998
21	1.3222	3.0445	56	1.7482	4.0254	91	1.9590	4.5109
22	1.3424	3.0910	57	1.7559	4.0430	92	1.9638	4.5218
23	1.3617	3.1355	58	1.7634	4.0604	93	1.9685	4.5326
24	1.3802	3.1780	59	1.7709	4.0775	94	1.9731	4.5433
25	1.3979	3.2189	60	1.7782	4.0943	95	1.9777	4.5539
26	1.4150	3.2581	61	1.7853	4.1109	96	1.9823	4.5643
27	1.4314	3.2958	62	1.7924	4.1271	97	1.9868	4.5747
28	1.4472	3.3322	63	1.7993	4.1431	98	1.9912	4.5850
29	1.4624	3.3673	64	1.8062	4.1589	99	1.9956	4.5951
30	1.4771	3.4012	65	1.8129	4.1744	100	2.0000	4.6052
31	1.4914	3.4340	66	1.8195	4.1897	1000	3.0000	6.9078
32	1.5052	3.4657	67	1.8261	4.2047	10000	4.0000	9.2103
33	1.5185	3.4965	68	1.8325	4.2195			
34	1.5315	3.5264	69	1.8388	4.2341			
35	1.5441	3.5554	70	1.8451	4.2485			

Table 7   The Exponential Function

x	e^x	e^{-x}	x	e^x	e^{-x}	x	e^x	e^{-x}
0.00	1.0000	1.0000	1.50	4.4817	0.2231	3.00	20.086	0.04979
0.05	1.0513	0.9512	1.55	4.7115	0.2122	3.05	21.115	0.04736
0.10	1.1052	0.9048	1.60	4.9530	0.2019	3.10	22.198	0.04505
0.15	1.1618	0.8607	1.65	5.2070	0.1920	3.15	23.336	0.04285
0.20	1.2214	0.8187	1.70	5.4739	0.1827	3.20	24.533	0.04076
0.25	1.2840	0.7788	1.75	5.7546	0.1738	3.25	25.790	0.03877
0.30	1.3499	0.7408	1.80	6.0496	0.1653	3.30	27.113	0.03688
0.35	1.4191	0.7047	1.85	6.3598	0.1572	3.35	28.503	0.03508
0.40	1.4918	0.6703	1.90	6.6859	0.1496	3.40	29.964	0.03337
0.45	1.5683	0.6376	1.95	7.0287	0.1423	3.45	31.500	0.03175
0.50	1.6487	0.6065	2.00	7.3891	0.1354	3.50	33.115	0.03020
0.55	1.7333	0.5769	2.05	7.7679	0.1287	3.55	34.813	0.02872
0.60	1.8221	0.5488	2.10	8.1662	0.1225	3.60	36.598	0.02732
0.65	1.9155	0.5220	2.15	8.5849	0.1165	3.65	38.473	0.02599
0.70	2.0138	0.4966	2.20	9.0250	0.1108	3.70	40.447	0.02472
0.75	2.1170	0.4724	2.25	9.4877	0.1054	3.75	42.521	0.02352
0.80	2.2255	0.4493	2.30	9.9742	0.1003	3.80	44.701	0.02237
0.85	2.3396	0.4274	2.35	10.486	0.09537	3.85	46.993	0.02128
0.90	2.4596	0.4065	2.40	11.023	0.09072	3.90	49.402	0.02024
0.95	2.5857	0.3867	2.45	11.589	0.08629	3.95	51.935	0.01925
1.00	2.7183	0.3679	2.50	12.182	0.08209	4.00	54.598	0.01832
1.05	2.8577	0.3499	2.55	12.807	0.07808	5.00	148.41	0.006738
1.10	3.0042	0.3329	2.60	13.464	0.07427	6.00	403.43	0.002479
1.15	3.1582	0.3166	2.65	14.154	0.07065	7.00	1096.6	0.000912
1.20	3.3201	0.3012	2.70	14.880	0.06721	8.00	2981.0	0.000335
						9.00	8103.1	0.000123
1.25	3.4903	0.2865	2.75	15.643	0.06393	10.00	22026.5	0.000045
1.30	3.6693	0.2725	2.80	16.445	0.06081			
1.35	3.8574	0.2592	2.85	17.288	0.05784			
1.40	4.0552	0.2466	2.90	18.174	0.05502			
1.45	4.2631	0.2346	2.95	19.106	0.05234			

Table 8   Trigonometric Functions

With degrees or radians at the left side of the table, use the function headings at the top of each page; with degrees or radians at the right, use the function headings at the bottom of each page. Thus, the cotangent of 20.5 degrees, or 0.3578 radians, is found to be 2.6746, while the same tabular entry gives the tangent of 69.5° as well. Linear interpolation will be accurate anywhere in the sine and cosine columns, in the tangent and cotangent columns for function values less than 1, and the secant and cosecant columns for function values less than about 1.2.

x (degrees)	x (radians)	$\sin x$	$\cos x$	$\tan x$	$\cot x$	$\sec x$	$\csc x$		
0.0	0.0000	0.0000	1.0000	0.0000	Infinity	1.0000	Infinity	90.0	1.5708
0.5	0.0087	0.0087	1.0000	0.0087	114.59	1.0000	114.59	89.5	1.5621
1.0	0.0175	0.0175	0.9998	0.0175	57.290	1.0002	57.299	89.0	1.5533
1.5	0.0262	0.0262	0.9997	0.0262	38.188	1.0003	38.202	88.5	1.5446
2.0	0.0349	0.0349	0.9994	0.0349	28.636	1.0006	28.654	88.0	1.5359
2.5	0.0436	0.0436	0.9990	0.0437	22.904	1.0010	22.926	87.5	1.5272
3.0	0.0524	0.0523	0.9986	0.0524	19.081	1.0014	19.107	87.0	1.5184
3.5	0.0611	0.0610	0.9981	0.0612	16.350	1.0019	16.380	86.5	1.5097
4.0	0.0698	0.0697	0.9976	0.0699	14.301	1.0024	14.338	86.0	1.5010
4.5	0.0785	0.0785	0.9969	0.0787	12.706	1.0031	12.745	85.5	1.4923
5.0	0.0873	0.0872	0.9962	0.0875	11.430	1.0038	11.474	85.0	1.4835
5.5	0.0960	0.0958	0.9954	0.0963	10.385	1.0046	10.433	84.5	1.4748
6.0	0.1047	0.1045	0.9945	0.1051	9.5144	1.0055	9.5668	84.0	1.4661
6.5	0.1134	0.1132	0.9936	0.1139	8.7769	1.0065	8.8337	83.5	1.4573
7.0	0.1222	0.1219	0.9925	0.1228	8.1443	1.0075	8.2055	83.0	1.4486
7.5	0.1309	0.1305	0.9914	0.1317	7.5958	1.0086	7.6613	82.5	1.4399
8.0	0.1396	0.1392	0.9903	0.1405	7.1154	1.0098	7.1853	82.0	1.4312
8.5	0.1484	0.1478	0.9890	0.1495	6.6912	1.0111	6.7655	81.5	1.4224
9.0	0.1571	0.1564	0.9877	0.1584	6.3138	1.0125	6.3925	81.0	1.4137
9.5	0.1658	0.1650	0.9863	0.1673	5.9758	1.0139	6.0589	80.5	1.4050
10.0	0.1745	0.1736	0.9848	0.1763	5.6713	1.0154	5.7588	80.0	1.3963
10.5	0.1833	0.1822	0.9833	0.1853	5.3955	1.0170	5.4874	79.5	1.3875
11.0	0.1920	0.1908	0.9816	0.1944	5.1446	1.0187	5.2408	79.0	1.3788
11.5	0.2007	0.1994	0.9799	0.2035	4.9152	1.0205	5.0159	78.5	1.3701
12.0	0.2094	0.2079	0.9781	0.2126	4.7046	1.0223	4.8097	78.0	1.3614
12.5	0.2182	0.2164	0.9763	0.2217	4.5107	1.0243	4.6202	77.5	1.3526
13.0	0.2269	0.2250	0.9744	0.2309	4.3315	1.0263	4.4454	77.0	1.3439
13.5	0.2356	0.2334	0.9724	0.2401	4.1653	1.0284	4.2837	76.5	1.3352
14.0	0.2443	0.2419	0.9703	0.2493	4.0108	1.0306	4.1336	76.0	1.3265
14.5	0.2531	0.2504	0.9681	0.2586	3.8667	1.0329	3.9939	75.5	1.3177
15.0	0.2618	0.2588	0.9659	0.2679	3.7321	1.0353	3.8637	75.0	1.3090
		$\cos x$	$\sin x$	$\cot x$	$\tan x$	$\csc x$	$\sec x$	x (degrees)	x (radians)

Table 8 (cont.)

x (degrees)	x (radians)	$\sin x$	$\cos x$	$\tan x$	$\cot x$	$\sec x$	$\csc x$		
15.0	0.2618	0.2588	0.9659	0.2679	3.7321	1.0353	3.8637	75.0	1.3090
15.5	0.2705	0.2672	0.9636	0.2773	3.6059	1.0377	3.7420	74.5	1.3003
16.0	0.2793	0.2756	0.9613	0.2867	3.4874	1.0403	3.6289	74.0	1.2915
16.5	0.2880	0.2840	0.9588	0.2962	3.3759	1.0429	3.5209	73.5	1.2828
17.0	0.2967	0.2924	0.9563	0.3057	3.2709	1.0457	3.4203	73.0	1.2741
17.5	0.3054	0.3007	0.9537	0.3153	3.1718	1.0485	3.3255	72.5	1.2654
18.0	0.3142	0.3090	0.9511	0.3249	3.0777	1.0515	3.2361	72.0	1.2566
18.5	0.3229	0.3173	0.9483	0.3346	2.9887	1.0545	3.1515	71.5	1.2479
19.0	0.3316	0.3256	0.9455	0.3443	2.9042	1.0576	3.0716	71.0	1.2392
19.5	0.3403	0.3338	0.9426	0.3541	2.8239	1.0608	2.9957	70.5	1.2305
20.0	0.3491	0.3420	0.9397	0.3640	2.7475	1.0642	2.9238	70.0	1.2217
20.5	0.3578	0.3502	0.9367	0.3739	2.6746	1.0676	2.8555	69.5	1.2130
21.0	0.3665	0.3584	0.9336	0.3839	2.6051	1.0711	2.7904	69.0	1.2043
21.5	0.3753	0.3665	0.9304	0.3939	2.5386	1.0748	2.7285	68.5	1.1956
22.0	0.3840	0.3746	0.9272	0.4040	2.4751	1.0785	2.6695	68.0	1.1868
22.5	0.3927	0.3827	0.9239	0.4142	2.4142	1.0824	2.6131	67.5	1.1781
23.0	0.4014	0.3907	0.9205	0.4245	2.3559	1.0864	2.5593	67.0	1.1694
23.5	0.4102	0.3987	0.9171	0.4348	2.2998	1.0904	2.5078	66.5	1.1606
24.0	0.4189	0.4067	0.9135	0.4452	2.2460	1.0946	2.4586	66.0	1.1519
24.5	0.4276	0.4147	0.9100	0.4557	2.1943	1.0989	2.4114	65.5	1.1432
25.0	0.4363	0.4226	0.9063	0.4663	2.1445	1.1034	2.3662	65.0	1.1345
25.5	0.4451	0.4305	0.9026	0.4770	2.0965	1.1079	2.3228	64.5	1.1257
26.0	0.4538	0.4384	0.8988	0.4877	2.0503	1.1126	2.2812	64.0	1.1170
26.5	0.4625	0.4462	0.8949	0.4986	2.0057	1.1174	2.2412	63.5	1.1083
27.0	0.4712	0.4540	0.8910	0.5095	1.9626	1.1223	2.2027	63.0	1.0996
27.5	0.4800	0.4617	0.8870	0.5206	1.9210	1.1274	2.1657	62.5	1.0908
28.0	0.4887	0.4695	0.8829	0.5317	1.8807	1.1326	2.1301	62.0	1.0821
28.5	0.4974	0.4772	0.8788	0.5430	1.8418	1.1379	2.0957	61.5	1.0734
29.0	0.5061	0.4848	0.8746	0.5543	1.8040	1.1434	2.0627	61.0	1.0647
29.5	0.5149	0.4924	0.8704	0.5658	1.7675	1.1490	2.0308	60.5	1.0559
30.0	0.5236	0.5000	0.8660	0.5774	1.7321	1.1547	2.0000	60.0	1.0472
		$\cos x$	$\sin x$	$\cot x$	$\tan x$	$\csc x$	$\sec x$	x (degrees)	x (radians)

Table 8 (cont.)

x (degrees)	x (radians)	$\sin x$	$\cos x$	$\tan x$	$\cot x$	$\sec x$	$\csc x$		
30.0	0.5236	0.5000	0.8660	0.5774	1.7321	1.1547	2.0000	60.0	1.0472
30.5	0.5323	0.5075	0.8616	0.5890	1.6977	1.1606	1.9703	59.5	1.0385
31.0	0.5411	0.5150	0.8572	0.6009	1.6643	1.1666	1.9416	59.0	1.0297
31.5	0.5498	0.5225	0.8526	0.6128	1.6319	1.1728	1.9139	58.5	1.0210
32.0	0.5585	0.5299	0.8480	0.6249	1.6003	1.1792	1.8871	58.0	1.0123
32.5	0.5672	0.5373	0.8434	0.6371	1.5697	1.1857	1.8612	57.5	1.0036
33.0	0.5760	0.5446	0.8387	0.6494	1.5399	1.1924	1.8361	57.0	0.9948
33.5	0.5847	0.5519	0.8339	0.6619	1.5108	1.1993	1.8118	56.5	0.9861
34.0	0.5934	0.5592	0.8290	0.6745	1.4826	1.2062	1.7883	56.0	0.9774
34.5	0.6021	0.5664	0.8241	0.6873	1.4550	1.2134	1.7655	55.5	0.9687
35.0	0.6109	0.5736	0.8192	0.7002	1.4281	1.2208	1.7434	55.0	0.9599
35.5	0.6196	0.5807	0.8141	0.7133	1.4019	1.2283	1.7221	54.5	0.9512
36.0	0.6283	0.5878	0.8090	0.7265	1.3764	1.2361	1.7014	54.0	0.9425
36.5	0.6370	0.5948	0.8039	0.7400	1.3514	1.2440	1.6812	53.5	0.9338
37.0	0.6458	0.6018	0.7986	0.7536	1.3270	1.2521	1.6616	53.0	0.9250
37.5	0.6545	0.6088	0.7934	0.7673	1.3032	1.2605	1.6427	52.5	0.9163
38.0	0.6632	0.6157	0.7880	0.7813	1.2799	1.2690	1.6243	52.0	0.9076
38.5	0.6720	0.6225	0.7826	0.7954	1.2572	1.2778	1.6064	51.5	0.8988
39.0	0.6807	0.6293	0.7771	0.8098	1.2349	1.2868	1.5890	51.0	0.8901
39.5	0.6894	0.6361	0.7716	0.8243	1.2131	1.2960	1.5721	50.5	0.8814
40.0	0.6981	0.6428	0.7660	0.8391	1.1918	1.3054	1.5557	50.0	0.8727
40.5	0.7069	0.6494	0.7604	0.8541	1.1708	1.3151	1.5398	49.5	0.8639
41.0	0.7156	0.6561	0.7547	0.8693	1.1504	1.3250	1.5243	49.0	0.8552
41.5	0.7243	0.6626	0.7490	0.8847	1.1303	1.3352	1.5092	48.5	0.8465
42.0	0.7330	0.6691	0.7431	0.9004	1.1106	1.3456	1.4945	48.0	0.8378
42.5	0.7418	0.6756	0.7373	0.9163	1.0913	1.3563	1.4802	47.5	0.8290
43.0	0.7505	0.6820	0.7314	0.9325	1.0724	1.3673	1.4663	47.0	0.8203
43.5	0.7592	0.6884	0.7254	0.9490	1.0538	1.3786	1.4527	46.5	0.8116
44.0	0.7679	0.6947	0.7193	0.9657	1.0355	1.3902	1.4396	46.0	0.8029
44.5	0.7767	0.7009	0.7133	0.9827	1.0176	1.4020	1.4267	45.5	0.7941
45.0	0.7854	0.7071	0.7071	1.0000	1.0000	1.4142	1.4142	45.0	0.7854
		$\cos x$	$\sin x$	$\cot x$	$\tan x$	$\csc x$	$\sec x$	x (degrees)	x (radians)

Table 9   Derivatives

In this table, u and v represent arbitrary differentiable functions of the independent variable x, while c and n represent constants. Roman numerals in parentheses after a formula refer to the formulas derived in Secs. 7.3 and 7.5.

1. $\dfrac{d(u+v)}{dx} = \dfrac{du}{dx} + \dfrac{dv}{dx}$.   (II)

2. $\dfrac{d(cu)}{dx} = c\dfrac{du}{dx}$.   (III)

3. $\dfrac{dc}{dx} = 0$.   (IV)

4. $\dfrac{d(u^n)}{dx} = nu^{n-1}\dfrac{du}{dx}$.   (I)

5. $\dfrac{d(\sqrt{u})}{dx} = \dfrac{1}{2\sqrt{u}}\dfrac{du}{dx}$.   (VIII)

6. $\dfrac{d(uv)}{dx} = u\dfrac{dv}{dx} + v\dfrac{du}{dx}$.   (V)

7. $\dfrac{d\left(\frac{u}{v}\right)}{dx} = \dfrac{v\frac{du}{dx} - u\frac{dv}{dx}}{v^2}$.   (VI)

8. $\dfrac{d(u(v(x)))}{dx} = \dfrac{du}{dy}\bigg|_{y=v(x)} \cdot \dfrac{dv}{dx}$.   (VII)

9. $\dfrac{dx}{du} = \dfrac{1}{du/dx}$.   (IX)

10. $\dfrac{d(u^v)}{dx} = vu^{v-1}\dfrac{du}{dx} + u^v \ln u \dfrac{dv}{dx}$.   (XIV)

11. $\dfrac{d(\sin u)}{dx} = \cos u \dfrac{du}{dx}$.   (X)

12. $\dfrac{d(\cos u)}{dx} = -\sin u \dfrac{du}{dx}$.   (XI)

13. $\dfrac{d(\tan u)}{dx} = \sec^2 u \dfrac{du}{dx}$.

14. $\dfrac{d(\cot u)}{dx} = -\csc^2 u \dfrac{du}{dx}$.

15. $\dfrac{d(\sec u)}{dx} = \sec u \tan u \dfrac{du}{dx}$.

16. $\dfrac{d(\csc u)}{dx} = -\csc u \cot u \dfrac{du}{dx}$.

17. $\dfrac{d(\sin^{-1} u)}{dx} = \dfrac{1}{\sqrt{1-u^2}}\dfrac{du}{dx}$.

18. $\dfrac{d(\cos^{-1} u)}{dx} = -\dfrac{1}{\sqrt{1-u^2}}\dfrac{du}{dx}$.

19. $\dfrac{d(\tan^{-1} u)}{dx} = \dfrac{1}{u^2+1}\dfrac{du}{dx}$.

20. $\dfrac{d\left(\cot^{-1}u\right)}{dx} = -\dfrac{1}{u^2+1}\dfrac{du}{dx}.$

21.* $\dfrac{d\left(\sec^{-1}u\right)}{dx} = \dfrac{1}{u\sqrt{u^2-1}}\dfrac{du}{dx}.$

22.* $\dfrac{d\left(\csc^{-1}u\right)}{dx} = -\dfrac{1}{u\sqrt{u^2-1}}\dfrac{du}{dx}.$

23. $\dfrac{d\left(e^u\right)}{dx} = e^u\dfrac{du}{dx}.$ (XIII)

24. $\dfrac{d\left(c^u\right)}{dx} = c^u \ln c\dfrac{du}{dx}.$

25. $\dfrac{d\left(\ln u\right)}{dx} = \dfrac{1}{u}\dfrac{du}{dx}.$ (XII)

26. $\dfrac{d\left(\log_c u\right)}{dx} = \dfrac{1}{u\ln c}\dfrac{du}{dx}.$

27. $\dfrac{d}{dx}\left(\int_c^u f(t)dt\right) = f(u)\dfrac{du}{dx}.$

* These formulas may have the absolute value of u instead of just u in their denominators, depending on the definitions of sec⁻¹ and csc⁻¹ for negative values of the argument, on which there is no universal agreement among mathematicians. If sec⁻¹ x is defined for negative x to lie between $\frac{\pi}{2}$ and π radians, the absolute value sign is required, but if it is defined to lie between $-\pi$ and $-\frac{\pi}{2}$, no absolute value sign is needed. Similarly, if csc⁻¹ x is defined for negative x to lie between $-\frac{\pi}{2}$ and 0, the absolute value sign is needed, whereas if it is defined to lie between $-\pi$ and $-\frac{\pi}{2}$ (or between π and $\frac{3\pi}{2}$), no absolute value sign is required. The opinions of mathematicians are slowly turning in favor of using the absolute value-requiring definitions.

Table 10   Integrals

In what follows, u and v are arbitrary integrable functions of x, while a, b, and n are constants. All logarithms are for the absolute values of their arguments.

1. $\int cu(x)dx = c\int u(x)dx.$

2. $\int [u(x) + v(x)]dx = \int u(x)dx + \int v(x)dx.$

3. $\int u(x)\frac{dv(x)}{dx}dx = u(x)\cdot v(x) - \int v(x)\frac{du(x)}{dx}dx.$

4. $\int u(v(x))\frac{dv(x)}{dx}dx = \int u(y)dy$ ($y = v(x)$ after integrating $u(y)\,dy$).

5. $\int x^n dx = \frac{x^{n+1}}{n+1} + C.$ ($n \neq -1$)

6. $\int (ax + b)^n dx = \frac{1}{a(n+1)}(ax + b)^{n+1} + C.$ ($n \neq -1$)

7. $\int \frac{dx}{x} = \ln x + C.$

8. $\int \frac{dx}{ax+b} = \frac{1}{a}\ln(ax + b) + C.$

9. $\int x(ax + b)^n dx = \frac{(ax+b)^{n+2}}{a^2(n+2)} - \frac{b(ax+b)^{n+1}}{a^2(n+1)} + C.$ ($n \neq -1, -2$)

10. $\int \frac{x\,dx}{ax+b} = \frac{x}{a} - \frac{b}{a^2}\ln(ax + b) + C.$

11. $\int \frac{x\,dx}{(ax+b)^2} = \frac{1}{a^2}\left[\ln(ax + b) + \frac{b}{ax+b}\right] + C.$

12. $\int x\sqrt{ax + b}\,dx = \frac{6ax-4b}{15a^2}\sqrt{(ax + b)^5} + C.$

13a. $\int \frac{dx}{x\sqrt{ax+b}} = \frac{1}{\sqrt{b}}\ln\left(\frac{\sqrt{ax+b}-\sqrt{b}}{\sqrt{ax+b}+\sqrt{b}}\right) + C.$ ($b > 0$)

13b. $\int \frac{dx}{x\sqrt{ax+b}} = \frac{2}{\sqrt{-b}}\tan^{-1}\sqrt{\frac{ax+b}{-b}} + C.$ ($b < 0$)

14. $\int \frac{dx}{a^2+x^2} = \frac{1}{a}\tan^{-1}\frac{x}{a} + C.$

15. $\int \frac{dx}{x^2-a^2} = \frac{1}{2a}\ln\frac{x+a}{x-a} + C.$

16. $\int \sqrt{a^2 + x^2}\,dx = \frac{x}{2}\sqrt{a^2 + x^2} + \frac{a^2}{2}\ln\left(x + \sqrt{a^2 + x^2}\right) + C.$

17. $\int \sqrt{a^2 - x^2}\,dx = \frac{x}{2}\sqrt{a^2 - x^2} + \frac{a^2}{2}\sin^{-1}\frac{x}{a} + C.$

18. $\int \sqrt{x^2 - a^2}\,dx = \frac{x}{2}\sqrt{x^2 - a^2} - \frac{a^2}{2}\ln(x + \sqrt{x^2 - a^2}) + C.$

19. $\int \frac{dx}{\sqrt{x^2+a^2}} = \ln\left(x + \sqrt{x^2 + a^2}\right) + C.$

20. $\int \frac{dx}{\sqrt{x^2-a^2}} = \ln\left(x + \sqrt{x^2 - a^2}\right) + C.$

21. $\int \frac{dx}{\sqrt{a^2-x^2}} = \sin^{-1}\frac{x}{a} + C.$

22. $\int \sin x \, dx = -\cos x + C.$

23. $\int \cos x \, dx = -\sin x + C.$

24. $\int \tan x \, dx = \ln \sec x + C.$

25. $\int \cot x \, dx = \ln \sin x + C.$

26. $\int \sec x \, dx = \ln(\tan x + \sec x) + C.$

27. $\int \csc x \, dx = -\ln(\cot x + \csc x) + C.$

28. $\int e^x \, dx = e^x + C.$

29. $\int a^x \, dx = \frac{a^x}{\ln a} + C.$

30. $\int \ln x \, dx = x \ln x - x + C.$

31. $\int e^x \sin x \, dx = \frac{e^x}{2}(\sin x - \cos x) + C.$

32. $\int e^x \cos x \, dx = \frac{e^x}{2}(\cos x + \sin x) + C.$

33. $\int_0^\infty x^n e^{-x} \, dx = n!$ (n an integer ≥ 0)

Table 11   Binomial Coefficients

This table contains the values of $\binom{n}{r}$, where $n \le 15$ and $r \le 10$.

$n \backslash r$	0	1	2	3	4	5	6	7	8	9	10
0	1										
1	1	1									
2	1	2	1								
3	1	3	3	1							
4	1	4	6	4	1						
5	1	5	10	10	5	1					
6	1	6	15	20	15	6	1				
7	1	7	21	35	35	21	7	1			
8	1	8	28	56	70	56	28	8	1		
9	1	9	36	84	126	126	84	36	9	1	
10	1	10	45	120	210	252	210	120	45	10	1
11	1	11	55	165	330	462	462	330	165	55	11
12	1	12	66	220	495	792	924	792	495	220	66
13	1	13	78	286	715	1287	1716	1716	1287	715	286
14	1	14	91	364	1001	2002	3003	3432	3003	2002	1001
15	1	15	105	455	1365	3003	5005	6435	6435	5005	3003

If $r > 10$, use the formula $\binom{n}{r} = \binom{n}{n-r}$; e. g., $\binom{15}{12} = \binom{15}{3} = 455$.

Table 12   Stirling Numbers

In these tables, the s_{ni} are Stirling numbers of the first kind, and the t_{ni} are Stirling numbers of the second kind. Thus, $s_{4,3} = -6$ and $t_{5,2} = 15$.

n	$s_{n,1}$	$s_{n,2}$	$s_{n,3}$	$s_{n,4}$	$s_{n,5}$	$s_{n,6}$	$s_{n,7}$	$s_{n,8}$
1	1							
2	-1	1						
3	2	-3	1					
4	-6	11	-6	1				
5	24	-50	35	-10	1			
6	-120	274	-225	85	-15	1		
7	720	-1764	1624	-735	175	-21	1	
8	-5040	13068	-13132	6769	-1960	322	-28	1

n	$t_{n,1}$	$t_{n,2}$	$t_{n,3}$	$t_{n,4}$	$t_{n,5}$	$t_{n,6}$	$t_{n,7}$	$t_{n,8}$
1	1							
2	1	1						
3	1	3	1					
4	1	7	6	1				
5	1	15	25	10	1			
6	1	31	90	65	15	1		
7	1	63	301	350	140	21	1	
8	1	127	966	1701	1050	266	28	1